Approximation Theory X

Abstract and Classical Analysis

Approximation Theory X

Abstract and Classical Analysis

Edited by

Charles K. Chui
Department of Mathematics
University of Missouri–St. Louis
St. Louis, Missouri

Larry L. Schumaker
Department of Mathematics
Vanderbilt University
Nashville, Tennessee

Joachim Stöckler
Department of Mathematics
University of Dortmund
Dortmund, Germany

VANDERBILT UNIVERSITY PRESS
Nashville

© 2002 Vanderbilt University Press
All Rights Reserved
First Edition 2002
Printed on acid-free paper.
Manufactured in the United States of America.

ISBN 0-8265-1415-4 (alk. paper)

INNOVATIONS IN APPLIED MATHEMATICS
An international series devoted to the latest research in
modern areas of mathematics, with significant applications
in engineering, medicine, and the sciences.

SERIES EDITOR
Larry L. Schumaker
Stevenson Professor of Mathematics
Vanderbilt University

CONTENTS

Preface vii

Contributors viii

Rate of Convergence of Bounded Linear Operators
George Anastassiou 1

The Metric Average of 1D Compact Sets
Robert Baier, Nira Dyn and Elza Farkhi 9

Uniform Strong Unicity of Order 2 for Generalized Haar Sets
M. W. Bartelt and J. J. Swetits 23

Markov-type Inequalities for Polynomials and Splines
Borislav Bojanov 31

Nonlinear Approximation by Refinable Functions and Piecewise
Polynomials
Yu. Brudnyi 91

On Spaces Admitting Minimal Projections Which are Orthogonal
Bruce L. Chalmers and Boris Shekhtman 113

Hilbert Transforms and Orthonormal Expansions for Exponential
Weights
S. B. Damelin 117

On the Separation of Logarithmic Points on the Sphere
P. D. Dragnev 137

Best One–Sided L^1–Approximation by Blending Functions
and Approximate Cubature
Dimiter Dryanov, Werner Haußmann and Petar Petrov 145

Polynomials with Littlewood-Type Coefficient Constraints
Tamás Erdélyi 153

Parameterization of Manifold Triangulations
Michael S. Floater, Kai Hormann, and Martin Reimers 197

Limit Theorems for Polynomial Approximations with Hermite
and Freud Weights
Michael I. Ganzburg 211

Complex Haar Spaces Generated by Shifts Applied to a Single
Function
Walter Hengartner and Gerhard Opfer 223

Optimal Location with Polyhedral Constraints
  R. Huotari, M. P. Prophet, and J. Ribando 239

Non-degenerate Rational Approximation
  Franklin Kemp 247

L_p Markov–Bernstein Inequalities on All Arcs of the Circle for
Generalized Polynomials
  C. K. Kobindarajah 267

Positive Definite Kernel Interpolation on Manifolds: Convergence
Rates
  Jeremy Levesley and David L. Ragozin 277

Asymptotic Expressions for Multivariate Positive Linear
Operators
  A.-J. López-Moreno, J. Martínez-Moreno and F. J. Muñoz-Delgado 287

Approximation by Penalized Least Squares
  José L. Martínez-Morales 309

Conditions of Regularity for Bernstein–Rogosinski-type Means of
Double Fourier Series
  Yuri Nosenko 325

Some Remarks on Zero-Increasing Transformations
  Allan Pinkus 333

The Aitken–Neville Scheme in Several Variables
  Thomas Sauer and Yuesheng Xu 353

Interpolation by Polynomials in Several Variables
  Boris Shekhtman 367

Nonlinear Approximation with Regard to Bases
  V. N. Temlyakov 373

PREFACE

The *Tenth International Symposium on Approximation Theory* was held March 26 – 29, 2001 at the Sheraton Conference Center in Westport Plaza, St. Louis. Previous conferences in this series were held in 1973, 1976, 1980, and 1992 in Austin; 1983, 1986, 1989, and 1995 in College Station; and 1998 in Nashville. The conference was attended by 124 active participants in addition to over twenty professional mathematicians and students from the local St. Louis area. Twenty-two different countries were represented.

The program included seven invited one-hour survey talks and the one-hour presentation of the 2001 Popov Prize winner, Emmanuel Candès. In addition to a large number of contributed talks, there was a special poster session on interdisciplinary topics of current research. Due to the large number of submitted papers, these proceedings appear in two separate volumes. The first, subtitled *Abstract and Classical Analysis*, covers several areas of classical and modern approximation theory, while the second, subtitled *Wavelets, Splines, and Applications*, is devoted to recent applied and computational developments.

The conference began with opening remarks by George Lorentz which included an overview of the history of the previous nine meetings, a review of the impact of the meetings, especially on international cooperation, and a call for everyone in the field to continue this cooperation by attending each other's meetings and publishing in each other's journals.

This conference was sponsored by the National Science Foundation and the U.S. Army Research Office. We are also indebted to the Department of Mathematics and Computer Science of the University of Missouri – St. Louis for their support. In particular, we would like to thank Kai Bittner for assistance in the preparation of the conference program, Michael Schulte and Jan Stöckler for creating and managing the web site, and Deloris Licklider for her assistance in registration and other logistics.

We would also like to acknowledge the support of the Departments of Mathematics at Vanderbilt University and the University of Dortmund in connection with the preparation of these proceedings volumes. Thanks are also due to the invited speakers and other presenters, as well as everyone who attended for making the conference a success. Finally, we would like to thank our reviewers who helped select articles for the proceedings, and also Gerda Schumaker who assisted in the preparation of these volumes.

Charles K. Chui   March 1, 2002
Larry L. Schumaker
Joachim Stöckler

CONTRIBUTORS

Numbers in parentheses indicate pages on which the author's contribution(s) begin. Page numbers marked with * refer to this volume, while those marked with a ** refer to the companion volume *Approximation Theory X: Wavelets, Splines, and Applications.*

GEORGE A. ANASTASSIOU (1*), *Department of Mathematical Sciences, The University of Memphis, Memphis, TN 38152* [ganastss@memphis.edu]

GERARD AWANOU (1**), *Department of Mathematics, University of Georgia, Athens, Ga 30602* [gawanou@math.uga.edu]

ROBERT BAIER (9*), *Chair of Applied Mathematics, University of Bayreuth, D-95440 Bayreuth, Germany* [Robert.Baier@uni-bayreuth.de]

M. W. BARTELT (23*), *Department of Mathematics, Christopher Newport University, Newport News, VA 23606* [mbartelt@pcs.cnu.edu]

R. K. BEATSON (17**), *Department of Mathematics and Statistics, University of Canterbury, Christchurch, New Zealand* [R.Beatson@math.canterbury.ac.nz]

AURELIAN BEJANCU (27**), *Department of Applied Mathematics, University of Leeds, Leeds LS2 9JT, UK* [A.Bejancu@amsta.leeds.ac.uk]

KAI BITTNER (41**), *Department of Mathematics and Computer Science, University of Missouri – St. Louis, St. Louis, MO 63121* [bittner@math.umsl.edu]

BORISLAV BOJANOV (31*), *Dept. of Mathematics, University of Sofia, Blvd. James Boucher 5, 1164 Sofia, Bulgaria* [boris@fmi.uni-sofia.bg]

LASSE BORUP (51**), *Dept. of Mathematical Sciences, Aalborg University, Aalborg, DK-9220, Denmark* [lasse@math.auc.dk]

MARCIN BOWNIK (63**), *Department of Mathematics, University of Michigan, 525 East University Ave., Ann Arbor, MI 48109–1109* [marbow@umich.edu]

YU. BRUDNYI (91*), *Department of Mathematics, Technion – Israel Institute of Technology, 32000 Haifa, Israel* [ybrudnyi@math.technion.ac.il]

EMMANUEL J. CANDÈS (87**), *Applied and Computational Mathematics, California Institute of Technology, Pasadena, California 91125* [emmanuel@acm.caltech.edu]

BRUCE CHALMERS (113*), *Department of Mathematics, University of California, Riverside, Riverside, California 92521* [blc@math.ucr.edu]

B. CHANANE (155**), *Department of Mathematical Sciences, K. F. U. P. M., Dhahran 31261, Saudi Arabia* [chanane@kfupm.edu.sa]

MARGARET CHENEY (167**), *Department of Mathematical Sciences, Rensselaer Polytechnic Institute, Troy, NY 12180 USA* [cheney@rpi.edu]

HOI LING CHEUNG (179**), *Department of Mathematics, City University of Hong Kong, Kowloon, Hong Kong* [96473695@plink.cityu.edu.hk]

CHARLES K. CHUI (41**,187**,207**), *Department of Mathematics and Computer Science, University of Missouri – St. Louis, St. Louis, MO 63121* [cchui@stat.stanford.edu]

STEVEN B. DAMELIN (117*), *Department of Mathematics and Computer Science, Georgia Southern University, Post Box 8093, Statesboro, GA 30460-8093, USA* [damelin@gasou.edu]

OLEG DAVYDOV (231**), *Mathematisches Institut, Justus-Liebig Universität, D-35392 Giessen, Germany* [oleg.davydov@math.uni-giessen.de]

UWE DEPCZYNSKI (241**), *FACT Unternehmensberatung GmbH, Goetheplatz 5, D-60313 Franfurt/Main, Germany* [u.depczynski@fact.de]

ASEN L. DONTCHEV (261**), *Mathematical Review, Ann Arbor, MI 48107, USA* [ald@ams.org]

P. D. DRAGNEV (137*), *Department of Mathematics, Indiana-Purdue University, Fort Wayne, IN 46825, USA* [dragnevp@ipfw.edu]

DIMITER DRYANOV (145*), *Département des Mathématiques et de Statistique, Université de Montréal, C. P. 6128, Montréal, Québec H3C 3J7, Canada* [dryanovd@dms.umontreal.ca]

NIRA DYN (9*), *School of Mathematical Sciences, Sackler Faculty of Exact Sciences, Tel Aviv University, Tel Aviv 69978, Israel* [niradyn@post.tau.ac.il]

TAMÁS ERDÉLYI (153*), *Department of Mathematics, Texas A & M University, College Station, Texas 77843, USA* [terdelyi@math.tamu.edu]

ELZA FARKHI (9*), *School of Mathematical Sciences, Sackler Faculty of Exact Sciences, Tel Aviv University, Tel Aviv 69978, Israel* [elza@post.tau.ac.il]

GREG FASSHAUER (271**), *Department of Applied Mathematics, Illinois Institute of Technology, Chicago, IL 60616* [fass@amadeus.math.iit.edu]

MICHAEL S. FLOATER (197*), *SINTEF, Post Box 124, Blindern, N-0314 Oslo, Norway* [mif@math.sintef.no]

MICHAEL I. GANZBURG (211*), *Dept. of Mathematics, Hampton University, Hampton, VA 23668, USA* [michael.ganzburg@hamptonu.edu]

DAVID E. GILSINN (283**), *National Institute of Standards and Technology, 100 Bureau Drive, Stop 8910, Gaithersburg, MD 20899-8910* [dgilsinn@nist.gov]

MANFRED V. GOLITSCHEK (295**), *Institut für Angewandte Mathematik und Statistik, Universität Würzburg, 97074 Würzburg, Germany* [goli@mathematik.uni-wuerzburg.de]

WERNER HAUSSMANN (145*), *Department of Mathematics, Gerhard–Mercator–University, 47048 Duisburg, Germany* [haussmann@math.uni-duisburg.de]

TIAN-XIAO HE (305**), *Department of Mathematics and Computer Science, Illinois Wesleyan University, Bloomington, IL 61702-2900* [the@sun.iwu.edu]

WENJIE HE (187**), *Department of Mathematics and Computer Science, University of Missouri–St. Louis, St. Louis, MO 63121-4499* [he@neptune.cs.umsl.edu]

WALTER HENGARTNER (223*), *Université de Laval, Département de Mathématique, Québec G1K 7P4, Canada* [walheng@mat.ulaval.ca]

KAI HORMANN (197*), *Computer Graphics Group, University of Erlangen-Nürnberg, Am Weichselgarten 9, D-91058 Erlangen, Germany* [hormann@cs.fau.de]

ROBERT HUOTARI (239*), *Mathematics Department, Glendale Community College, Glendale AZ 85302* [r_huotari@yahoo.com]

FRANKLIN KEMP (247*), *3004 Pataula Lane, Plano, TX 75074-8765* [lfkemp@attbi.com]

BRUCE KESSLER (323**), *1 Big Red Way, Department of Mathematics, Western Kentucky University, Bowling Green, KY 42101* [bruce.kessler@wku.edu]

C. K. KOBINDARAJAH (267*), *Department of Mathematics, Eastern University, Chenkalady, Sri Lanka* [kshant@yahoo.com]

OGNYAN KOUNCHEV (333**), *Institute of Mathematics, Bulgarian Akademy of Sciences, Bonchev Str. 9, 1143 Sofia, Bulgaria* [kounchev@bas.bg]

BORIS I. KVASOV (355**), *School of Mathematics, Suranaree University of Technology, 111 University Avenue, Nakhon Ratchasima 30000, Thailand* [boris@math.sut.ac.th]

MING-JUN LAI (1**,369**,385**), *Mathematics Department, University of Georgia, Athens, GA 30602* [mjlai@math.uga.edu]

JOHN E. LAVERY (283**), *Computing and Information Sciences Division, Army Research Office, Army Research Laboratory, P. O. Box 12211, Research Triangle Park, NC 27709-2211* [lavery@arl.aro.army.mil]

J. LEVESLEY (17**,277*), *Department of Mathematics and Computer Science, University of Leicester, Leicester LE1 7RH, UK* [jl1@mcs.le.ac.uk]

JIAN-AO LIAN (207**), *Dept. of Mathematics, Prairie View A&M University, Prairie View, TX 77446* [unurbs@yahoo.com]

A.-J. LÓPEZ-MORENO (287*), *Departamento de Matemáticas, Universidad de Jaén, Spain* [ajlopez@ujaen.es]

JOSÉ L. MARTÍNEZ-MORALES (309*), *Instituto de Matemáticas, Universidad Nacional Autónoma de México, Ap. Post. 6-60, Cuernavaca, Mor. 62131, Mexico* [martinez@matcuer.unam.mx]

J. MARTÍNEZ-MORENO (287*), *Departamento de Matemáticas, Universidad de Jaén, Spain* [jmmoreno@ujaen.es]

F. J. MUÑOZ-DELGADO (287*), *Departamento de Matemáticas, Universidad de Jaén, Spain* [fdelgado@ujaen.es]

MORTEN NIELSEN (51**), *Dept. of Mathematics, University of South Carolina, Columbia, SC 29208* [nielsen@math.sc.edu]

YURI L. NOSENKO (325*), *Dept. of Mathematics, Donetsk National Technical University, 58, Artema str., Donetsk, 83000, Ukraine* [yuri@nosenko.dgtu.donetsk.ua]

GÜNTHER NÜRNBERGER (405**), *Institut für Mathematik, Universität Mannheim, D-618131 Mannheim, Germany* [nuern@euklid.math.uni-mannheim.de]

GERHARD OPFER (223*), *University of Hamburg, Department of Mathematics, Bundesstr. 55, 20146 Hamburg, Germany* [opfer@math.uni-hamburg.de]

PETAR PETROV (145*), *Department of Mathematics, University of Sofia, 1164 Sofia, Bulgaria* [peynov@fmi.uni-sofia.bg]

ALEXANDER PETUKHOV (425**), *Department of Mathematics, University of South Carolina, Columbia, SC 29208* [petukhov@math.sc.edu]

ALLAN PINKUS (333*), *Dept. of Mathematics, Technion – Israel Institute of Technology, Haifa, 32000, Israel* [pinkus@tx.technion.ac.il]

MICHAEL P. PROPHET (241*), *Department of Mathematics, University of Northern Iowa, Cedar Falls, IA 50614* [prophet@math.uni.edu]

HOUDUO QI (261**), *Department of Applied Mathematics, The Hong Kong Polytechnic University, Hung Hom, Kowloon, Hong Kong* [mahdqi@polyu.edu.hk]

LIQUN QI (261**), *Department of Applied Mathematics, The Hong Kong Polytechnic University, Hung Hom, Kowloon, Hong Kong* [maqilq@polyu.edu.hk]

DAVID L. RAGOZIN (277*), *University of Washington, Department of Mathematics, Seattle, WA 98195-4350, USA* [rag@math.washington.edu]

MARTIN REIMERS (197*), *Institutt for informatikk, Universitetet i Oslo, Post Box 1080, Blindern, N-0316 Oslo, Norway* [martinre@ifi.uio.no]

HERMANN RENDER (333**), *Fachbereich Mathematik, Gerhard-Mercator Universität Duisburg, D- 47048 Duisburg, Germany* [render@math.uni-duisburg.de]

JASON RIBANDO (239*), *Department of Mathematics, University of Northern Iowa, Cedar Falls, IA 50614* [jribando@math.uni.edu]

DAVID W. ROACH (369**), *Mathematics and Statistics Department, Murray State University, Murray, KY 42071* [david.roach@murraystate.edu]

THOMAS SAUER (353*), *Lehrstuhl für Numerische Mathematik, Justus–Liebig–Universität Gießen, Heinrich–Buff–Ring 44, D–35092 Gießen, Germany* [Tomas.Sauer@math.uni-giessen.de]

R. SCHABACK (433**), *Institut für Numerische und Angewandte Mathematik, Zentrum für Informatik, Lotzestraße 16–18, D-37083 Göttingen, Germany* [schaback@math.uni-goettingen.de]

LARRY L. SCHUMAKER (405**), *Center for Constructive Approximation, Department of Mathematics, Vanderbilt University, Nashville, TN 37205, USA* [s@mars.cas.vanderbilt.edu]

BORIS SHEKHTMAN (113*,367*), *Department of Mathematics, University of South Florida, Tampa, FL 33620-5700, USA* [boris@math.usf.edu]

DARRIN SPEEGLE (63**), *Department of Mathematics, Saint Louis University, 221 N. Grand Boulevard, Saint Louis, MO 63103* [speegled@slu.edu]

JOACHIM STÖCKLER (187**), *Universität Dortmund, Fachbereich Mathematik, 44221 Dortmund, Germany* [joachim.stoeckler@math.uni-dortmund.de]

HANS STRAUSS (441**), *Institut für Angewandte Mathematik, Universität Erlangen–Nürnberg, Martensstr. 3, D–91058 Erlangen, Germany* [strauss@am.uni-erlangen.de]

JOHN J. SWETITS (23*), *Department of Mathematics and Statistics, Old Dominion University, Norfolk, VA 23529* [jswetits@odu.edu]

CANQIN TANG (179**), *Department of Mathematics, Changde Teacher's College, Hunan, P. R. China* [tangcq2000@yahoo.com.cn]

V. N. TEMLYAKOV (373*), *Department of Mathematics, University of South Carolina, Columbia, SC 29208, USA* [temlyak@math.sc.edu]

JIANZHONG WANG (453**), *Sam Houston State University, Huntsville, TX 77341-2206, USA* [mth_jxw@shsu.edu]

HOLGER WENDLAND (473**), *Institut für Numerische und Angewandte Mathematik, Universität Göttingen, Lotzestr. 16-18, 37083 Göttingen, Germany* [wendland@math.uni-goettingen.de]

PAUL WENSTON (385**), *Department of Mathematics, The University of Georgia, Athens, GA 30602* [paul@math.uga.edu]

YUESHENG XU (353*), *Department of Mathematics, West Virginia University, Morgantown, WV 26506, USA* [yxu@math.wvu.edu]

LUNG-AN YING (385**), *School of Mathematical Sciences, Peking University, Beijing, China* [yingla@pku.edu.cn]

FRANK ZEILFELDER (405**), *Institut für Mathematik, Universität Mannheim, D-618131 Mannheim, Germany* [zeilfeld@euklid.math.uni-mannheim.de]

Rate of Convergence of Bounded Linear Operators

George Anastassiou

Abstract. This is a quantitative study for the rate of pointwise convergence of a sequence of bounded linear operators to an arbitrary operator in a very general setting involving the modulus of continuity. This is achieved through the Riesz representation theorem and the weak convergence of the corresponding signed measures to zero, studied quantitatively in various important cases.

§1. Introduction

Our work has been greatly motivated by the following result, see [2] and [3], which solves a problem of P. Lévy.

Theorem A. *Let* $\{\mu_a\}$ *be a bounded net (or sequence) of signed Borel measures on* $[0,1]$; *i.e., there is a number* $M > 0$ *such that* $|\int f d\mu_a| \leq M\|f\|_\infty$ *for* $f \in C[0,1]$. *Define* $K_a(x) = \mu_a[0,x]$ *for* $0 \leq x \leq 1$. *Then the following are equivalent:*

(i) $\lim\limits_a \int f d\mu_a = 0$ *for each* $f \in C[0,1]$ *(i.e.,* $\{\mu_a\}$ *converges weakly to zero).*

(ii) $\lim\limits_a \left(\int |K_a| dx + |K_a(1)| \right) = 0$.

Here dx stands for the Lebesgue measure on $[0,1]$.

Approximation Theory X: Abstract and Classical Analysis
Charles K. Chui, Larry L. Schumaker, and Joachim Stöckler (eds.), pp. 1–7.
Copyright © 2002 by Vanderbilt University Press, Nashville, TN.
ISBN 0-8265-1415-4.

§2. Results

We present our first result:

Theorem 1. *Let $f \in C[-1, 1]$ and $\{\mu_m\}_{m \in \mathbf{N}}$ be a sequence of nontrivial finite Borel signed measures on $[-1, 1]$. Set $M_m := \mu_m[-1, 1]$ and write $|\mu_m| = \mu_m^+ + \mu_m^-$, where μ_m^+, μ_m^- are the positive and negative parts, respectively, in the Jordan–Hahn decomposition of $\mu_m = \mu^+ - \mu^-$. Then*

$$\left| \int_{-1}^{1} f d\mu_m \right| \le |f(0)| |M_m|$$

$$+ \{1 + |\mu_m|[-1, 1]\} \omega_1 \left(f, \int_{-1}^{1} |t| d|\mu_m| \right), \tag{1}$$

where ω_1 stands for the first modulus of continuity. Moreover,

$$0 < \int_{-1}^{1} |t| d|\mu_m| < +\infty.$$

If $M_m \to 0$ and $\int_{-1}^{1} |t| d|\mu_m| \to 0$, as $m \to +\infty$, then $\{\mu_m\}_{m \in \mathbf{N}}$ converges weakly to zero.

Proof: Let $f \in C[-1, 1]$. Then

$$\int_{-1}^{1} f(t) d\mu_m(t) = \int_{-1}^{1} f(t) d(\mu_m^+ - \mu_m^-)(t)$$

$$= \int_{-1}^{1} f(t) d\mu_m^+(t) - \int_{-1}^{1} f(t) d\mu_m^-(t),$$

and

$$\left| \int_{-1}^{1} f(t) d\mu_m(t) \right| \le \int_{-1}^{1} |f(t)| d|\mu_m|(t).$$

We observe that

$$\int_{-1}^{1} f d\mu_m = \int_{-1}^{1} (f - f(0)) d\mu_m + f(0) \mu_m[-1, 1].$$

Thus,

$$\left| \int_{-1}^{1} f d\mu_m \right| \le \left| \int_{-1}^{1} (f - f(0)) d\mu_m \right| + |f(0)| |M_m|$$

$$\le \int_{-1}^{1} |f - f(0)| d|\mu_m|(t) + |f(0)| |M_m|$$

(by Corollary 7.1.1, p. 209, [1])

$$\le \omega_1(f, h_m) \int_{-1}^{1} \left\lceil \frac{|t|}{h_m} \right\rceil d|\mu_m|(t) + |f(0)| |M_m|$$

($\lceil \cdot \rceil$ is the ceiling of the number)

$$\leq \omega_1(f, h_m) \left\{ |\mu_m|[-1,1] + \frac{1}{h_m} \int_{-1}^{1} |t| d|\mu_m| \right\} + |f(0)| |M_m|$$

(by choosing $h_m = \int_{-1}^{1} |t| d|\mu_m|$)

$$= \omega_1 \left(f, \int_{-1}^{1} |t| d|\mu_m| \right) \{1 + |\mu_m|[-1,1]\} + |f(0)| |M_m|.$$

We have established (1). If $M_m \to 0$ and $\int_{-1}^{1} |t| d|\mu_m| \to 0$, as $m \to \infty$, we obtain $\int_{-1}^{1} f d\mu_m \to 0$, as $m \to \infty$. $\square$

Theorem 2. *Let* $f \in C^1[-1,1]$ *and let* $\{\mu_m\}_{m\in\mathbb{N}}$ *be a sequence of nontrivial Borel signed measures on* $[-1,1]$. *We assume that each* μ_m *is bounded, and set* $M_m := \mu_m[-1,1]$. *Define* $K_m(x) := \mu_m[-1,x]$, *$-1 \leq x \leq 1$, K_m is of bounded variation. Then*

$$\left| \int_{-1}^{1} f d\mu_m \right| \leq |f(1)| |M_m| + |f'(0)| \left| \int_{-1}^{1} K_m(x) dx \right| + \left(1 + \int_{-1}^{1} |K_m(x)| dx \right)$$

$$\cdot \omega_1 \left(f', \int_{-1}^{1} |K_m(x)| |x| dx \right), \ \forall m \in \mathbb{N}. \tag{2}$$

Here ω_1 is the first modulus of continuity. If $M_m \to 0$, $\int_{-1}^{1} |K_m(x)| dx \to 0$, as $m \to \infty$, then $\{\mu_m\}_{m\in\mathbb{N}}$ converges weakly to zero.

Proof: By integration by parts,

$$\int_{-1}^{1} f d\mu_m = \int_{-1}^{1} f dK_m + f(-1) K_m(-1)$$

$$= - \int_{-1}^{1} K_m df + f(1) K_m(1) - f(-1) K_m(-1)$$

$$\quad + f(-1) K_m(-1)$$

$$= - \int_{-1}^{1} K_m(x) f'(x) dx + f(1) M_m$$

$$= - \left[\int_{-1}^{1} K_m(x) (f'(x) - f'(0)) dx + f'(0) \int_{-1}^{1} K_m(x) dx \right]$$

$$\quad + f(1) M_m.$$

Thus,

$$\left| \int_{-1}^{1} f d\mu_m \right| \leq |f(1)| \, |M_m| + |f'(0)| \left| \int_{-1}^{1} K_m(x) dx \right|$$

$$+ \int_{-1}^{1} |K_m(x)| \, |f'(x) - f'(0)| dx$$

$$\leq |f(1)| \, |M_m| + |f'(0)| \left| \int_{-1}^{1} K_m(x) dx \right|$$

$$+ \omega_1(f', h_m) \cdot \int_{-1}^{1} |K_m(x)| \left\lceil \frac{|x|}{h_m} \right\rceil dx.$$

Here $\lceil \cdot \rceil$ is the ceiling of the number, and the last inequality comes from Corollary 7.1.1, p. 209, [1]. Since

$$\left\lceil \frac{|x|}{h_m} \right\rceil \leq 1 + \frac{|x|}{h_m},$$

we get that

$$\int_{-1}^{1} |K_m(x)| \left\lceil \frac{|x|}{h_m} \right\rceil dx \leq \int_{-1}^{1} |K_m(x)| \left(1 + \frac{|x|}{h_m} \right) dx$$

$$= \int_{-1}^{1} |K_m(x)| dx + \frac{1}{h_m} \int_{-1}^{1} |K_m(x)| \, |x| dx.$$

Clearly $K_m \not\equiv 0$, by $K_m(x) \neq 0$ and may be 0 only at some points. Here we choose

$$h_m = \int_{-1}^{1} |K_m(x)| \, |x| dx.$$

It is obvious that $h_m > 0$, and h_m is finite by $\|K_m\|_\infty < \infty$. Putting this together, we have established the validity of (2). Under the special assumptions $M_m \to 0$ and $\int_{-1}^{1} |K_m(x)| dx \to 0$, as $m \to \infty$, we obtain that $\int_{-1}^{1} f d\mu_m \to 0$, as $m \to \infty$. $\square$

Finally we have the following theorem which for $n = 1$ implies Theorem 2.

Theorem 3. *Let $f \in C^n[-1, 1]$, $n \geq 1$, and $\{\mu_m\}_{m \in \mathbf{N}}$ be a sequence of nontrivial Borel signed measures on $[-1, 1]$. Assume that each μ_m is bounded, and set $M_m := \mu_m[-1, 1]$. Define $K_m(x) := \mu_m[-1, x]$, $-1 \leq x \leq 1$. Then*

$$\left| \int_{-1}^{1} f d\mu_m \right| \leq |f(1)| \, |M_m| + \sum_{k=0}^{n-1} \frac{|f^{(k+1)}(0)|}{k!} \left| \int_{-1}^{1} K_m(x) x^k dx \right|$$

$$+ \left(\frac{1}{(n-1)!} \int_{-1}^{1} |K_m(x)| \, |x|^{n-1} dx + \frac{1}{n!} \right) \omega_1 \left(f^{(n)}, \int_{-1}^{1} |K_m(x)| \, |x|^n dx \right).$$

$$(3)$$

If $M_m \to 0$, $\int_{-1}^{1} |K_m(x)| dx \to 0$ with $m \to \infty$, then $\{\mu_m\}_{m \in \mathbb{N}}$ converges weakly to zero.

Proof: Let $f \in C^n[-1, 1]$, $n \geq 1$. We have that

$$f'(x) = \sum_{k=0}^{n-1} \frac{f^{(k+1)}(0)}{k!} x^k + \int_0^x (f^{(n)}(t) - f^{(n)}(0)) \frac{(x-t)^{n-2}}{(n-2)!} dt, \qquad (4)$$

for any $-1 \leq x \leq 1$. We also have again

$$\int_{-1}^{1} f d\mu_m = -\int_{-1}^{1} K_m(x) f'(x) dx + f(1) M_m. \qquad (5)$$

So combining (4) and (5), we get that

$$\int_{-1}^{1} f d\mu_m = -\sum_{k=0}^{n-1} \frac{f^{(k+1)}(0)}{k!} \int_{-1}^{1} K_m(x) x^k dx$$

$$- \int_{-1}^{1} K_m(x) \left(\int_0^x (f^{(n)}(t) - f^{(n)}(0)) \frac{(x-t)^{n-2}}{(n-2)!} dt \right) dx$$

$$+ f(1) M_m.$$

In particular, we see that

$$\left| \int_0^x (f^{(n)}(t) - f^{(n)}(0)) \frac{(x-t)^{n-2}}{(n-2)!} dt \right|$$

$$\leq \omega_1(f^{(n)}, h_m) \left| \int_0^x \left\lceil \frac{|t|}{h_m} \right\rceil \frac{|x-t|^{n-2}}{(n-2)!} dt \right| = \omega_1(f^{(n)}, h_m) \phi_{n-1}(|x|).$$

Here

$$\phi_{n-1}(|x|) := \int_0^{|x|} \left\lceil \frac{t}{h_m} \right\rceil \frac{(|x|-t)^{n-2}}{(n-2)!} dt, \quad x \in \mathbb{R}$$

(see (7.1.13) in Remark 7.1.3, p. 210, [1]). Therefore,

$$\left| \int_{-1}^{1} f d\mu_m \right| = |f(1)| |M_m| + \sum_{k=0}^{n-1} \frac{|f^{(k+1)}(0)|}{k!} \left| \int_{-1}^{1} K_m(x) x^k dx \right|$$

$$+ \omega_1(f^{(n)}, h_m) \cdot \int_{-1}^{1} |K_m(x)| \phi_{n-1}(|x|) dx. \qquad (6)$$

By (7.2.9), p. 217, [1], we have

$$\phi_{n-1}(|x|) \leq \frac{|x|^{n-1}}{(n-1)!} \left(1 + \frac{|x|}{n h_m} \right) = \frac{|x|^{n-1}}{(n-1)!} + \frac{|x|^n}{n! h_m}, \quad x \in \mathbb{R}.$$

Hence

$$|K_m(x)||\phi_{n-1}(|x|) \le |K_m(x)|\frac{|x|^{n-1}}{(n-1)!} + \frac{|K_m(x)|\,|x|^n}{n!h_m}.$$

Consequently, we get

$$\int_{-1}^{1} |K_m(x)||\phi_{n-1}(|x|)dx \le \frac{1}{(n-1)!} \int_{-1}^{1} |K_m(x)|\,|x|^{n-1}dx$$

$$+ \frac{1}{n!h_m} \int_{-1}^{1} |K_m(x)|\,|x|^n dx$$

$$\left(\text{by picking } h_m = \int_{-1}^{1} |K_m(x)|\,|x|^n dx\right)$$

$$= \frac{1}{(n-1)!} \int_{-1}^{1} |K_m(x)|\,|x|^{n-1}dx + \frac{1}{n!},$$

i.e.,

$$\int_{-1}^{1} |K_m(x)||\phi_{n-1}(|x|)dx \le \frac{1}{(n-1)!} \int_{-1}^{1} |K_m(x)|\,|x|^{n-1}dx + \frac{1}{n!}. \qquad (7)$$

So by combining (6) and (7), we obtain (3). If $\int_{-1}^{1} |K_m(x)|dx \to 0$, as $m \to \infty$, then

$$\int_{-1}^{1} |K_m(x)|\,|x|^N dx \to 0, \quad \text{as } m \to \infty, \qquad \text{for any } N \in \mathbb{N}.$$

Assuming also $M_m \to 0$, as $m \to \infty$ we get $\int_{-1}^{1} fd\mu_m \to 0$. $\square$

Remark 1. Let $f \in C[-1,1]$ and suppose L_m, T are bounded linear operators from $C[-1,1]$ into itself such that $L_m(f) \to T(f)$ uniformly as $m \to \infty$, i.e., $K_m(f) = (L_m - T)(f) \to 0$, uniformly, as $m \to \infty$. By the Riesz representation theorem, we have that

$$(K_m(f))(x_0) = \int_{-1}^{1} f(t)\mu_{mx_0}(dt), \quad x_0 \in [-1,1], \quad \forall f \in C[-1,1],$$

where μ_{mx_0} is a unique finite Baire signed measure, see [4], p. 310, Theorem 8. Here $\mu_{mx_0}[-1,1] =: M_{mx_0} \in \mathbb{R}$. So the pointwise convergence $(K_m(f))(x_0) \to 0$ as $m \to \infty$ is equivalent to the weak convergence of μ_{mx_0} to zero. The last implies $M_{mx_0} \to 0$ as $m \to \infty$. Clearly, Theorems 1, 2, and 3 provide estimates and rates of pointwise convergence to zero for the sequence K_m. Equivalently, this paper presents a quantitative study of the pointwise convergence of operators L_m to T as $m \to +\infty$. This treatment is very general, and as such is the first to our knowledge. Due to lack of space we do not give applications here. Their list is expected to be very long and interesting, especially in Probability.

References

1. Anastassiou, George A., *Moments in Probability and Approximation Theory*, Longman Scientific and Technical, Pitman Research Notes in Mathematics Series, No. 287, England, 1993.

2. Hognas, G., Characterization of weak convergence of signed measures on $[0, 1]$, Math. Scand. **41** (1977), 175–184.

3. Johnson, J., An elementary characterization of weak convergence of measures, The American Math. Monthly **92**, No. 2 (1985), 136–137.

4. Royden, H. L., *Real Analysis*, second edition, Macmillan, New York, 1968.

George A. Anastassiou
Department of Mathematical Sciences
The University of Memphis
Memphis, TN 38152
ganastss@memphis.edu

The Metric Average of 1D Compact Sets

Robert Baier, Nira Dyn and Elza Farkhi

Abstract. We study properties of a binary operation between two compact sets depending on a weight in $[0, 1]$, termed metric average. The metric average is used in spline subdivision schemes for compact sets in $\mathbb{R}^n$, instead of the Minkowski convex combination of sets, to retain non-convexity [3]. Some properties of the metric average of sets in $\mathbb{R}$, like the cancellation property, and the linear behavior of the Lebesgue measure of the metric average with respect to the weight, are proven. We present an algorithm for computing the metric average of two compact sets in $\mathbb{R}$ which are finite unions of intervals, as well as an algorithm for reconstructing one of the metric average's operands, given the second operand, the metric average and the weight.

§1. Introduction

In this paper we study properties of a binary operation, termed metric average, between two compact sets $A, B \subset \mathbb{R}$. For two compact sets A, B in $\mathbb{R}^n$ and a weight $t \in [0, 1]$, the **metric average of A and B with weight t** is given by

$$A \oplus_t B = \big\{ t\{a\} + (1-t)\Pi_B(a) \ : \ a \in A \big\} \cup \big\{ t\Pi_A(b) + (1-t)\{b\} \ : \ b \in B \big\},$$

where $\Pi_A(b)$ is the set of all closest points to b from the set A, and the addition above is the Minkowski addition of sets.

The metric average is introduced in [1] for piecewise linear approximation of set-valued functions. It is used in spline subdivision schemes for compact sets to replace the average between numbers [3]. With this binary average, the limit set-valued function of a spline subdivision scheme, operating on initial data consisting of samples of a univariate Lipschitz continuous set-valued function, approximates the sampled function with error of the order of $O(h)$ for samples h distance apart [3]. Thus, the limit set-valued functions of the spline subdivision schemes retain the

Approximation Theory X: Abstract and Classical Analysis   9
Charles K. Chui, Larry L. Schumaker, and Joachim Stöckler (eds.), pp. 9–22.
Copyright © 2002 by Vanderbilt University Press, Nashville, TN.
ISBN 0-8265-1415-4.

non-convexity nature of the approximated set-valued functions, while if we use the Minkowski average instead of the metric average, any limit is convex, and the spline subdivision schemes fail to approximate set-valued functions with non-convex images [4].

The metric average has many important properties [1,3]. It is a subset of the Minkowski average $tA + (1 - t)B$, generally non-convex, recovering the set A for $t = 1$, and B for $t = 0$. Here we consider the metric average as an operation between compact sets in $\mathbb{R}$. In this setting, the metric average has several more important properties, such as the cancellation property which guarantees that for a given weight t, if the metric average and one of its operands are known, then the second operand is determined uniquely. For Minkowski sums, such a property is valid only for convex sets, and not for non-convex ones. While redundant convexifying parts may appear in the Minkowski average of non-convex sets, there are no redundancies in the metric average of compacts in $\mathbb{R}$. In this sense the metric average of sets in $\mathbb{R}$ is optimal.

We also show that the computation of the metric average is not costly and does not require the computation of distances. By presenting an algorithm for the calculation of the metric average, we prove that the number of operations required is linear in the sum of the numbers of closed intervals in the two operands, independently of the weight parameter.

The metric average for sets in $\mathbb{R}$ is important in the reconstruction of 2D sets from their cross-sections, and more generally, in approximating set-valued functions with images in $\mathbb{R}$.

Here is an outline of the paper: Definitions and notation are presented in Section 2. Properties of the metric average for 1D sets are presented in Section 3, without proofs. An algorithm for calculating the metric average is given in Section 4. The cancellation property is derived from an algorithm for the reconstruction of the set A from the sets B, C and the weight $t \in (0, 1)$ when $C = A \oplus_t B$. This is done in Section 5, where a central theorem for the validity of the cancellation algorithm is stated. The main proofs are postponed to the last section.

§2. Definitions and Notation

Denote by $\mathcal{K}(\mathbb{R}^n)$ the set of all compact, nonempty subsets of $\mathbb{R}^n$, by $\mathcal{C}(\mathbb{R}^n)$ the set of all compact, convex, nonempty subsets of $\mathbb{R}^n$, and by $\mathcal{K}_\mathcal{F}(\mathbb{R})$ the set of all compact, nonempty subsets of $\mathbb{R}$ which are finite unions of nonempty intervals.

The Lebesgue measure of the set A is denoted by $\mu(A)$. The Hausdorff distance between the sets $A, B \in \mathcal{K}(\mathbb{R}^n)$ is $\text{haus}(A, B)$. The Euclidean distance from a point a to a set $B \in \mathcal{K}(\mathbb{R}^n)$ is $\text{dist}(a, B) = \inf_{b \in B} \|a - b\|_2$. The set of all projections of $a \in \mathbb{R}^n$ on the set $B \in \mathcal{K}(\mathbb{R}^n)$ is denoted by

$$\Pi_B(a) := \{b \in B : \|a - b\|_2 = \text{dist}(a, B)\}.$$

The set difference of $A, B \in \mathcal{K}(\mathbb{R}^n)$ is $A \setminus B = \{ a \ : \ a \in A, \ a \notin B \}$. A linear Minkowski combination of two sets A and B is

$$\lambda A + \nu B = \{\lambda a + \nu b \ : \ a \in A, \ b \in B\},$$

for $A, B \in \mathcal{K}(\mathbb{R}^n)$ and $\lambda, \nu \in \mathbb{R}$.

The Minkowski sum $A + B$ corresponds to a linear Minkowski combination with $\lambda = \nu = 1$. The linear Minkowski combination with $\lambda, \nu \in [0, 1]$, $\lambda + \nu = 1$, is termed Minkowski average or Minkowski convex combination. A segment is denoted by $[c, d] = \{\lambda c + (1 - \lambda)d \ : \ 0 \leq \lambda \leq 1\}$, for $c, d \in \mathbb{R}^n$.

Definition 1. *Let $A, B \in \mathcal{K}(\mathbb{R}^n)$ and $0 \leq t \leq 1$. The t-weighted metric average of A and B is*

$$A \oplus_t B = \{t\{a\} + (1-t)\Pi_B(a) \ : \ a \in A\} \cup \{t\Pi_A(b) + (1-t)\{b\} \ : \ b \in B\}, \tag{1}$$

where the linear combinations in (1) are in the Minkowski sense.

The sets $A, B \in \mathcal{K}_{\mathcal{F}}(\mathbb{R})$ are given as

$$A = \bigcup_{i=1}^{M} [a_i^l, a_i^r], \qquad B = \bigcup_{j=1}^{N} [b_j^l, b_j^r]. \tag{2}$$

Each interval is proper, i.e. the left endpoint is not bigger than the right one, and equality is possible which stands for a so-called point (or degenerate) interval. The intervals are ordered in an increasing order, i.e. $a_i^r < a_{i+1}^l$ and $b_j^r < b_{j+1}^l$ for all relevant i, j.

We extend the representation (2) to a common closed interval containing the considered sets by adding point intervals to the left and to the right of the sets:

$$A = \bigcup_{i=0}^{M+1} [a_i^l, a_i^r], \qquad B = \bigcup_{j=0}^{N+1} [b_j^l, b_j^r], \tag{3}$$

where

$$x_{\min} = a_0^l = a_0^r = b_0^l = b_0^r < \min\{a_1^l, b_1^l\} - |a_1^l - b_1^l|, \tag{4}$$

$$x_{\max} = a_{M+1}^l = a_{M+1}^r = b_{N+1}^l = b_{N+1}^r > \max\{a_M^r, b_N^r\} + |a_M^r - b_N^r|. \tag{5}$$

This choice guarantees that $C = A \oplus_t B$ is changed only by the addition of the point intervals $\{x_{\min}\}$ and $\{x_{\max}\}$ for any $t \in [0, 1]$. Denote $X = [x_{\min}, x_{\max}]$.

The "holes" of each set, namely the maximal open intervals in X, which do not intersect the set, play an important role, as well as their centers. Denote the holes by

$$H_i^A := (a_i^r, a_{i+1}^l), \quad i = 0, \ldots, M, \qquad H_j^B := (b_j^r, b_{j+1}^l), \quad j = 0, \ldots, N, \tag{6}$$

and their centers by

$$a_i^* := \frac{a_i^r + a_{i+1}^l}{2}, \quad i = 0, \ldots, M, \qquad b_j^* := \frac{b_j^r + b_{j+1}^l}{2}, \quad j = 0, \ldots, N. \tag{7}$$

The dual representation of A and B is:

$$A = X \setminus \bar{A}, \quad \text{where} \quad \bar{A} = \bigcup_{i=0}^{M} H_i^A, \tag{8}$$

$$B = X \setminus \bar{B}, \quad \text{where} \quad \bar{B} = \bigcup_{j=0}^{N} H_j^B. \tag{9}$$

The t-weighted metric average $A \oplus_t B$ is denoted by C,

$$C = A \oplus_t B = \bigcup_{k=0}^{L+1} [c_k^l, c_k^r], \tag{10}$$

where $c_0^l = c_0^r = x_{\min}$, $c_{L+1}^l = c_{L+1}^r = x_{\max}$, and the center of the k-th hole $H_k^C = (c_k^r, c_{k+1}^l)$ is denoted by c_k^*.

§3. Properties of the Metric Average

Let $A, B, C \in \mathcal{K}(\mathbb{R}^n)$ and $0 \le t \le 1$, $0 \le s \le 1$. Then the following properties of the metric average are known [1,3]:

1) $A \oplus_0 B = B$, $\qquad A \oplus_1 B = A$, $\qquad A \oplus_t B = B \oplus_{1-t} A$.
2) $A \oplus_t A = A$.
3) $A \cap B \subseteq A \oplus_t B \subseteq tA + (1-t)B \subseteq co(A \cup B)$.
4) $\text{haus}(A \oplus_t B, A \oplus_s B) = |t - s| \text{haus}(A, B)$.

The following properties are valid for sets in $\mathbb{R}$.

Proposition 2. Let $A, B \in \mathcal{K}(\mathbb{R})$, $C, D \in \mathcal{C}(\mathbb{R})$, $t \in [0, 1]$. Then

(a) $C \oplus_t D = tC + (1-t)D$,
(b) $\mu(A \oplus_t B) = t\mu(A) + (1-t)\mu(B)$.
(c) $\mu(\overline{A \oplus_t B}) = t\mu(\bar{A}) + (1-t)\mu(\bar{B})$.

The proof of the first assertion follows trivially from the definition. The third assertion is proven in the last section. The second one follows directly from the third. In the following we present two properties of the metric average which are valid for sets in $\mathcal{K}_{\mathcal{F}}(\mathbb{R})$.

Proposition 3. *Let $A, B \in \mathcal{K}_{\mathcal{F}}(\mathbb{R})$, and let $H(A)$ denote the number of holes of A. Then for every $t \in (0,1)$*

(a) $H(A \oplus_t B) \leq H(A) + H(B)$

(b) *The number of operations necessary for the calculation of $A \oplus_t B$ is $\mathcal{O}(H(A) + H(B))$.*

The first assertion is proven in the last section. The second one follows from the algorithm presented in the sequel. The cancellation property of the metric average is

Proposition 4. *Let $A', A'', B \in \mathcal{K}_{\mathcal{F}}(\mathbb{R})$. Then for any $t \in (0,1)$,*

$$A' \oplus_t B = A'' \oplus_t B \Longrightarrow A' = A''. \tag{11}$$

The proof of this claim follows from the considerations in Section 5. It can be extended to sets in $\mathcal{K}(\mathbb{R})$, since all relevant statements are valid for sets consisting of an infinite number of compact segments.

To understand the nature of the metric average of two sets A and B, we distinguish four types of holes in A with respect to B, and vice versa:

Definition 5. *Let H_i^A be a hole of A. According to its position with respect to B, H_i^A is said to be:*

1) paired with a hole of B, *if there is a hole H_j^B of B, such that $a_i^* \in H_j^B$ and $b_j^* \in H_i^A$,*

2) paired with a point in B, *if the center $a_i^* \in B$,*

3) left shadow of a hole of B, *if there is a hole H_j^B of B, such that $a_i^* \in H_j^B$, and $b_j^* \geq a_{i+1}^l$,*

4) right shadow of a hole of B, *if there is a hole H_j^B of B, such that $a_i^* \in H_j^B$, and $b_j^* \leq a_i^r$.*

Clearly, each hole of A belongs to exactly one of the above categories of holes with respect to B, and vice versa. Note that H_0^A is paired with H_0^B by the choice of $x_{\min}$, and similarly, H_M^A is paired with H_N^B by the choice of $x_{\max}$. When A is averaged with B and $t \in (0,1]$, each hole of A creates a "child" hole of C, which inherits the type of its parent with respect to B, as stated below.

Proposition 6. *Let H_i^A and H_j^B be holes of A and B, respectively, $t \in [0,1]$ and $C = A \oplus_t B$.*

1) *If H_i^A and H_j^B are paired, then the interval*

$$H^C = t H_i^A + (1-t) H_j^B$$

is a hole of C, paired with both H_j^B and H_i^A.

2) If H_i^A is paired with a point of B, then for $t > 0$ the interval

$$H^C = tH_i^A + (1 - t)\{a_i^*\}$$

is a hole of C, paired with the point $a_i^* \in B$.

3) If H_i^A is a left shadow of H_j^B, then for $t > 0$ the interval

$$H^C = tH_i^A + (1 - t)\{b_j^r\}$$

is a hole of C, and a left shadow of H_j^B.

4) If H_i^A is a right shadow of H_j^B, then for $t > 0$ the interval

$$H^C = tH_i^A + (1 - t)\{b_{j+1}^l\}$$

is a hole of C, and a right shadow of H_j^B.

The proof of this proposition is postponed to the last section. Interchanging the roles of A and B and replacing t with $1 - t$ in the proposition, we get that for $t \in (0, 1)$ some holes of C are generated by holes of A or by holes of B, by the four ways presented above. The following result, proved in the last section, states that every hole of C has this property.

Proposition 7. Let $H^C = (c', c'')$ be a hole of $C = A \oplus_t B$, $t \in (0, 1)$. Then H^C is obtained either from a hole of A, by one of the four ways presented in Proposition 6, or from a hole of B, in a symmetric way.

In the example below we have plotted the one-dimensional sets A, B and the set $C_t = A \oplus_t B$ in one picture, giving B at the y-coordinate 0, A at $y = 1$, and C_t at $y = t$ for $t = \frac{1}{4}, \frac{1}{2}, \frac{3}{4}$ (see Figure 1).

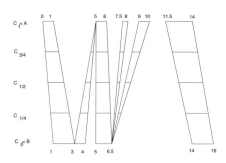

Fig. 1. The sets A, B and C_t of Example 8.

The lines connecting the boundary points of A to points of B and vice versa, show which holes of A are connected with which holes or points of B, according to their type with respect to B, and similarly for the holes of B. These lines give the holes of C_t when crossed with the line $y = t$.

Example 8. *Consider the two sets*

$$A = [0,1] \cup [5,6] \cup [7.5,8] \cup [9,10] \cup [11.5,14],$$
$$B = [1,4] \cup [5,6.5] \cup [14,16].$$

For these two sets a possible X is $[-2,20]$. The metric average $C_t = A \oplus_t B$ is $C_t = X \setminus \bar{C}_t$, where

$$\bar{C}_t = \big(t(-2,0) + (1-t)(-2,1)\big) \cup \big(t(1,5) + (1-t)\{3\}\big)$$
$$\cup \big(t\{5\} + (1-t)(4,5)\big) \cup \big(t(6,7.5) + (1-t)\{6.5\}\big)$$
$$\cup \big(t(8,9) + (1-t)\{6.5\}\big)$$
$$\cup \big(t(10,11.5) + (1-t)(6.5,14)\big) \cup \big(t(14,20) + (1-t)(16,20)\big).$$

The end points of X and of A, B, C_t, $x_{\min} = -2$, $x_{\max} = 20$, are not present in the figure. The holes of the set A are related to the set B as follows:

The hole $(1,5)$ is paired with a point in B, each of the holes $(6,7.5)$, $(8,9)$ is a left shadow of a hole of B, the hole $(10,11.5)$ is paired with a hole of B.

The holes of the set B are related to the set A as follows:

The hole $(4,5)$ is a right shadow of a hole of A and the hole $(6.5,14)$ is paired with a hole of A.

Figure 1 shows how these holes induce holes in $C_t = A \oplus_t B$.

§4. Algorithm for Computing the Metric Average

In this section we propose an algorithm for calculating the metric average $C = A \oplus_t B$, of two given sets $A, B \in \mathcal{K}_{\mathcal{F}}(\mathbb{R})$, and $t \in (0,1)$.

Relying on Propositions 6 and 7, we construct the holes of the set C considering the generating holes of A and B, in an order from left to right, determining the type of each hole.

Algorithm for calculating $C = A \oplus_t B$. Given $t \in (0,1)$, let $A, B \in \mathcal{K}_{\mathcal{F}}(\mathbb{R})$ be of the form (3).

1) $H_0^C := tH_0^A + (1-t)H_0^B$, $\quad i := 1, \quad j := 1, \quad k := 1$.
2) While $i \leq M$ and $j \leq N$,
   (a) If H_i^A is a right shadow of H_{j-1}^B, then
   $$H_k^C := tH_i^A + (1-t)\{b_j^l\}, \quad k := k+1, \ i := i+1.$$

(b) Else, if H_j^B is a right shadow of H_{i-1}^A, then
$$H_k^C := t\{a_i^l\} + (1-t)H_j^B, \quad k := k+1, \ j := j+1.$$

(c) Else, if H_i^A is a left shadow of H_j^B, then
$$H_k^C := tH_i^A + (1-t)\{b_j^r\}, \quad k := k+1, \ i := i+1.$$

(d) Else, if H_j^B is a left shadow of H_i^A, then
$$H_k^C := t\{a_i^r\} + (1-t)H_j^B, \quad k := k+1, \ j := j+1.$$

(e) Else, if $a_i^* < b_j^r$, then
$$H_k^C := tH_i^A + (1-t)\{a_i^*\}, \quad k := k+1, \ i := i+1.$$

(f) Else, if $b_j^* < a_i^r$, then
$$H_k^C := t\{b_j^*\} + (1-t)H_j^B, \quad k := k+1, \ j := j+1.$$

(g) Else (H_i^A and H_j^B are paired)
$$H_k^C := tH_i^A + (1-t)H_j^B, \quad k := k+1, \ i := i+1, \ j := j+1.$$
End of the loop.

3) $L := k - 1, \qquad C = X \setminus \left(\bigcup_{k=0}^{L} H_k^C \right).$

Each hole of A, B belongs exactly to one of the cases described in Step 2 of the algorithm. A hole which is a shadow hole, or paired with a point of the other set, is connected to a single point of the other set to generate a hole of C (cases (a)–(f) of Step 2). Note that the condition (e) (resp. (f)) checked after the condition (a) (resp.(b)) yields that the hole H_i^A (resp. H_j^B) is paired with a point of B (A).

Note also that the order of the holes from the left to the right yields that all the right shadow holes of a given hole are considered after it. That is why in the cases (a),(b) of Step 2, we check for right shadows of the previously considered holes H_{i-1}^A, H_{j-1}^B.

§5. Cancellation Property

To prove the cancellation property (11), we present an algorithm which computes the set A, if $t \in (0,1)$, B and $C(= A \oplus_t B)$ are given. The following proposition is the basis for our cancellation algorithm.

Proposition 9. *Given are* $t \in (0,1)$, $C = A \oplus_t B$ *with holes* H_k^C ($0 \le k \le L$), *and* B *with holes* H_j^B ($0 \le j \le N$).

1) *Let* H_k^C *and* H_j^B *be paired and define* $a' = a_k' = \frac{1}{t}c_k^r + (1 - \frac{1}{t})b_j^r$, $a'' = a_k'' = \frac{1}{t}c_{k+1}^l + (1 - \frac{1}{t})b_{j+1}^l$. *If* $a' < a''$, *then* $(a', a'') \subset X \setminus A$.

2) *Let* H_k^C *be paired with a point of* B, *and define* $a' = a_k' = \frac{1}{t}c_k^r + (1 - \frac{1}{t})c_k^*$, $a'' = a_k'' = \frac{1}{t}c_{k+1}^l + (1 - \frac{1}{t})c_k^*$. *Then* $(a', a'') \subset X \setminus A$.

3) *Let* H_k^C *be a left shadow of the hole* H_j^B, *and define* $a' = a_k' = \frac{1}{t}c_k^r + (1 - \frac{1}{t})b_j^r$, $a'' = a_k'' = \frac{1}{t}c_{k+1}^l + (1 - \frac{1}{t})b_j^r$. *Then* $(a', a'') \subset X \setminus A$.

4) Let H_k^C be a right shadow hole of H_j^B, and define $a' = a_k' = \frac{1}{t}c_k^r + (1 - \frac{1}{t})b_{j+1}^l$, $a'' = a_k'' = \frac{1}{t}c_{k+1}^l + (1 - \frac{1}{t})b_{j+1}^l$. Then $(a', a'') \subset X \setminus A$.

Definition 10. A hypothetic hole (a', a'') of A is any proper open interval (a', a'') constructed in one of the four ways described in the above proposition.

Let $C = A \oplus_t B$, $t \in (0, 1)$. By Proposition 9, every hypothetic hole of A is a subset of some (real) hole of A. Thus the union of all hypothetic holes is contained in the union of all the holes of A.

On the other hand, by Proposition 6, every hole of A generates a "child" hole of C of the same type with respect to B. The procedure described in Proposition 9 guarantees that every hole of A will be recovered by its "child" hole of C. Thus the set of all holes of A is contained in the set of all hypothetic holes of A constructed from the holes of C. Therefore, the union of all the holes of A is equal to the union of all the hypothetic holes of A.

Theorem 11. Let $J = \{k : 0 \le k \le L, a_k' < a_k''\}$, where a_k', a_k'' are defined in Proposition 9. Then $A = X \setminus \left(\bigcup_{k \in J} (a_k', a_k'') \right)$.

Note that Propositions 6 and 9 and Theorem 11 remain true when B and C are infinite unions of compact segments, since their proofs do not use essentially the finite number of segments. Thus the cancellation property is true for sets in $\mathcal{K}(\mathbb{R})$.

Given two sets $B, C \in \mathcal{K}_\mathcal{F}(\mathbb{R})$, and a weight $t \in (0, 1)$, we propose the following algorithm for reconstructing $A \in \mathcal{K}_\mathcal{F}(\mathbb{R})$, if $C = A \oplus_t B$.

Cancellation Algorithm

1) $J := \emptyset$, $k := 0$.
2) While $k \le L$,
   (a) Compute a_k', a_k'' according to Proposition 9.
   (b) If $a_k' \ge a_k''$, then $J := J \cup \{k\}$.
3) $A = [b_0^l, b_{N+1}^r] \setminus \left(\bigcup_{\substack{k=0 \\ k \notin J}}^{L} (a_k', a_k'') \right)$.

§6. Proofs

First we prove Propositions 6 and 7, which are then used in the proof of Propositions 2(c) and 3(a).

Proof of Proposition 6:

Statement 1: Let H_i^A be paired with H_j^B, and let $c' = ta_i^r + (1 - t)b_j^r$, $c'' = ta_{i+1}^l + (1 - t)b_{j+1}^l$. To prove that $H^C = (c', c'') = tH_i^A + (1 - t)H_j^B$ is a hole of $C = A \oplus_t B$, we first prove that $H^C \cap C = \emptyset$. Suppose that there is $c \in H^C \cap C$. Then $c = ta' + (1-t)b'$, where either $a' \in A, b' \in \Pi_B(a')$, or $a' \in \Pi_A(b'), b' \in B$. Suppose that $b' \in \Pi_B(a')$, where $a' \in A$, and $a' \le a_i^r$. Then since $a' < b_j^*$, it follows that $b' \le b_j^r$, and $c = ta' + (1-t)b' \le ta_i^r + (1-t)b_j^r = c'$, i.e. $c \notin H^C$, a contradiction. Similarly one gets contradictions if $a' \ge a_{i+1}^l, b' \in \Pi_B(a')$, or if $a' \in \Pi_A(b')$, where $b' \in B$ satisfies $b \le b_j^r$ or $b \ge b_{j+1}^l$.

Thus we have proven that $H^C \subset X \setminus C$. To verify that H^C is a hole of C, we have to prove that its end points are elements of C. This follows trivially from the definition of H^C and the fact that, for the left end points, either $b_j^r \in \Pi_B(a_i^r)$, or $a_i^r \in \Pi_A(b_j^r)$, and similarly, for the right end points, either $b_{j+1}^l \in \Pi_B(a_{i+1}^l)$, or $a_{i+1}^l \in \Pi_A(b_{j+1}^l)$. The proof that H^C is paired with A and B is trivial and follows from the relation $\frac{c'+c''}{2} = ta_i^* + (1-t)b_j^* \in H_i^A \cap H_j^B$.

Statement 2: Let H_i^A be paired with a point of B, i.e. $a_i^* \in B$. Then $a_i^r, a_{i+1}^l \in \Pi_A(a_i^*)$, hence $c' = ta_i^r + (1-t)a_i^* \in C$, $c'' = ta_{i+1}^l + (1-t)a_i^* \in C$. To prove that $H^C = (c', c'')$ is a hole of C, we have to prove that $H^C \cap C = \emptyset$. If there is $c \in H^C \cap C$, then $c = ta' + (1-t)b'$, where either $a' \in A, b' \in \Pi_B(a')$, or $a' \in \Pi_A(b'), b' \in B$. Suppose first that $a' \in A, b' \in \Pi_B(a')$ and $a' \le a_i^r$. Then since $a_i^* \in B$, it follows that $b' \le a_i^*$. Thus $c = ta' + (1-t)b' \le ta_i^r + (1-t)a_i^* = c'$, i.e. $c \notin H^C$, a contradiction. Similarly one proves the other three cases. Thus (c', c'') is a hole of C with $\frac{c'+c''}{2} = a_i^* \in B$, and hence H^C is paired with a point of B.

Statement 3: Let H_i^A be a left shadow of H_j^B, and let $c' = ta_i^r + (1-t)b_j^r$, $c'' = ta_{i+1}^l + (1-t)b_j^r$, and $H^C = (c', c'')$. Clearly, $b_j^r \in \Pi_B(a_{i+1}^l)$, and either $b_j^r \in \Pi_B(a_i^r)$, or $a_i^r \in \Pi_A(b_j^r)$. Thus $c', c'' \in C$. If $H^C \cap C = \emptyset$, it is trivial to show that H^C is a left shadow of H_j^B.

It remains to show that $H^C \cap C = \emptyset$. Suppose that $c \in H^C \cap C$, i.e. $c = ta' + (1-t)b'$, where either $a' \in A, b' \in \Pi_B(a')$, or $a' \in \Pi_A(b'), b' \in B$. Suppose first that $a' \in A, b' \in \Pi_B(a')$ and $a' \le a_i^r$. Then since $a_i^r < b_j^*$, it follows that $b' \le b_j^r$. Hence $c = ta' + (1-t)b' \le ta_i^r + (1-t)b_j^r = c'$, i.e. $c \notin H^C$, a contradiction. The other three cases are proven similarly.

Statement 4: The case that H_i^A is a right shadow of H_j^B is symmetric to the previous case and we omit the proof. $\square$

Proof of Proposition 7:

Let $c' = ta' + (1-t)b'$, where either $b' \in \Pi_B(a')$ for some $a' \in A$, or $a' \in \Pi_A(b')$ for some $b' \in B$, and let $c'' = ta'' + (1-t)b''$, where either $b'' \in \Pi_B(a'')$ for some $a'' \in A$, or $a'' \in \Pi_A(b'')$ for some $b'' \in B$.

First we prove that both inequalities $a' \leq a''$, $b' \leq b''$ hold and at least one of them is strict. Clearly, if $a' \geq a''$ and $b' \geq b''$, then $c' \geq c''$, which is impossible. Next we show that $a' > a''$, $b' < b''$ is impossible. We use the inequality

$$\max\{|b' - a'|, |b'' - a''|\} > \max\{|b' - a''|, |b'' - a'|\}, \qquad (12)$$

which is proven at the end of the present proof.

Suppose, e.g. that $|b' - a'| = \max\{|b' - a'|, |b'' - a''|\}$. It follows from (12) that $a' \notin \Pi_A(b')$ and $b' \notin \Pi_B(a')$, a contradiction. Similarly we get a contradiction if $|b' - a'| \leq |b'' - a''|$.

Thus $a' \leq a''$, $b' \leq b''$ and at least one of these inequalities is strict. To prove that $(a', a'') \subset X \setminus A$, suppose that there exists $a \in (a', a'') \cap A$. Then a belongs to one of the following ranges (some of them might be empty):

1) If $b' \leq a \leq b''$, there is $b \in \Pi_B(a) \cap [b', b'']$, hence $ta + (1-t)b \in (c', c'')$, a contradiction.

2) If $a < b'$, then there exists $a(b') \in \Pi_A(b') \cap (a', a'') \cap (a', b')$, such that $ta(b') + (1 - t)b' \in (c', c'')$, a contradiction.

3) The case $b'' < a$ is symmetric to the previous one.

Thus $(a', a'') \subset X \setminus A$. Similarly one proves that $(b', b'') \subset X \setminus B$.

Next, we prove that the intervals (a', a''), (b', b'') satisfy the conditions of one of the four cases of Proposition 6.

Assume that (a', a''), (b', b'') are non-degenerate, i.e. $a' < a''$ and $b' < b''$. We will prove that they are paired. If $a^* = \frac{a' + a''}{2} \notin (b', b'')$, for instance $a^* \leq b'$, then either $a^* \leq b' \leq a''$, implying that $a'' \in \Pi_A(b')$ and $c' < ta'' + (1 - t)b' < c''$, a contradiction, or $a'' < b'$, implying that $a' \notin \Pi_A(b')$, and therefore $b' \in \Pi_B(a')$, from which it is concluded that in the interval $(a' - (b' - a'), a' + (b' - a'))$ there are no points of B. Thus $b' \in \Pi_B(a'')$ and $c' < ta'' + (1 - t)b' < c''$, a contradiction. The case $a^* \geq b''$ is symmetric. Similarly one proves that $b^* = \frac{b' + b''}{2} \in (a', a'')$. Therefore (a', a''), (b', b'') are paired.

Let one of (a', a''), (b', b'') be degenerate, for instance, $b' = b''$. If $b'' < a^*$, then $a'' \notin \Pi_A(b'')$ and $b'' \in \Pi_B(a'')$. Hence in the interval $(a'' - |b'' - a''|, a'' + |b'' - a''|)$ there are no points of B. Let $b''' = \min\{b \in B, \ b \geq a''\}$, then (a', a'') is a left shadow of (b'', b'''). Similarly, if $b'' > a^*$, we get that (a', a'') is a right shadow of a hole of B.

If $b'' = a^*$, then obviously (a', a'') is paired with $a^* = b'' \in B$. In a similar way, if $a' = a''$, then (b', b'') is a shadow of a hole in A, or paired with $b^* = a'' \in A$. $\square$

Proof of (12):

The inequality (12) follows easily from the Pythagorean theorem and the fact that in the trapezoid with vertices $(a'', 0)$, $(a', 0)$, $(b'', 1)$, $(b', 1)$, the large diagonal is longer than the sides. To prove this geometric fact, it is sufficient to prove that if one of the sides BC, AD of the trapezoid $ABCD$ ($AB \| CD$), is not less than one of the diagonals of $ABCD$, then it is less than the other diagonal. Suppose, for instance, that $BC \geq BD$. We prove that $BC < AC$. In the triangle BCD, the inequality of the sides yields inequality of the angles, $\angle BCD \leq \angle BDC$. Continuing to compare the angles, since $AB \| CD$, it follows that $\angle BAC = \angle ACD < \angle BCD$. On the other hand, $\angle BDC = \angle ABD < \angle ABC$. Thus we get $\angle BAC < \angle ABC$, hence, in the triangle ABC, $BC < AC$, which completes the geometric proof of (12). $\square$

Proof of Proposition 2(c) and Proposition 3(a):

As was proven in Proposition 7, every hole in C is generated either by a hole of A connected to a single point of B (if the hole of A is a shadow of some hole of B or is paired with a point of B), or, symmetrically, by a hole of B connected to a single point of A, or by two paired holes of A and B. By Proposition 6, different holes of A (or of B) produce different holes in C, and the only case when two holes, one of A and one of B, produce one hole of C is the case of paired holes. This yields the claim of Proposition 3(a).

Denote by I_A (respectively I_B) the set of indices of holes in A (resp. B) which are connected to a single point in B (resp. A). Since for every $i \notin I_A$ there exists a unique $j(i) \notin I_B$ such that H_i^A is paired with $H_{j(i)}^B$, Proposition 6 implies

$$\mu(\bar{C}) = \sum_{i \in I_A} t\mu(H_i^A) + \sum_{j \in I_B} (1-t)\mu(H_j^B)$$

$$+ \sum_{i \notin I_A} \left(t\mu(H_i^A) + (1-t)\mu(H_{j(i)}^B) \right) = t\mu(\bar{A}) + (1-t)\mu(\bar{B}) . \quad \square$$

Proof of Proposition 9:

Statement 1: Let H_k^C be paired with H_j^B. Since $C = A \oplus_t B$, by Propositions 6, 7, the only possibility for H_k^C is that $H_k^C = tH_i^A + (1-t)H_j^B$, where H_i^A is a hole of A, paired with H_j^B. Then clearly $H_i^A = (a', a'')$.

Statement 2: Let H_k^C be paired with a point of B, i.e. $c_k^* \in B$. Then $a' = \frac{1}{t}c_k^r + (1-\frac{1}{t})c_k^*$, $a'' = \frac{1}{t}c_{k+1}^l + (1-\frac{1}{t})c_k^*$. Suppose that $a \in A \cap (a', a'')$. Then, since a is closer to c_k^* than a' and a'', there is a point $a_0 \in (a', a'') \cap \Pi_A(c_k^*)$. Hence $ta_0 + (1-t)c_k^* \in C \cap H_k^C$, a contradiction.

Statement 3: Let H_k^C be a left shadow of H_j^B, $a' = \frac{1}{t}c_k^r + (1 - \frac{1}{t})b_j^r$, $a'' = \frac{1}{t}c_{k+1}^l + (1 - \frac{1}{t})b_j^r$. Define also the point $a_s' \geq b_j^r$ such that $|a_s' - b_j^r| = |a' - b_j^r|$ (possibly $a_s' = a'$). We perform the proof in several steps.

Step 1. First we prove that $(a', a_s') \cap A = \emptyset$. Assume $(a', a_s') \neq \emptyset$, i.e. $a' < b_j^r < a_s'$, and suppose that there is $a \in A \cap (a', a_s')$. Then for b_j^r there is $a(b_j^r) \in \Pi_A(b_j^r) \cap (a', a_s')$. Thus $ta(b_j^r) + (1 - t)b_j^r \in (c_k^r, c_{k+1}^l) \cap C$, which is a contradiction.

Step 2. We prove now that $(a', a'') \cap [a_s', b_j^*] \cap A = \emptyset$. Suppose that there is $a \in A \cap (a', a'') \cap [a_s', b_j^*]$. Then since $a \leq b_j^*$, it follows that $b_j^r \in \Pi_B(a)$. Hence $ta + (1 - t)b_j^r < ta'' + (1 - t)b_j^r = c_{k+1}^l$. On the other hand, $ta + (1 - t)b_j^r > ta' + (1 - t)b_j^r = c_k^r$. Thus $ta + (1 - t)b_j^r \in (c_k^r, c_{k+1}^l) \cap C$, which is a contradiction.

Clearly, if $a'' \leq b_j^*$, the proof is completed. In all next steps we suppose that $b_j^* < a''$. Note that the point $c_{k+1}^l \in C$ is obtained either by

$$c_{k+1}^l = ta(b) + (1 - t)b, \qquad \text{where } b \in B, \ a(b) \in \Pi_A(b), \qquad (13)$$

or by

$$c_{k+1}^l = ta + (1 - t)b(a), \qquad \text{where } a \in A, \ b(a) \in \Pi_B(a''). \qquad (14)$$

In Steps 3 and 4 we suppose that (13) holds and prove that $[b_j^*, a'') \cap A = \emptyset$, which implies $(a', a'') \cap A = \emptyset$. In Step 5 we show that (14) is impossible when $b_j^* < a''$.

Step 3. We prove that $[b_j^*, a'') \cap A = \emptyset$, in case $b \leq b_j^r$ in (13). Indeed, since $\frac{1}{t} > 1$, then $a(b) = \frac{1}{t}c_{k+1}^l + (1 - \frac{1}{t})b \geq \frac{1}{t}c_{k+1}^l + (1 - \frac{1}{t})b_j^r = a''$. This yields that there are no elements of A in the interval $[b, a'') \supset [b_j^*, a'')$.

Step 4. We prove that $[b_j^*, a'') \cap A = \emptyset$, in case $b \geq b_{j+1}^l$ in (13). Assume $b \geq b_{j+1}^l$. Define $b_j^s > c_{k+1}^l$ such that $b_j^s - c_{k+1}^l = c_{k+1}^l - b_j^r$. Such a point exists since H_k^C is a left shadow of H_j^B, i.e. $b_j^r < c_{k+1}^l < b_j^s < b_{j+1}^l$.

Denote $a_s'' = \frac{1}{t}c_{k+1}^l + (1 - \frac{1}{t})b_j^s$. Then by the definition of a'' we get $a'' - c_{k+1}^l = c_{k+1}^l - a_s'' = (\frac{1}{t} - 1)(c_{k+1}^l - b_j^r)$. Since $b \geq b_{j+1}^l > b_j^s$ and $a(b) = \frac{1}{t}c_{k+1}^l + (1 - \frac{1}{t})b$, it follows that $a(b) < a_s''$. Since $a(b)$ is a projection of b, it follows that there are no points of A in the interval $I = [b - (b - a_s''), b + (b - a_s'')] \supset [b_j^s - (b_j^s - a_s''), b_j^s + (b_j^s - a_s'')] = I'$. Since $a_s'' < c_{k+1}^l < b_j^*$ and $b_j^s + (b_j^s - a_s'') = 2c_{k+1}^l - b_j^r + \frac{1}{t}(c_{k+1}^l - b_j^r) = a'' + (b_j^s - b_j^r) > a''$, it is easy to see that $[b_j^*, a'') \subset I' \subset I$. Thus there are no points of A in $[b_j^*, a'')$.

Step 5. Let $c_{k+1}^l = ta + (1 - t)b(a)$, where $a \in A$, $b(a) \in \Pi_B(a)$. Since $a \in A$, it follows by Steps 1,2 that $a \notin (a', b_j^*] \cap (a', a'')$, hence either $a \geq a''$, or $a > b_j^*$, or $a \leq a'$.

If $a > a''$, then since $a'' > c_{k+1}^l > b_j^r$, it follows that $b(a) \geq b_j^r$ and $ta + (1 - t)b(a) > ta'' + (1 - t)b_j^r = c_{k+1}^l$, a contradiction. If $a > b_j^*$, then $b(a) \geq b_{j+1}^l$ and $ta + (1 - t)b(a) > tb_j^* + (1 - t)b_{j+1}^l > b_j^*$, a contradiction. The case $b_j^* < a \leq a'$ is identical to the previous one.

If $a \leq \min\{a', b_j^*\}$, then since $a \leq b_j^*$, $b(a) \leq b_j^r$. Hence $c_{k+1}^l \leq ta' + (1 - t)b_j^r = c_k^r < c_{k+1}^l$, a contradiction. Thus the only possibility for $a \in A$ is $a = a''$, hence $b(a) = b_j^r$. But, since $b_j^* < a''$, we obtain that $b_j^r \notin \Pi_B(a'')$, which is a contradiction. This completes the proof of 3.

Statement 4: The case that H_k^C is a right shadow of H_j^B is symmetric to the previous case and is proven similarly. $\square$

Acknowledgments. The second and third authors were partially supported by the Israel Science Foundation – Center of Excellence Program and by the Hermann Minkowski Center for Geometry at Tel Aviv University.

References

1. Artstein, Z., Piecewise linear approximations of set-valued maps, Journal of Approx. Theory **56** (1989), 41–47.

2. Dyn, N., Subdivision schemes in Computer-Aided Geometric Design, in *Advances in Numerical Analysis*, W. Light (ed.), Vol. II, Wavelets, Subdivision Algorithms and Radial Basis Functions, Clarendon Press, Oxford, 1992, 36–104.

3. Dyn, N. and E. Farkhi, Spline subdivision schemes for compact sets with metric averages, in *Trends in Approximation Theory*, K. Kopotun, T. Lyche and M. Neamtu (eds.), Vanderbilt Univ. Press, Nashville, TN, USA, 2001, 95–104.

4. Dyn, N. and E. Farkhi, Convexification rate in Minkowski averaging processes, preprint.

Robert Baier
Chair of Applied Mathematics
University of Bayreuth
D-95440 Bayreuth, Germany
Robert.Baier@uni-bayreuth.de

Nira Dyn and Elza Farkhi
School of Mathematical Sciences
Sackler Faculty of Exact Sciences
Tel Aviv University
Tel Aviv 69978, Israel
niradyn@post.tau.ac.il,   elza@post.tau.ac.il

Uniform Strong Unicity of Order 2
for Generalized Haar Sets

M. W. Bartelt and J. J. Swetits

Abstract. Zukhovitskii and Stechken introduced the concept of a generalized Haar space and showed that in vector valued uniform approximation, G is a generalized Haar subspace if and only if the best approximation to every function is unique. It is known that for real valued uniform approximation on a finite set from a Haar set there is a uniform strong uniqueness constant of order 1. It is shown that for vector valued approximation on a finite set from a generalized Haar set there is a uniform strong uniqueness constant of order 2.

Let $I = \{1, ..., m\}$ be a finite set with the discrete topology and $C(I, \mathbb{R}^r)$ be the space of vector-valued functions from the index set I to the r-dimensional Euclidean space $\mathbb{R}^r$. For f in $C(I, \mathbb{R}^r)$, let

$$\|f\| := \max_{i \in I} \|f(i)\|_2 = \max_{i \in I} \left(\sum_{j=1}^{r} (f_j(i))^2 \right)^{\frac{1}{2}},$$

where $f(i) = (f_1(i), ..., f_r(i))$. Let G be an n-dimensional subspace of $C(I, \mathbb{R}^r)$ with $n \geq 1$. Zukhovitskii and Stechken [21] proved the following generalization of the classical Haar theorem.

Theorem 1. (Zukhovitskii and Stechken). *An n-dimensional subspace G of $C(I, \mathbb{R}^r)$ is a generalized Haar subspace if and only if $P_G(f)$ is a singleton for every f in $C(I, \mathbb{R}^r)$.*

A generalized Haar space is defined as follows.

Approximation Theory X: Abstract and Classical Analysis
Charles K. Chui, Larry L. Schumaker, and Joachim Stöckler (eds.), pp. 23–30.
Copyright ℗ 2002 by Vanderbilt University Press, Nashville, TN.
ISBN 0-8265-1415-4.

Definition 1. *An n-dimensional subspace G of $C\left(I, \mathbb{R}^r\right)$ is called a gen-eralized Haar subspace if the following two conditions hold:*

1) *every nonzero g in G has at most s zeros;*

2) *for any s distinct points $i \in I$ and any s vectors $\{x_1, ..., x_s\}$ in $\mathbb{R}^r$, there is a vector-valued function p in G such that $p(i) = x_i$ for $1 \le i \le s$, where s is the maximum integer less than $\frac{n}{r}$, i.e. $sr < n \le (s+1)r$.*

It was shown in [5] that for a finite dimensional subspace G of $C\left(I, \mathbb{R}^r\right)$, G is a generalized Haar subspace if and only if $P_G(f)$ is strongly unique of order 2 at each f in $C\left(I, \mathbb{R}^r\right)$.

Uniform strong unicity has been the subject of many papers. It is well known [2,9] that when approximating from a Haar subspace in $C\left(I, \mathbb{R}^1\right)$ there is a uniform strong unicity constant. We know that [3] when approximating from any n-dimensional subspace G in $C\left(I, \mathbb{R}^r\right)$ for $r \ge 2$ there will be a uniform Hausdorff strong unicity constant of order 1 if and only if G is interpolation on its support. However, this note shows that if G in $C\left(I, \mathbb{R}^r\right)$ for $r \ge 2$ is a generalized Haar set, then there is in fact a uniform strong unicity constant of order 2.

One important reason for studying approximation in $C\left(I, \mathbb{R}^r\right)$ for $r \ge 2$ is that if G is a Haar set of dimension n in $C\left(I, \mathbb{C}\right)$, where $\mathbb{C}$ denotes the complex numbers, then the associated set of functions $\hat{G}$ in $C\left(I, \mathbb{R}^2\right)$ is a generalized Haar set, where $f_1 + if_2$ in $C\left(I, \mathbb{C}\right)$ is identi-fied with (f_1, f_2) in $C\left(I, \mathbb{R}^2\right)$. The design of digital filters is an important practical problem (cf. [1,8,14,18,19,20]) which involves the complex Cheby-shev approximation problem on a set which is sometimes finite. Typically algorithms are presented in papers on digital filter design, but often a proof of convergence is not given. Cromme [10] showed that strong unic-ity is useful in proving the convergence of algorithms, particularly the Remez algorithm, variations of which are sometimes used in digital filter design. Thus the results herein on strong unicity may lead to convergence proofs for some algorithms in digital filter design. The specific complex approximation problem often encountered in digital filter design is that of approximation from the set $\left\{1, z^{-1}, ..., z^{-n+1}\right\}$ on the unit circle of the complex plane. The following proposition describes part of the relation-ship between complex Haar sets, generalized Haar sets and the specific set of complex functions just mentioned. In the proposition the set I need not be finite.

Proposition 1. *Let I be a locally compact Hausdorff space which is not necessarily finite.*

(i) *If the subspace $G \subseteq C\left(I, \mathbb{C}\right)$ has dimension n and basis $\{f_k + ig_k\}_{k=1}^n$, then the associated subspace $\hat{G} \subseteq C\left(I, \mathbb{R}^2\right)$ has di-mension $2n$ and basis $\{(f_k, g_k)\}_{k=1}^n \cup \{(-g_k, f_k)\}_{k=1}^n$.*

(ii) If the subspace $G \subseteq C(I, \mathbb{C})$ is a Haar set of dimension n, then the associated subspace $\acute{G} \subseteq C(I, \mathbb{R}^2)$ is a generalized Haar set of dimension $2n$.

(iii) There are generalized Haar sets in $C(I, \mathbb{R}^2)$, in general, which do not arise from Haar sets in $C(I, \mathbb{C})$.

(iv) The set of functions $\{1, z^{-1}, ..., z^{-n+1}\}$ is a complex Haar set of dimension n on any subset of the complex plane not containing 0.

Proof: The proofs of (i) and (ii) are straightforward and hence omitted. For (iii), let $\acute{G} = (x, x^2) \in C([1, 2], \mathbb{R}^2)$. Then it is easy to verify that $\acute{G}$ is a generalized Haar set of dimension 1 but it cannot arise from a Haar set G in $C([1, 2], \mathbb{C})$ since then by (i) its dimension would be even. By (ii) then, in fact, no generalized Haar set of odd dimension arises in this manner from a complex Haar set.

Part (iv) follows because if $P(z) = a_0 + a_1 z^{-1} + \cdots + a_{n-1} z^{-n+1}$, with not all a_i equal zero, then $z^{n-1} P(z)$ is a polynomial of degree less than or equal to $n - 1$ which thus has at most $n - 1$ distinct zeros. Thus $P(z)$ itself has at most $n - 1$ distinct zeros since 0 is not in the domain of $P(z)$. $\square$

Before we state the theorem, we need to define annihilator and state a relevant result of F. Deutsch [11].

Definition 2. We say that μ is an annihilator of G on J if $J = supp(\mu) \subseteq supp(G)$, $\mu \neq 0$, $\mu \in C(I, \mathbb{R}^r)$, and

$$\sum_{i \in J} \langle \mu(i), g(i) \rangle = 0, \qquad g \in G.$$

Proposition 2. [11] Let G be an n-dimensional subspace of the space $C_0(T, \mathbb{R}^r)$. Let $f \in C_0(T, \mathbb{R}^r) \backslash G$ and g^* in G. Then g^* is a best approximation to f (i.e., $g^* \in P_G(f)$) if and only if there exists an annihilator σ of G such that $supp(\sigma) \subset E(f - g^*)$ with $card(supp(\sigma)) \leq n + 1$ and sgn $\sigma(t) = sgn(f(t) - g^*(t))$ for $t \in supp(\sigma)$. We say that the annihilator in Proposition 2 corresponds to f and $P_G(f)$.

It is known [6] that if G is a generalized Haar subspace of $C(I, \mathbb{R}^r)$ and μ is an annihilator of G, then

$$\|g\|_\mu := \left(\sum_{i \in supp(\mu)} \|g(i)\|_2^2 \right)^{\frac{1}{2}}$$

is a norm on G.

Theorem 2. *Let G be a generalized Haar set in $C(I, \mathbb{R}^r)$. Then there exists a uniform Hausdorff strong unicity constant of order 2 for all f in $C(I, \mathbb{R}^r)$, i.e., there exists a positive constant γ such that*

$$\|f - g\|^2 \geq dist(f, G)^2 + \gamma \cdot \|g - P_G(f)\|^2, \qquad g \in G.$$

Proof: We can assume without loss that $\|f\| = 1$ and $P_G(f) = 0$. Observe that if μ is an annihilator of G, then if

$$\|g\|_{\text{supp}(\mu)} = \max \{\|g(i)\|_2 : i \in \text{supp}(\mu)\} = 0 \tag{1}$$

then $g = 0$ because $\|g\|_\mu$ is a norm on G. Hence $\|g\|_{\text{supp}(\mu)}$ is also a norm on G.

First we claim that there exists a positive constant C such that

$$\|g\|_\mu \geq C \|g\|, \qquad g \in G, \tag{2}$$

for all annihilators μ corresponding to functions in $C(I, \mathbb{R}^r)$, where as before we assume $\|f\| = 1$ and $P_G(f) = 0$. If there were no such positive constant C then there would exist a sequence of annihilators $\{\mu_k\}$ and a sequence $\{g_k\}$ of non-zero functions in G such that

$$\|g_k\|_{\mu_k} \leq \frac{1}{k} \|g_k\|. \tag{3}$$

Since I is a finite set, some subset A of I must be repeated infinitely often as the support of μ_k, so without loss of generality, by using a subsequence of $\{\mu_k\}$, assume that $\text{supp}(\mu_k) = A \subseteq I$, $k \geq 1$. Now from (3) we have

$$\left\|\frac{g_k}{\|g_k\|}\right\|_{\mu_k} = \sum_{i \in A} \left\|\frac{g_k(i)}{\|g_k\|}\right\|_2^2 \leq \frac{1}{k}. \tag{4}$$

We can assume that $\left\{\frac{g_k}{\|g_k\|}\right\}$ converges to a function $\bar{g}$ in G with $\|\bar{g}\| = 1$. Then letting $k \to \infty$ in (4), we obtain

$$\sum_{i \in A} \|\bar{g}(i)\|_2^2 = 0, \tag{5}$$

and hence $\|\bar{g}\|_{\mu_k} = 0$ for any k. Since $\|g\|_{\mu_k}$ is a norm on G, it follows that $\bar{g} = 0$ which contradicts the fact that $\|\bar{g}\| = 1$.

Secondly, we claim there exists a positive constant Λ such that

$$\sum_{i \in \text{supp}(\mu)} \lambda_\mu(i) \|g(i)\|_2^2 \geq \Lambda \|g\|_\mu \tag{6}$$

for all annihilators μ of G corresponding to functions f in $C\left(I, \mathbb{R}^{r}\right)$ with $\|f\| = 1$ and $P_G(f) = 0$ and all functions g in G, where $\mu(i) = \lambda_\mu(i)f(i)$ for $i \in \text{supp}(\mu)$. If there were no such Λ, then there would exist a sequence of annihilators $\{\mu_k\}$ and a sequence of non-zero functions $\{g_k\}$ of functions in G such that

$$\sum_{i \in \text{supp}(\mu_n)} \lambda_{\mu_k}(i)\|g_k(i)\|_2^2 \leq \frac{1}{k}\|g_k\|_{\mu_k}, \tag{7}$$

where

$$\mu_k(i) = \lambda_k(i)f_k(i), \quad i \in \text{supp}(\mu_k),$$

$$1 \geq \lambda_k(i) > 0,$$

$$\sum_{i \in \text{supp}(\mu_k)} \lambda_k(i) = 1, \tag{8}$$

$$\sum_{i \in \text{supp}(\mu_k)} \langle \mu_k(i), g(i) \rangle = 0, \quad g \in G.$$

As before we can assume that $\text{supp}(\mu_k) = A \subseteq I$ for all k. Since $g_k \neq 0$, $\|g_k\|_A \neq 0$ for all k. By dividing both sides of (7) by $\|g_k\|_A$ we can assume that $\|g_k\|_A = 1$.

Now since $\|f_k\| = 1$, using a subsequence, we can assume that $\{f_k\} \to \bar{f}$ for some function $\bar{f}$ in $C\left(I, \mathbb{R}^{r}\right)$. From (8) we see that we can assume $\{\lambda_k(i)\} \to \lambda(i)$ with $\sum_{i \in A} \lambda(i) = 1$ for some constants $\lambda(i)$ with $1 \geq \lambda(i) \geq 0$. Thus letting $k \to \infty$ in (8), we obtain

$$\sum_{i \in A} \langle \lambda(i)\overline{f}(i), g(i) \rangle = 0, \quad g \in G,$$

and so

$$\mu(i) = \begin{cases} \lambda(i)\bar{f}(i), & i \in A, \\ 0, & i \notin A \end{cases}$$

is an annihilator of G corresponding to $\bar{f}$ where $\|\overline{f}\| = 1$ and $P_G\left(\bar{f}\right) = 0$. Note that we only claim that $\text{supp}(\mu) \subseteq A$.

Since all norms on a finite dimensional space are equivalent, there exist positive constants c_1 and c_2 such that

$$c_1\|g\|_A \leq \|g\| \leq c_2\|g\|_A, \quad g \in G.$$

Hence $c_1 \leq \|g_k\| \leq c_2$ because $\|g_k\|_A = 1$, and so we can assume that $\{g_k\} \to \bar{g} \neq 0$ in G. Now by (7)

$$\sum_{i \in A} \lambda_k(i)\|g_k(i)\|_2^2 \leq \frac{1}{k} \cdot \sum_{i \in A} \|g_k(i)\|_2^2, \tag{9}$$

and letting $k \to \infty$ in (9) we obtain

$$\sum_{i \in A} \lambda(i) \|\bar{g}(i)\|_2^2 = 0. \tag{10}$$

But then

$$\sum_{i \in A} \lambda(i) \|\bar{g}(i)\|_2^2 \geq \sum_{i \in \text{supp}(\mu)} \lambda(i) \|\bar{g}(i)\|_2^2 = \|\bar{g}\|_\mu \,,$$

and so by (10), $\|\bar{g}\|_\mu = 0$. Since $\|g\|_\mu$ is a norm on G, then $\bar{g} = 0$, but this contradicts $\bar{g} \neq 0$. Hence we have shown that (6) holds.

Now to prove the theorem, let f in $C(I, \mathbb{R}^r)$, and assume as before that $\|f\| = 1$ and $0 = P_G(f)$. Let μ be an annihilator of G corresponding to f so that $\mu(i) = \lambda(i) f(i)$, $i \in \text{supp}(\mu)$, $1 \geq \lambda(i) > 0$, and $\sum_{i \in \text{supp}(\mu)} \lambda(i) = 1$. Then for a function g in G and $i \in I$, we see that

$$\begin{aligned}
\|f - g\|^2 &\geq \|f(i) - g(i)\|_2^2 \\
&= \langle f(i) - g(i), \, f(i) - g(i) \rangle \\
&= \|f(i)\|_2^2 - 2 \langle f(i), g(i) \rangle + \|g(i)\|_2^2 \,.
\end{aligned} \tag{11}$$

Now multiplying both sides of (11) by $\lambda(i)$ and summing over $i \in \text{supp}(\mu)$, we obtain

$$\begin{aligned}
\sum_{i \in \text{supp}(\mu)} \lambda(i) \|f - g\|^2 &\geq \sum_{i \in \text{supp}(\mu)} \lambda(i) \|f(i)\|_2^2 - 2 \sum_{i \in \text{supp}(\mu)} \lambda(i) \langle f(i), g(i) \rangle \\
&\quad + \sum_{i \in \text{supp}(\mu)} \lambda(i) \|g(i)\|_2^2 \,.
\end{aligned} \tag{12}$$

Since $\sum_{i \in \text{supp}(\mu)} \lambda(i) = 1$ and μ is an annihilator corresponding to f, it follows that $\sum_{i \in \text{supp}(\mu)} \lambda(i) \langle f(i), g(i) \rangle = 0$ and (12) implies that

$$\|f - g\|^2 \geq \sum_{i \in \text{supp}(\mu)} \lambda(i) \|f(i)\|_2^2 + \sum_{i \in \text{supp}(\mu)} \lambda(i) \|g(i)\|_2^2 \,. \tag{13}$$

Since $\text{supp}(\mu) \subseteq E(f), \|f(i)\|_2 = \|f\|$ if $i \in \text{supp}(\mu)$. Then from (2), (6) and (13) we obtain

$$\begin{aligned}
\|f - g\|^2 &\geq \|f\|^2 + \Lambda \|g\|_\mu^2 \\
&\geq \|f\|^2 + \Lambda C \|g\|^2 , \, g \in G,
\end{aligned}$$

and the proof is complete. $\square$

References

1. Alkhairy, A. K. Christian and J. Lim, Design of FIR filters by complex Chebyshev approximation, IEEE Trans. on Signal Processing **41** (1993), 559–572.

2. Bartelt, M., On Lipschitz conditions, strong unicity, and a theorem of A. K. Cline, J. Approx. Theory **14** (1975), 245–250.

3. Bartelt, M., Hausdorff strong unicity in vector-valued Chebyshev approximation on finite sets, in *Trends in Approximation Theory*, K. Kopotun, T. Lyche and M. Neamtu, (eds), Vanderbilt University Press, Nashville (2001), 31–38.

4. Bartelt, M. and J. Swetits, Uniform strong unicity constants for subsets of C(X), J. Approx. Theory **55** (1988), 304–317.

5. Bartelt, M. and W. Li, Abadie's constraint qualification, Hoffman's error bounds and Hausdorff strong unicity, J. Approx. Theory **97** (1999), 140–157.

6. Bartelt, M. and W. Li, Haar theory in vector-valued continuous function spaces, in *Approximation Theory VIII, Vol. 1: Approximation and Interpolation*, Charles K. Chui and Larry L. Schumaker (eds.), World Scientific Publishing Co., Inc., Singapore, 1995, 39–46.

7. Bartelt, M. and W. Li, Exact order of Hoffman's error bounds for elliptic quadratic inequalities derived from vector-valued Chebyshev approximation, Math. Programming, Series B, (2) **88** (2000), 223–254.

8. Chit, N. N. and J. S. Mason, Complex Chebyshev approximation for FIR digital filters, IEEE Transactions on Signal Processing **39** (1991), 49–54.

9. Cline, A. K., Lipschitz conditions on uniform approximation operators, J. Approx. Theory **8** (1973), 160–172.

10. Cromme, L., A far-reaching criterion for the convergence analysis of iterative processes, Numer. Math **29** (1978), 179–193.

11. Deutsch, F., Best approximation in the space of continuous vector-valued functions, J. Approx. Theory **53** (1988), 112–116.

12. Gutknecht, M., Non-strong uniqueness in real and complex Chebyshev approximation, J. Approx. Theory **23** (1978), 204–213.

13. Henry, M. S., D. Schmidt and J. J. Swetits, Uniform strong unicity for rational approximation, J. Approx. Theory **33** (1981), 131–146.

14. Ikehara, M., M. Funaishi and H. Kuroda, Design of all-pass networks using Remez algorithm, IEEE International Symposium on Circuits and Systems, XLVIII+3177, Vol. 1 (1991), 364–367.

15. Li, W., Strong uniqueness and Lipschitz continuity of metric projections: A generalization of the classical Haar theory, J. Approx. Theory **56** (1989), 164–184.

16. Parks, S.H., Uniform Hausdorff strong uniqueness, J. Approx. Theory **55** (1989), 78–89.

17. Paur, S. O. and J. A. Roulier, Uniform Lipschitz and strong unicity constants on subintervals, J. Approx. Theory **32** (1981), 247–255.

18. Preuss, K., On the design of FIR filters by complex Chebyshev approximation, IEEE Transactions on Acoustics, Speech and Signal Processing **37** (1989), 702–712.

19. Schulist, M., FIR filter design with additional constraints using complex Chebyshev approximation, Signal Processing, 33, no. 1, (1993), 111–119.

20. Spagl, C., On complex valued functions with strongly unique best Chebyshev approximation, J. Approx. Theory **74**, no. 1 (1993), 16–27.

21. Zukhovitskii, S. I. and S. B. Stechkin, On approximation of abstract functions with values in Banach space, Dokl. Adad. Nauk, SSSR **106** (1956), 773–776.

M. W. Bartelt
Department of Mathematics
Christopher Newport University
Newport News, VA 23606
mbartelt@pcs.cnu.edu

John J. Swetits
Department of Mathematics and Statistics
Old Dominion University
Norfolk, VA 23529
jswetits@odu.edu

Markov-type Inequalities for
Polynomials and Splines

Borislav Bojanov

Abstract. We shall discuss various extremal problems concerning estimation of a derivative of a function from a given class on the basis of bounds for its uniform norm on the domain or on a subset of points. The Vladimir Markov inequality for polynomials and the extension given by Duffin and Schaeffer, as well as the Landau-Kolmogorov inequality, are typical examples of results of this kind. We shall present new methods leading to exact inequalities for the L_p-norm of the derivatives, and to a complete characterization of the extremal function. Finally, we shall formulate open problems.

§1. Introduction

In this survey we discuss certain extremal problems in classes of functions, mainly polynomials and splines, which lead to inequalities between the norms of the derivative and the function itself. The story of this subject is more than 100 years old and goes back to the genesis of Approximation Theory. The beginning was a question, a simple question raised by the famous Russian chemist Dmitrii Ivanovich Mendeleev. In 1887, in his paper "Study of water solutions in terms of their specific gravity", Mendeleev asked:

> What is the maximum of $|a_k|$, a coefficient of the second degree polynomial $p(x) = a_0 + a_1 x + a_2 x^2$, provided the deviation from zero of this polynomial on a given interval $[a, b]$ does not exceed a certain fixed number M?

This is a great question. It gives rise to a class of interesting and difficult extremal problems. Notice that $a_k = p^{(k)}(0)/k!$, and thus the question is actually about the maximum of $|p^{(k)}(0)|$ under the condition

Approximation Theory X: Abstract and Classical Analysis
Charles K. Chui, Larry L. Schumaker, and Joachim Stöckler (eds.), pp. 31–90.
Copyright ℗ 2002 by Vanderbilt University Press, Nashville, TN.
ISBN 0-8265-1415-4.

that the value of $|p(x)|$ is bounded by M on $[a, b]$. Since the interval $[a, b]$ is arbitrary, it could contain or not contain the zero. Instead of fixing the point 0 and considering any interval $[a, b]$, we can fix the interval to be $[-1, 1]$ and consider the maximum of $|p^{(k)}(\xi)|$ at any fixed point ξ. Further, most mathematicians would rather formulate the question directly for polynomials of any degree n. Thus, one naturally arrives at the following general extremal problem:

Problem 1.1. *Suppose that the polynomial*

$$P(x) = a_0 + a_1 x + \cdots + a_n x^n$$

satisfies

$$|P(x)| \leq 1 \quad \text{on} \quad [-1, 1].$$

For any fixed point ξ and integer $0 \leq k \leq n$, find $\max |P^{(k)}(\xi)|$.

The solution of this problem depends on the location of the point ξ. If ξ is on the real line $\mathbb{R}$ and $|\xi| \geq 1$, then for any k the corresponding extremal function to the above problem is the Tchebycheff polynomial $T_n(x)$, defined on $[-1, 1]$ by

$$T_n(x) = \cos n \arccos x.$$

This was noticed by Tchebycheff (for $k = 0$) and A. A. Markov (for $1 \leq k \leq n$). The case $-1 < \xi < 1$ is really a hard problem. It was studied by Vladimir Markov. In his celebrated work [82], he proved the inequality

$$\|P^{(k)}\| \leq T_n^{(k)}(1) \|P\| \quad \text{for } k = 1, \ldots, n, \quad \forall P \in \pi_n,$$

where $\|P\|$ is the uniform norm of P on $[-1, 1]$ and π_n denotes the set of all real algebraic polynomials of degree less than or equal to n.

If ξ is a complex number lying outside the unit disk $D := \{z : |z| < 1\}$, the corresponding problem admits a beautiful solution which was given by S. N. Bernstein and rediscovered later by P. Erdős. The extremal polynomial is again T_n. The case $\xi \in D$ is still open. The material related to these classical results is covered in Section 2 and Section 3. We include there also a new, shorter proof of the Vladimir Markov inequality.

There exists a huge literature on Markov-type inequalities (see, e.g., [36,69,86,98,99,102,125], and the references therein). In this paper, we concentrate only on those results that produce exact inequalities, in which the extremal function is completely characterized. There are only a few such examples. Among them we should first distinguish the result of Duffin and Schaeffer [47]. They proved that

$$\|P^{(k)}\| \leq T_n^{(k)}(1) \quad \text{for} \quad k = 1, \ldots, n, \quad \forall P \in \pi_n,$$

under the condition that $|P(x)|$ is bounded by 1 only at the extremal points $-1 = \eta_0 < \eta_1 < \cdots < \eta_{n-1} < \eta_n = 1$ of the Tchebycheff polynomial $T_n(x)$ in $[-1, 1]$. Recently Nikolov [92] proved that the same result holds for $k = 1$ even if $|P(x)|$ is bounded by $|T_n(x)|$ on any other system of points $\{t_j\}_0^n$ which interlace with the zeros of T_n. These and other results of Duffin and Schaeffer type are presented in Section 3.

Section 4 is devoted to L_p extensions of the Markov inequality. We discuss the progress in proving the conjecture

$$\|P^{(k)}\|_p \leq \|T_n^{(k)}\|_p \|P\|, \quad \forall P \in \pi_n, \quad 1 \leq p < \infty.$$

For example, it is true for $k = 1$. Also, it was proved for each k, under the additional restriction that the polynomial P has all its zeros in $[-1, 1]$. We present the method of proof which is based on new interesting facts about polynomials with only real zeros, in particular, on inheritance theorems that assert that a certain relation between two polynomials P and Q is inherited by their derivatives P' and Q'.

Section 5 contains a review of the most recent results on estimating the derivative $f^{(k)}$ of a function f in terms of the norms of the function and a higher derivative $f^{(n)}$ on a given bounded interval. These are results related to Kolmogorov-type inequalities for a finite interval, one of the oldest open problems in Approximation Theory. Complete solutions are given in the case of twice and three times differentiable functions. Markov-type inequalities for perfect splines are discussed. As is believed, the perfect spline would play an essential role in the proof of a Kolmogorov-type inequality for a finite interval. Finally, in Section 6 we formulate some open problems.

In this paper $\|f\|$ denotes the uniform norm of f on $[-1, 1]$ and

$$\|f\|_p := \left(\int_{-1}^1 |f(x)|^p \, dx \right)^{1/p}, \quad 1 \leq p \leq \infty,$$

with the only exception when f is a trigonometric polynomial — then the standard interval is $[0, 2\pi]$. Whenever the norm is taken over any other interval, we shall always indicate it explicitly.

§2. Andrei Markov Inequality

Inspired by the question of Mendeleev, the Russian mathematician Andrei Andreevich Markov (1856–1922) published his memoir [81], where the maximum of $|a_0|$, $|a_1|$, and $|a_n|$ was found for polynomials $P(x) = a_0 + a_1 x + \cdots + a_n x^n$ of any given degree n such that $|P(x)| \leq 1$ on $[-1, 1]$. In the same paper, Markov considers also other extremal problems for polynomials, and proves the inequality

$$\|P'\| \leq n^2 \|P\| \tag{2.1}$$

for every algebraic polynomial P of degree n, where the equality sign is attained only for the Tchebycheff polynomial T_n (up to any constant factor). The proof of (2.1) can be found in most of the books on approximation theory.

A. Markov's paper marks the start of an exciting body of research in polynomial inequalities, containing wonderful results. It is a curious fact that the corresponding inequality for trigonometric polynomials

$$s_n(\theta) = a_0 + \sum_{j=1}^{n}(a_j \cos j\theta + b_j \sin j\theta)$$

was proved much later, although the proof in the trigonometric case is quite easy. Namely, Bernstein [13] was the first to show that

$$|s_n'(\theta)| \le n \max_{t\in[0,2\pi]} |s_n(t)|, \quad \forall \theta \in [0, 2\pi]. \tag{2.2}$$

Inequality (2.2) follows immediately from a remarkable extremal property of $\cos n\theta$, the discovery of which can be attributed to de La Vallée Poussin (see [121]). The following is true:

Theorem 2.1. *Let α be any number in $(-1, 1)$ and let s_n be any trigonometric polynomial of degree n satisfying $\|s_n\| \le 1$. Let ξ and η be any two points for which*

$$s_n(\xi) = \cos n\eta = \alpha.$$

Then

$$|s_n'(\xi)| \le \left|\left\{\cos n\theta\right\}'_{\theta=\eta}\right| = n|\sin n\eta|.$$

For $|\alpha| < 1$, the equality is attained only for $s_n(\theta) = \cos n\theta$ and its translates.

Proof: Using a translation of $s_n(\theta)$, if necessary, we can assume that $\xi = \eta$ and both $s_n'(\xi)$ and $\sin n\eta$ are of the same sign (or both are zero). If we assume now that

$$|s_n'(\xi)| > n|\sin n\eta|$$

and take into account that the graph of $\cos n\theta$ equioscillates $2n$ times in $[0, 2\pi)$, we conclude that $s_n(\theta) - \cos n\theta$ would have at least $2n+1$ zeros in $[0, 2\pi)$, a contradiction to the known fact that any trigonometric polynomial of degree n has at most $2n$ zeros in $[0, 2\pi)$, counting the multiplicities. The proof is complete. $\square$

In the above, we have actually estimated the first derivative $s'(\theta)$ of any trigonometric polynomial s of degree n in terms of its value $s(\theta)$ at the same point θ. Indeed, assume for simplicity that $\|s\| = 1$. For each $\theta \in [0, 2\pi]$, there is a point $\eta \in [0, 2\pi]$ such that

$$s(\theta) = \cos n\eta.$$

Then, by the above property of $\cos nt$,

$$|s'(\theta)| \leq n|\sin n\eta| = n\sqrt{1 - \cos^2 n\eta} = n\sqrt{1 - s^2(\theta)}.$$

This yields

$$s'^2(\theta) + n^2 s^2(\theta) \leq n^2, \quad \forall \theta,$$

which is the Szegö inequality [113] (see also [42]).

Because of the relation between the trigonometric and algebraic polynomials, the Szegö inequality yields a similar result for algebraic polynomials. Let $P \in \pi_n$ and $\|P\| = 1$. Then $P(\cos \theta)$ is a trigonometric polynomial of degree n, and by Szegö's inequality,

$$P'^2(\cos \theta) \sin^2 \theta + n^2 P^2(\cos \theta) \leq n^2.$$

Returning to the variable $x = \cos \theta$, we can rewrite the above inequality as

$$P'^2(x)(1 - x^2) \leq n^2(1 - P^2(x)), \quad \forall x \in [-1, 1]. \tag{2.3}$$

Now one can easily derive

$$|P'(x)| \leq \frac{n}{\sqrt{1 - x^2}}\|P\|, \quad \forall x \in (-1, 1),$$

which is known as Bernstein's inequality, although, according to Bernstein (see [17], Volume 1, p.18), it was given implicitly in Markov's paper.

Note that (2.3) turns into an identity for $P = \pm T_n$, that is, we have

$$T_n'^2(x)(1 - x^2) = n^2(1 - T_n^2(x)), \quad \forall x.$$

Another consequence of Theorem 2.1 is the following observation concerning algebraic polynomials (see Lemma 1 in [21]).

Lemma 2.2. *Given* $-1 \leq \alpha \leq 1$, *let* P *be an arbitrary polynomial from* π_n, *normalized by* $\max_{-1 \leq x \leq 1} |P(x)| = 1$. *If*

$$P(\eta) = T_n(\xi) = \alpha$$

for certain points $0 \leq \eta \leq \xi \leq 1$, *then*

$$|P'(\eta)| \leq |T_n'(\xi)|.$$

For $-1 < \alpha < 1$ *the equality is attained only for* $P = \pm T_n$.

Proof: Given ξ, η as in the assumptions, we have

$$P'^2(\eta)(1 - \eta^2) \leq n^2(1 - P^2(\eta))$$
$$= n^2(1 - T_n^2(\xi)) = T_n'^{\,2}(\xi)(1 - \xi^2).$$

Since $1 - \eta^2 \geq 1 - \xi^2$, we conclude that $P'^2(\eta) \leq T_n'^{\,2}(\xi)$ which was to be shown. The proof is complete. □

Lemma 2.2 easily implies A.A. Markov's inequality.

Corollary 2.3 (A. A. Markov inequality). *For every real algebraic polynomial $P(x)$ of degree n and every finite interval $[a, b]$, we have*

$$|P'(x)| \leq \frac{2n^2}{b - a} \max_{a \leq x \leq b} |P(x)|, \qquad \forall x \in [a, b].$$

Proof: It suffices to prove the inequality for $[a, b] = [-1, 1]$ and polynomials P with uniform norm on $[-1, 1]$ equal to 1. Let P_* be the polynomial in this class of maximal uniform norm of its first derivative. Clearly $|P_*(1)| = |P_*(-1)| = 1$, since otherwise a compression of P_* would produce a polynomial of bigger norm of the first derivative. Let $\|P_*'\| = |P_*'(\xi)|$. Assume without loss of generality that $0 \leq \xi$ and $P_*(1) = 1$. If there is a point η such that $\xi \leq \eta$ and $P_*(\xi) = T_n(\eta)$, then Lemma 2.2 leads to $P_* = \pm T_n$. If there is no such point, then taking into account the condition $P_*(1) = T_n(1)$ and the alternating property of T_n, we deduce that $P_* - T_n$ has at least $n + 1$ zeros, thus $P_* = T_n$. This completes the proof. □

§3. Vladimir Markov Inequality

Clearly, the inequality (2.1) is equivalently related to the extremal problem of characterizing the polynomial P_* that has a maximal uniform norm of its first derivative on $[-1, 1]$ among all real algebraic polynomials of degree n that are bounded by 1 on $[-1, 1]$. A. A. Markov has proved that the Tchebycheff polynomial is the unique (up to a factor -1) extremal polynomial for this problem.

A real challenge was to characterize the extremal polynomials for higher derivatives. And this was done by Vladimir Andreevich Markov (May 7, 1871– January 18, 1897), a much younger brother of Andrei. In his celebrated paper [82] he proved that for any k, $1 \leq k \leq n$, the Tchebycheff polynomial T_n has a maximal uniform norm of its k-th derivative among all real algebraic polynomials of degree n that are bounded by 1 on $[-1, 1]$, i.e.,

$$\|P^{(k)}\| \leq \|T_n^{(k)}\| \, \|P\|, \qquad \forall P \in \pi_n, \, 1 < k \leq n.$$

Fig. 1. Vladimir Markov.

This is a deep result and the original proof goes on for 110 pages, exploiting in full the techniques of classical analysis. A more detailed examination could notice in this work even elements of a new mathematics that was developed much later and flourished in the Lvov mathematical school as "Functional Analysis". Several bright ideas have been combined to finally get this lovely result. Among them, the following observation concerning a property of algebraic polynomials is indeed a small gem. It is well-known in the literature as the Markov interlacing property.

Lemma 3.1. *Assume that the zeros of the polynomials*

$$P(x) := (x - x_1) \cdots (x - x_n) \quad \text{and} \quad Q(x) := (x - y_1) \cdots (x - y_n)$$

interlace, that is,

$$x_1 \leq y_1 \leq x_2 \leq \cdots \leq x_n \leq y_n \leq \infty.$$

Then the zeros of P' and Q' also interlace. The case $y_n = \infty$ corresponds to $Q(x) := (x - y_1) \cdots (x - y_{n-1})$.

Moreover, if $x_i < y_i$ for at least one i, then the zeros of P' and Q' strictly interlace.

A simple proof of this lemma, similar to that given by V. Markov, can be seen in Rivlin's book [99] (Lemma 2.7.1, p. 125). As was shown in [27], the next observation (which is of its own interest) provides a new proof of Lemma 3.1 which admits extensions to other systems of polynomial-like functions (T-systems and perfect splines).

Lemma 3.2. *Each zero η of the derivative of an algebraic polynomial $P(x) := (x - x_1) \cdots (x - x_n)$ is a strictly increasing function of x_k in the domain $x_1 < \cdots < x_n$.*

Proof: Differentiating the identity

$$0 = \frac{P'(\eta)}{P(\eta)} = \sum_{i=1}^{n} \frac{1}{\eta - x_i}$$

with respect to x_k, we get

$$0 = - \left(\sum_{i=1}^{n} \frac{1}{(\eta - x_i)^2} \right) \frac{\partial \eta}{\partial x_k} + \frac{1}{(\eta - x_k)^2}.$$

This yields

$$\frac{\partial \eta}{\partial x_k} > 0$$

and thus, the strict monotonicity of η. $\square$

The monotonicity, described in Lemma 3.2, is known. Actually, it follows immediately from Markov's interlacing property: If we move a certain zero x_k of P forward to a position $x_k + \varepsilon$ and denote the resulting polynomial by P_ε, then it is clear that the zeros of P and P_ε interlace. By Lemma 3.1, the zeros of P' and P'_ε also interlace (strictly) and therefore the zeros of P'_ε are to the right of the zeros of P, thus monotonicity.

The direct proof given in Lemma 3.2 is folklore in Approximation Theory. Similar reasoning can be found in the work [52] by Erdős and Szegő, in Dimitrov [45].

The inverse statement, that Markov's interlacing property follows from Lemma 3.2, can be easily verified (see [27], Theorem 1).

V. Markov uses frequently the interlacing property in his proof. For any given $\xi \in [-1, 1]$ and $0 < k \le n$ he studies the quantity

$$M_{nk}(\xi) := \sup\{|P^{(k)}(\xi)| \; : \; P \in \pi_n, \; \|P\| \le 1\}$$

and establishes first that $M_{nk}(\xi) = |T_n^{(k)}(\xi)|$ if ξ belongs to certain subintervals $\{[\alpha_j, \beta_j]\}$ of $[-1, 1]$ (let us call them Tchebycheff subintervals). The study of $M_{nk}(\xi)$ on the remaining subintervals of $[-1, 1]$ (call them Zolotorev subintervals) is the difficult part of his proof. The supremum value $M_{nk}(\xi)$ is attained there by the Zolotorev polynomials. He applies fine analysis to show that the maximal value of $M_{nk}(\xi)$ on the Zolotorev subintervals $[\beta_j, \alpha_{j+1}]$ is attained at the end-points and therefore

$$\max_{\xi \in [-1, 1]} M_{nk}(\xi) = \max_{\xi \in [-1, 1]} |T_n^{(k)}(\xi)|.$$

There are several other proofs of Vladimir Markov's inequality. Firstly, Bernstein [16] gave a shorter but still too elaborate variational proof. Then Duffin and Schaeffer [103] found another shorter proof. They gave also a wonderful extension of V. Markov's inequality in [46]. Shadrin [108] simplified the original idea of V. Markov and gave the currently simplest proof of the inequality. Tikhomirov demonstrated in [117] that V. Markov's inequality can be proved using standard methods of the variational calculus. It is a pity that many text books on Approximation Theory do not supply a proof of this important inequality. For a book with a proof, see Schönhage [106].

Vladimir Markov died from tuberculosis too young, at the age of 26. A shortened version of his fundamental paper [82] was translated to German and published in Mathematische Annalen in 1916 with a preface by Professor Bernstein.

3.1 Another proof of V. Markov's inequality

We shall present below a proof of Vladimir Markov's inequality that uses only classical real analysis and elementary properties of algebraic polynomials. It is a simplification of Tikhomirov's variational approach as outlined in [117] and demonstrated on various extremal problems in his book [116]. The Lagrange multiplier method, the Lagrange principle of first and second order, as well as Fermat's theorem, which are basic tools in Tikhomirov's approach, are avoided here.

Firstly, we give in the next lemma a differentiation formula for algebraic polynomials.

Lemma 3.3. *Given arbitrary points* $t_1 < \cdots < t_n$, *let* $\omega(x) = (x - t_1) \cdots (x - t_n)$. *If* $\omega^{(k)}(\xi) = 0$ *for some* ξ *and* k, $0 < k < n$, *then there exist nonzero constants* $A_1, \ldots, A_n$ *such that*

$$P^{(k)}(\xi) = \sum_{j=1}^{n} A_j P(t_j), \quad \forall P \in \pi_n. \tag{3.1}$$

Moreover, $A_j A_{j+1} < 0$ *for* $j = 1, \ldots, n - 1$.

Proof: Denote by $\{\ell_{nj}\}_1^n$ the Lagrange basic polynomials associated with the nodes $t_1, \ldots, t_n$. Then, by the Lagrange interpolation formula,

$$Q(x) = \sum_{j=1}^{n} Q(t_j)\ell_{nj}(x), \quad \forall Q \in \pi_{n-1}.$$

Since every polynomial P of degree n can be written in the form $P(x) = c\omega(x) + Q(x)$, where $Q \in \pi_{n-1}$ and c is a constant, we clearly have

$$P^{(k)}(\xi) = \sum_{j=1}^{n} Q(t_j)\ell_{nj}^{(k)}(\xi).$$

It remains to notice that $P(t_j) = Q(t_j)$. Then (3.1) follows with

$$A_j = \ell_{nj}^{(k)}(\xi), \quad j = 1, \ldots, n.$$

Since the zeros of $\omega(x)$ interlace with the zeros of $\ell_{nj}(x)$, then by Lemmas 3.1 and 3.2, the zeros $\xi_1 < \cdots < \xi_{n-k}$ of $\omega^{(k)}(x)$ and the zeros $y_{1,j} < \cdots < y_{n-1-k,j}$ of $l_{nj}^{(k)}(x)$ strictly interlace. Therefore,

$$\xi_m < y_{m,n} < y_{m,n-1} < \cdots < y_{m,1} < \xi_{m+1}, \quad m = 1, \ldots, n-1.$$

Now it is clear that $\ell_{nj}^{(k)}(\xi) \neq 0$ at any fixed zero $\xi = \xi_m$ of $\omega^{(k)}(x)$ and the values $\ell_{n1}^{(k)}(\xi), \ldots, \ell_{n,n}^{(k)}(\xi)$ alternate in sign. The proof is complete. □

Theorem 3.4 (Vladimir Markov inequality). *Let P be any real polynomial of degree less than or equal to n. Then*

$$\|P^{(k)}\| \leq T_n^{(k)}(1)\,\|P\|, \quad for \quad k = 1, \ldots, n.$$

Moreover, equality is attained only for polynomials of the form $P = cT_n$, where c is a nonzero constant.

Proof: As we have mentioned already, for fixed k, Markov's inequality is equivalent to the extremal problem

$$\|P^{(k)}\| \quad \to \sup \quad \text{over all} \quad P \in \pi_n, \ \|P\| \leq 1.$$

Assume that f is an extremal function to this problem. Let

$$M_k := \|f^{(k)}\| = |f^{(k)}(\xi)|.$$

In the case $\xi = 1$ (or $\xi = -1$) it can be easily seen that the Tchebycheff polynomial is the only extremal function. Indeed, by the Lagrange interpolation formula at the extremal points $-1 = \eta_0 < \cdots < \eta_n = 1$ of the Tchebycheff polynomial $T_n(x)$, we get

$$|f^{(k)}(1)| = \left| \sum_{j=0}^{n} f(\eta_j)\ell_{nj}^{(k)}(1) \right| \leq \sum_{j=0}^{n} |\ell_{nj}^{(k)}(1)|$$

$$= \left| \sum_{j=0}^{n} T_n(\eta_j)\ell_{nj}^{(k)}(1) \right| = T_n^{(k)}(1),$$

and the claim is proved.

Note that in the particular cases $k = n - 1$ and $k = n$, for any extremal polynomial f, the quantity $|f^{(k)}(x)|$ attains its maximal value at 1 (or -1). Thus, we may assume in what follows that $k \leq n - 2$.

Let $|f(x)|$ attain its maximal value 1 on $[-1,1]$ at the points $t_1 < \cdots < t_m$. Note that $t_1 = -1$ and $t_m = 1$, since otherwise an appropriate compression of f would increase $\|f^{(k)}\|$. Evidently $m \le n + 1$. We shall show that $m = n + 1$. This would imply that $f = \pm T_n$ and the V. Markov inequality will be proved.

Assume first that $m < n$. For any fixed extremal point ξ of $f^{(k)}(x)$ in $[-1,1]$, we construct the polynomial g of degree $m + 1$ which satisfies the interpolation conditions

$$g(t_j) = 0, \quad j = 1, \ldots, m, \quad g^{(k)}(\xi) = \text{sign } f^{(k)}(\xi), \quad g^{(k+1)}(\xi) = 0.$$

By the Atkinson-Sharma Theorem [7], this is a regular Birkhoff interpolation problem, and thus g exists and is uniquely defined. (The regularity can be proved directly: Assume that the corresponding homogeneous problem admits a non-zero solution $g \in \pi_{m+1}$. Then, by Rolle's theorem, $g^{(k)}$ would have $m - k$ sign changes and, in addition, a double zero at ξ, thus, totally $m + 2 - k$ zeros. Since $g^{(k)} \in \pi_{m+1-k}$, this implies $g^{(k)} \equiv 0$, and consequently, $g \equiv 0$, a contradiction.)

Having g, we consider the polynomial $f + \varepsilon g$ for sufficiently small $\varepsilon > 0$. It is not difficult to verify that

$$\|f + \varepsilon g\| = 1 + o(\varepsilon), \quad \text{while} \quad |f^{(k)}(\xi) + \varepsilon g^{(k)}(\xi)| = M_k + \varepsilon.$$

Thus $(f + \varepsilon g)/\|f + \varepsilon g\|$ has a larger k-th derivative at ξ, a contradiction with the assumed extremal property of f.

Suppose now that $m = n$. Then the polynomial $\omega(x) = (x-t_1) \cdots (x-t_m)$ is in π_n. If ξ is an end-point, we showed already by the Lagrange interpolation formula that $f = \pm T_n$.

Assume next that ξ is an interior point of $[-1,1]$. Then

$$f^{(k+1)}(\xi) = 0.$$

If $\omega^{(k)}(\xi) \ne 0$, then one of the polynomials $(f \pm \varepsilon\omega(x))/\|f \pm \varepsilon\omega\|$ will have larger k-th derivative than f, a contradiction. It remains to consider the case when

$$\omega^{(k)}(\xi) = 0. \tag{3.2}$$

Recall that

$$t_1 := -1 < t_2 < \cdots < t_{n-1} < 1 =: t_n$$

are the zeros of $\omega(x)$ and $|f(t_j)| = 1$, $j = 1, \ldots, n$. Notice that

$$f(t_j) = \sigma(-1)^j, \quad j = 1, \ldots, n, \tag{3.3}$$

where $\sigma = 1$ or $\sigma = -1$. Indeed, if say, $f(t_i)f(t_{i+1}) > 0$ for some i, then f' has a zero in (t_i, t_{i+1}). It is seen that the zeros $t_1, \ldots, t_n$ of ω interlace with the zeros of f', and by Lemma 3.1, the zeros of $\omega^{(k)}$ and $f^{(k+1)}$ interlace strictly. Since $f^{(k+1)}(\xi) = 0$, we then conclude that $\omega^{(k)}(\xi) \neq 0$, a contradiction to (3.2).

For every sufficiently small ε (in absolute value), we introduce the system of points $\mathbf{x}(\varepsilon) := \{x_1(\varepsilon), \ldots, x_n(\varepsilon)\}$ defined by

$$x_j(\varepsilon) := t_j - \frac{1}{2}\frac{\omega'(t_j)}{f''(t_j)}\varepsilon, \quad j = 2, \ldots n - 1,$$

$$x_1(\varepsilon) := -1, \quad x_n(\varepsilon) := 1,$$

and set

$$\omega(\mathbf{x}(\varepsilon); t) := (t - x_1(\varepsilon)) \cdots (t - x_n(\varepsilon)).$$

Consider the polynomial

$$f_\varepsilon(t) := f(t) + \varepsilon\,\omega(\mathbf{x}(\varepsilon); t).$$

Clearly $f_\varepsilon \in \pi_n$. We shall show that

$$\|f_\varepsilon\| = 1 + o(\varepsilon^2).$$

Indeed, since $x_j(\varepsilon) \to t_j$ as $\varepsilon \to 0$, then f_ε tends uniformly to f as $\varepsilon \to 0$. Consequently, the coefficients of f_ε approach the coefficients of f as $\varepsilon \to 0$. This implies that the zeros of $f_\varepsilon^{(i)}$ approach the zeros of $f^{(i)}$ as $\varepsilon \to 0$. Thus, for every sufficiently small $|\varepsilon|$, there exist points $t_2(\varepsilon), \ldots, t_{n-1}(\varepsilon)$ such that

$$f_\varepsilon'(t_j(\varepsilon)) = 0, \quad j = 2, \ldots, n - 1,$$

and $t_j(\varepsilon) \to t_j$ as $\varepsilon \to 0$. Set $t_1(\varepsilon) = -1$, $t_n(\varepsilon) = 1$. Clearly,

$$\max_{x \in [t_j - d, t_j + d]} |f_\varepsilon(x)| = |f_\varepsilon(t_j(\varepsilon))|$$

for some small $d > 0$ and $j = 2, \ldots, n - 1$. Moreover,

$$\|f_\varepsilon\| = \max_{1 \leq j \leq n} |f_\varepsilon(t_j(\varepsilon))|.$$

Consider the function

$$F_j(\varepsilon) := f_\varepsilon(t_j(\varepsilon)) = f(t_j(\varepsilon)) + \varepsilon\,\omega(\mathbf{x}(\varepsilon); t_j(\varepsilon))$$

for any fixed $j \in \{2, \ldots, n - 1\}$. We have

$$F_j(0) = f(t_j).$$

Further,

$$F_j'(\varepsilon) = f'(t_j(\varepsilon))\, t_j'(\varepsilon) + \omega(\mathbf{x}(\varepsilon); t_j(\varepsilon)) + \varepsilon\, \frac{\partial}{\partial \varepsilon} \omega(\mathbf{x}(\varepsilon); t_j(\varepsilon)).$$

From the definition of $t_j(\varepsilon)$,

$$f'(t_j(\varepsilon)) + \varepsilon\, \omega'(\mathbf{x}(\varepsilon); t_j(\varepsilon)) = 0$$

for each sufficiently small $|\varepsilon|$. Differentiating this identity with respect to ε at $\varepsilon = 0$, we obtain

$$f''(t_j(0))t_j'(0) + \omega'(\mathbf{x}(0); t_j(0)) = 0,$$

and thus

$$t_j'(0) = -\frac{\omega'(t_j)}{f''(t_j)}, \quad j = 2, \ldots, n-1.$$

Then we obtain

$$F_j'(0) = 0.$$

Let us compute $F_j''(0)$. We have

$$F_j''(\varepsilon) = f''(t_j(\varepsilon))[t_j'(\varepsilon)]^2 + f'(t_j(\varepsilon))t_j''(\varepsilon) + 2\frac{\partial}{\partial \varepsilon}\omega(\mathbf{x}(\varepsilon); t_j(\varepsilon))$$

$$+ \varepsilon\, \frac{\partial^2}{\partial \varepsilon^2}\, \omega(\mathbf{x}(\varepsilon); t_j(\varepsilon)).$$

Taking into account that

$$\frac{\partial}{\partial \varepsilon}\omega(\mathbf{x}(\varepsilon); t_j(\varepsilon)) = \omega'(\mathbf{x}(\varepsilon); t_j(\varepsilon))t_j'(\varepsilon) - \sum_{i=2}^{n-1} \frac{\omega(\mathbf{x}(\varepsilon); x)}{x - x_i(\varepsilon)}\bigg|_{x=t_j(\varepsilon)} \cdot x_i'(\varepsilon),$$

we obtain

$$\frac{\partial}{\partial \varepsilon}\omega(\mathbf{x}(\varepsilon); t_j(\varepsilon))\bigg|_{\varepsilon=0} = \omega'(t_j)\, t_j'(0) - \sum_{i=2}^{n-1} \omega'(t_j)x_i'(0) \cdot \delta_{ij}$$

and hence,

$$F_j''(0) = f''(t_j)[t_j'(0)]^2 + 2\omega'(t_j)t_j'(0) - 2x_j'(0)\omega'(t_j)$$

$$= -\frac{[\omega'(t_j)]^2}{f''(t_j)} - 2\left(-\frac{1}{2}\frac{\omega'(t_j)}{f''(t_j)}\right)\omega'(t_j) = 0.$$

Therefore

$$F_j(\varepsilon) = F_j(0) + F_j'''(0)\varepsilon^3 + o(\varepsilon^3) = f(t_j) + O(\varepsilon^3),$$

and consequently

$$\|f_\varepsilon\| = 1 + o(\varepsilon^2).$$

Similarly, we study the function

$$D(\varepsilon) := f^{(k)}(\eta(\varepsilon)) + \varepsilon\omega^{(k)}(\mathbf{x}(\varepsilon); \eta(\varepsilon)),$$

where $\eta(\varepsilon)$ is defined uniquely by $\eta(0) = \xi$ and

$$f^{(k+1)}(\eta(\varepsilon)) + \varepsilon\omega^{(k+1)}(\mathbf{x}(\varepsilon); \eta(\varepsilon)) = 0.$$

Differentiating the last identity with respect to ε, we obtain

$$\eta'(0) = -\frac{\omega^{(k+1)}(\xi)}{f^{(k+2)}(\xi)}.$$

Therefore,

$$D(0) = f^{(k)}(\xi), \quad D'(0) = 0,$$

$$D''(0) = f^{(k+2)}(\xi)\eta'^2(0) + 2\left[\omega^{(k+1)}(\xi)\,\eta'(0) - \sum_{j=2}^{n-1} \ell_{nj}^{(k)}(\xi)\omega'(t_j)\,x_j'(0)\right]$$

$$= -\frac{[\omega^{(k+1)}(\xi)]^2}{f^{(k+2)}(\xi)} + \sum_{j=2}^{n-1} A_j\frac{[\omega'(t_j)]^2}{f''(t_j)}.$$

Our next task is to show that $D''(0) \neq 0$ and sign $f^{(k)}(\xi) =$ sign $D''(0)$. Let us mention first that

$$A_j f(t_j) f^{(k)}(\xi) > 0, \quad j = 1, \ldots, n. \tag{3.4}$$

Indeed, by Lemma 3.3,

$$f^{(k)}(\xi) = \sum_{j=1}^{n} A_j f(t_j).$$

Then the relation (3.4) follows from the fact that all the numbers $A_1 f(t_1)$, $\ldots$, $A_n f(t_n)$ are of the same sign (since, in view of (3.1) and (3.3), $\{A_j\}_1^n$, as well as $\{f(t_j)\}_1^n$, alternate).

In addition, notice that the sign of $f(t_j)\omega'(t_j)$ does not depend on j, for $j = 1, \ldots, n$ (since $\{\omega'(t_j)\}_1^n$ alternate in sign). We may assume, without loss of generality, that

$$f(t_j)\omega'(t_j) > 0, \quad j = 1, \ldots, n.$$

In view of (3.4), this implies

$$A_j\omega'(t_j)f^{(k)}(\xi) > 0, \quad j = 1, \ldots, n. \tag{3.5}$$

Then, by Lemma 3.3,

$$\operatorname{sign} \omega^{(k+1)}(\xi) = \operatorname{sign} \sum_{j=1}^{n} A_j\omega'(t_j) = \operatorname{sign} f^{(k)}(\xi) = -\operatorname{sign} f^{(k+2)}(\xi).$$

Therefore,

$$\alpha := \frac{f^{(k+2)}(\xi)}{\omega^{(k+1)}(\xi)} < 0.$$

Introduce the polynomial

$$h(x) := f'(x) - \alpha\omega(x).$$

Clearly $h(t_j) = 0$ for $j = 2, \ldots, n-1$ and $h^{(k)}(\xi) = h^{(k+1)}(\xi) = 0$. Then, it follows by Rolle's theorem that h has no other zeros on the whole real line except $t_2, \ldots, t_{n-1}$. As a consequence, $h(x)$ changes sign alternatively at $t_2, \ldots, t_{n-1}$. Since $f''(x)$ also changes sign alternatively at these points, we conclude that the numbers

$$B_j := \frac{h'(t_j)}{f''(t_j)}, \quad j = 2, \ldots, n-1,$$

are of the same sign. But, as seen from the definition of h, $h(1) = f'(1)$ and $h(t_{n-1}) = f'(t_{n-1}) = 0$. Since $h(x)$ and $f'(x)$ do not vanish in $(t_{n-1}, 1)$, this implies $h'(t_{n-1})/f''(t_{n-1}) > 0$ and hence $B_{n-1} > 0$. Therefore, $B_j > 0$ for $j = 2, \ldots, n-1$. Set, in addition, $B_1 = B_n = 1$.

Next, by Lemma 3.3,

$$\omega^{(k+1)}(\xi) = A_1\omega'(-1) + A_n\omega'(1) + \sum_{j=2}^{n-1} A_j\omega'(t_j).$$

Then a straightforward computation and application of the observations made above show that

$$-\alpha D''(0) = \omega^{(k+1)}(\xi) - \sum_{j=2}^{n-1} A_j \frac{\omega'(t_j)}{f''(t_j)}[f''(t_j) - h'(t_j)] = \sum_{j=1}^{n} A_j B_j\omega'(t_j)$$

and thus, in view of (3.5), $-\alpha D''(0)f^{(k)}(\xi) > 0$. Taking into account that $\alpha < 0$, we conclude that sign $D''(0) = $ sign $f^{(k)}(\xi)$.

Hence

$$\|f_\varepsilon^{(k)}\| \geq |D(\varepsilon)| = |f^{(k)}(\xi) + \frac{D''(0)}{2}\varepsilon^2 + o(\varepsilon^2)| > M_k + \frac{|D''(0)|}{3}\varepsilon^2$$

for small ε, while $\|f_\varepsilon\| = 1 + o(\varepsilon^2)$. This contradicts the optimality of f. The proof is complete. $\square$

Markov-type inequalities for polynomials satisfying boundary conditions have been studied by Schur [107]. He proved that

$$\|P'\| \leq \|\tilde{T}_n'\|\,\|P\| = n\cos\frac{\pi}{2n}\|P\|$$

for every polynomial P from the subset

$$\pi_n(-1,1) := \{Q \in \pi_n : Q(-1) = Q(1) = 0\},$$

with equality holding only if $P = \pm\tilde{T}_n$, where $\tilde{T}_n$ denotes the Tchebycheff polynomial, stretched so that the first zero goes to -1 and the last one to 1.

Only recently, Milev and Nikolov [84] showed that Schur's inequality holds for higher derivatives.

Theorem 3.5. *(Milev and Nikolov [84]). For every polynomial* $P \in \pi_n(-1,1)$,

$$\|P^{(k)}\| \leq \|\tilde{T}_n^{(k)}\|\,\|P\|, \quad k = 1,\ldots,n.$$

Equality holds only if $P = \pm\tilde{T}_n$, *up to a constant multiplier.*

An interesting extension of V. Markov's inequality has been established by Videnskii [124]. He proved the following.

Theorem 3.6. *If the polynomial* $P \in \pi_n$ *satisfies the inequality*

$$|P(x)| \leq |\alpha x + i\sqrt{1 - x^2}|, \quad (\alpha \geq 0)$$

on $[-1,1]$, *then for each* $1 \leq k \leq n$,

$$|P^{(k)}(x)| \leq M_n^{(k)}(1) = \frac{\alpha+1}{2}T_n^{(k)}(1) + \frac{\alpha-1}{2}T_{n-2}^{(k)}(1), \quad -1 \leq x \leq 1, \tag{3.6}$$

where

$$M_n(x) := \frac{\alpha+1}{2}T_n(x) + \frac{\alpha-1}{2}T_{n-2}(x).$$

The equality sign is attained in (3.6) *only for the polynomials* $P(x) = cM_n(x)$, $|c| = 1$, *at the points* $x = \pm 1$.

In particular, taking $\alpha = 0$, we derive from Videnskii's theorem an exact bound for the derivatives of a polynomial $P \in \pi_n$ whose graph lies in the unit disk. This particular problem was solved independently by Rahman [96] in a study of a question raised by Paul Turan (see also [93]). Further extensions are given by Pierre and Rahman [93], Rahman and Schmeisser in [97].

Similar Markov type extremal problems for trigonometric polynomials admit more elegant treatment due to the periodicity. For example, just a repeat application of the Bernstein inequality

$$\|s'_n\| \le n\|s_n\|$$

for trigonometric polynomials of degree n leads to

$$\|s_n^{(k)}\| \le n^k\|s_n\|, \quad k = 0, 1, 2, \ldots$$

Equality is attained for the translates of $\cos n\theta$. Even more is known, namely

$$\|s_n^{(k)}\|_p \le n^k\|s_n\|_p \tag{3.7}$$

for each $0 \le p < \infty$. The case $1 \le p < \infty$ is due to Zygmund [127]. His proof is based on the M. Riesz interpolation formula

$$s'_n(\theta) = \sum_{j=1}^{2n}(-1)^{j+1}\lambda_j \cdot s_n(\theta + \theta_j), \quad \forall \theta, \tag{3.8}$$

where

$$\lambda_j := \frac{1}{4n\sin^2\frac{\theta_j}{2}}$$

and $\theta_j := \frac{2j-1}{2n}\pi$. Let us sketch it. Note that $\lambda_1 + \cdots + \lambda_{2n} = n$. Then, by the classical Jensen inequality, for every integrable convex and increasing function $\phi(t)$ on $[0, \infty)$ we have

$$\int_0^{2\pi} \phi(|s'_n(\theta)|)\, d\theta \le \int_0^{2\pi} \phi\left(\sum_{j=1}^{2n}\frac{\lambda_j}{n}n|s_n(\theta + \theta_j)|\right) d\theta$$

$$\le \int_0^{2\pi} \sum_{j=1}^{2n}\frac{\lambda_j}{n}\phi(n|s_n(\theta + \theta_j)|)\, d\theta$$

$$= \left(\sum_{j=1}^{2n}\frac{\lambda_j}{n}\right)\int_0^{2\pi} \phi(n|s_n(\theta)|)\, d\theta = \int_0^{2\pi} \phi(n|s_n(\theta)|)\, d\theta.$$

Now (3.7) follows as a particular case for $\phi(x) = x^p$, $1 \le p < \infty$.

For $0 \le p < 1$ inequality (3.7) stayed as an open problem for a long time. It was proved by Vitalii Arestov in an outstanding work [3] (see also [57] for a simplified presentation).

There is no sharp inequality known for algebraic polynomials in L_p norm on a finite interval for any $0 < p < \infty$. Even in the usually "easy" case $p = 2$ the exact constant $C(n,2)$ in

$$\left(\int_{-1}^{1} |P'(x)|^p \, dx \right)^{1/p} \le C(n,p) \left(\int_{-1}^{1} |P(x)|^p \, dx \right)^{1/p}$$

is not known. Milovanovic [85] described $C(n,2)$ as the biggest eigenvalue of a certain matrix. Special cases of weighted L_2 norms have been studied and sharp inequalities have been obtained for infinite intervals. Turan [119] and Schmidt [104] proved

$$\|P'\| \le \sqrt{2n}\|P\|, \quad \text{for} \quad \|P\| := \int_{\infty}^{\infty} e^{-x^2} P^2(x) \, dx,$$

and Turan [119] showed that

$$\|P'\| \le \left(2\sin\frac{\pi}{4n+2} \right)^{-1} \|P\|, \quad \text{for} \quad \|P\| := \int_{0}^{\infty} e^{-x} P^2(x) \, dx.$$

Sharp point-wise bounds for the derivative $P'(x)$ in $[-1,1]$, under the condition that P does not vanish in the open unit disk are obtained in [6]. Sharp inequalities for $\|P'\|_p$ in the same class are studied in [46].

3.2 Bounds for polynomials in the complex plane

An immediate consequence of the equioscillation of the Tchebycheff polynomial $T_n(x)$ is the property that T_n is the most rapidly growing polynomial outside the interval $[-1,1]$ among all polynomials from π_n bounded by 1 on $[-1,1]$. Later, answering the question raised by Mendeleev, A. Markov [81] showed that this is true even for any derivative of T_n, namely, for each real t such that $|t| \ge 1$, we have

$$|P^{(k)}(t)| \le |T_n^{(k)}(t)|, \quad k = 0, \ldots, n, \qquad P \in \pi_n,$$

provided $|P(x)| \le 1$ on $[-1,1]$. Bernstein [14] showed that this holds also for points lying outside the open unit disk D, even under weaker conditions. More precisely, if $P \in \pi_n$ and

$$|P(\eta_j)| \le 1, \quad j = 0, \ldots, n,$$

then for all z lying outside D

$$|P^{(k)}(z)| \le |T_n^{(k)}(z)|.$$

The proof is really nice: Set $v(z) := (z - \eta_0) \cdots (z - \eta_n)$. Since z lies outside the disk D, the segment $[-1, 1]$ is seen from z in an angle less than or equal to $\pi/2$. Then the vectors $\bar{z} - \eta_j$, $j = 0, \ldots, n$, lie in an angle not greater than $\pi/2$ and, consequently, the number

$$|P(z)| = \left| \sum_{j=0}^{n} P(\eta_j) \frac{T_n(z)}{(z - \eta_j)v'(\eta_j)} \right| = \left| T_n(z) \sum_{j=0}^{n} \frac{P(\eta_j)}{v'(\eta_j)} \frac{\bar{z} - \eta_j}{|z - \eta_j|^2} \right|$$

is maximal if $P(\eta_j) = \pm(-1)^j$, i.e., if $P = \pm T_n$. The same proof was rediscovered by Erdős [50].

The inequality was actually proved for $k = 0$. It follows for each $k \in \{1, \ldots, n\}$ from another observation due to Bernstein [15] (also de Bruijn [38]) which can be recognized as an example of an inheritance theorem:

Theorem 3.7. *Let $P(z)$ and $Q(z)$ be algebraic polynomials of degree n with complex coefficients. Assume that $|P(z)| \leq |Q(z)|$ for each z from the boundary ∂G of a given convex domain G. Then $|P'(z)| \leq |Q'(z)|$ on ∂G, provided Q has n zeros in G.*

The proof of this theorem is very short and can be seen in Rivlin's book ([99], Theorem 2.26).

To extend Bernstein's bound to derivatives of P, let us take any closed convex curve Γ (for instance, a circle) that passes through the point z and contains the disk D. By the result just proved, $|P(u)| \leq |T_n(u)|$ for every $u \in \Gamma$. Then the above inheritance theorem implies $|P'(u)| \leq |T_n'(u)|$ for each $u \in \Gamma$, and in particular for $u = z$. Thus, the inequality is proved for the first derivative. Then we proceed similarly for the higher derivatives.

In the same way one can show even more: Let $-1 = t_0 < \cdots < t_n = 1$ be any set of points and $h_0, \ldots, h_n$ a set of positive numbers. Let H be the polynomial of degree n defined by the interpolation conditions: $H(t_j) = (-1)^j h_j$, $j = 0, \ldots, n$. Assume that $|z| \geq 1$. Then

$$|P^{(k)}(z)| \leq |H^{(k)}(z)|$$

for every real polynomial P of degree n satisfying the conditions

$$|P(t_j)| \leq h_j, \quad j = 0, \ldots, n.$$

Theorem 3.7 is one of the most beautiful results in the analytic theory of algebraic polynomials. It was proved first by Bernstein [15] for the disk and extended to any convex domain by de Bruijn [38]. It implies the Bernstein inequality:

If P is a polynomial of degree n such that $|P(z)| \leq 1$ on the unit disk D, then $|P'(z)| \leq n$ on D.

The proof is immediate: Take $Q(z) = z^n$. We clearly have $|P(z)| \leq 1 = |Q(z)|$ on the boundary ∂D of D. Thus, by Theorem 3.7, $|P'(z)| \leq |Q'(z)| = n$ on ∂D.

According to an observation due to Szegö [113] (see also Malik [79] for a nice simple proof), if $|\Re P(z)| \leq 1$ for $|z| \leq 1$, then $|P'(z)| \leq n$ on D.

In case P does not vanish in D, it was conjectured by Erdös that

$$|P'(z)| \leq \frac{n}{2} \max_{|z|=1} |P(z)|.$$

The conjecture was proved by Lax [77]. The inequality is exact. An elegant proof based on Theorem 3.7 can be seen in Rivlin's book [99]. This result was extended by Malik [79] to the form:

If P is a polynomial of degree n and does not vanish in the disk $|z| < K$, $K \geq 1$, then

$$\max_{|z|=1} |P'(z)| \leq \frac{n}{1+K} \max_{|z|=1} |P(z)|.$$

The inequality is sharp.

Concerning estimates in L_p norm, the following was established by de Bruijn [38]:

Let $P(z) = a_n \prod_{j=1}^{n} (z - z_j)$, $a_n \neq 0$. If $|z_j| \geq 1$, $1 \leq j \leq n$, then for $p \geq 1$,

$$\left(\int_0^{2\pi} |P'(e^{i\theta})|^p \, d\theta \right)^{1/p} \leq n C_n \left(\int_0^{2\pi} |P(e^{i\theta})|^p \, d\theta \right)^{1/p},$$

where

$$C_n := \left\{ 2\pi / \int_0^{2\pi} |1 + e^{i\theta}|^p \, d\theta \right\}^{1/p}.$$

Equality holds for the polynomials $P(z) = a + bz^n$, $|a| = |b|$.

In dealing with L_p norms of polynomials the next property of subordination is a very useful tool (see [58], §8, Theorem 1). We say that the function f is **subordinate** to F (denoted by $f \prec F$) if $f(z) = F(\gamma(z))$, where $\gamma(z)$ is any analytic function on the disk $|z| < 1$ that satisfies the conditions $\gamma(0) = 0$, $|\gamma(z)| < 1$ for $|z| < 1$.

Theorem 3.8. If the functions f and F are analytic in $|z| < 1$, $f(0) = F(0) = 0$ and $f(z) \prec F(z)$ in $|z| < 1$, then for each $p > 0$ and $0 < r < 1$ we have

$$\int_0^{2\pi} |f(r\,e^{i\theta})|^p \, d\theta \leq \int_0^{2\pi} |F(r\,e^{i\theta})|^p \, d\theta.$$

An ingenious application of the property of subordination has led Saff and Sheil-Small [100] to the exact bound of $|P^{(k)}(0)|$ (for $k \neq n/2$) in terms of $M := \max_{|z|=1} |P(z)|$ for polynomials P having all zeros on the unit circle.

3.3 Duffin and Schaeffer type inequalities

A remarkable extension of V. Markov's inequality was given by Duffin and Schaeffer [47]. Making a step into the complex plane and using common tools from the analytic theory of polynomials, like the Gauss-Lucas Theorem and Rouché's Theorem, they proved the following.

Theorem 3.9 (Duffin-Schaeffer inequality). *Assume that $P \in \pi_n$ satisfies the conditions*

$$|P(\eta_j)| \leq 1, \quad j = 0, \ldots, n.$$

Then, for $-1 \leq x \leq 1$ and $-\infty < y < \infty$,

$$|P^{(k)}(x + iy)| \leq |T_n^{(k)}(1 + iy)|, \quad k = 1, \ldots, n,$$

with equality holding only if $P = \pm T_n$ and $x = \pm 1$.

The main ingredient of the proof is the following important proposition.

Theorem 3.10. *Let $q(x) := c(x-x_1) \cdots (x-x_n)$ where $x_1 < \cdots < x_n < 1$ and c is a nonzero constant. Assume that $f \in \pi_n$,*

$$|f'(x_j)| \leq |q'(x_j)|, \quad j = 1, \ldots, n, \tag{3.9}$$

and, for some $\xi < 1$,

$$|q(\xi + iy)| \leq |q(1 + iy)|, \quad -\infty < y < \infty. \tag{3.10}$$

Then $|f'(\xi + iy)| \leq |q'(1 + iy)|, \quad -\infty < y < \infty.$

To deduce the Duffin-Schaeffer inequality from Theorem 3.10, one need only show that (3.9) and (3.10) hold for $q = T_n$ and any polynomial $f \in \pi_n$ such that $|f(\eta_j)| \leq 1, j = 0, \ldots, n$.

The bound (3.9) is a known property of the Tchebycheff polynomial T_n. We shall derive it from the Riesz formula (3.8). Let us apply the formula for $t_n(\theta) := f(\cos \theta)$ at the point $\theta = \theta_j$ ($\xi_j := \cos \theta_j$). We have

$$t'_n(\theta_j) = -f'(\cos \theta_j) \sin \theta_j = \sum_{k=1}^{2n} (-1)^{k+1} \lambda_k f(\cos(\theta_j + \theta_k)).$$

Making use of the fact that $\lambda_1 + \cdots + \lambda_{2n} = n$ and the assumption $|f(\eta_j)| \leq 1$, we obtain

$$|f'(\xi_j) \sin \theta_j| = |f'(\cos \theta_j) \sin \theta_j| \leq \sum_{k=1}^{2n} \lambda_k = n.$$

Therefore
$$|f'(\xi_j)| \le \frac{n}{\sin\theta_j} = \frac{n}{\sqrt{1-\xi_j^2}} = |T_n'(\xi_j)|.$$

The equality is attained only if $f(\eta_j) = \pm(-1)^j$, $j = 0, \ldots n$, i.e., for $P = \pm T_n$.

The property (3.10) of the Tchebycheff polynomials $q = T_n$, for every $\xi \in [-1, 1]$, was discovered by Duffin and Schaeffer. Hence, as a corollary of Theorem 3.10, we obtain

$$|f'(\xi + iy)| \le |T_n'(1 + iy)|, \quad -\infty < y < \infty.$$

Now a repeated application of Theorem 3.7 finishes the proof of the Duffin-Schaeffer inequality.

Are there polynomials q, other than T_n and its derivatives, possessing the extremal property as described in Duffin-Schaeffer's theorem? In other words, is it true that if $q \in \pi_n$ and q has n distinct real zeros in $(-1, 1)$, then for each $k = 1, \ldots, n$,

$$|f^{(k)}(x)| \le \|q^{(k)}\|, \quad \forall x \in [-1, 1],$$

provided
$$|f(t_j)| \le |q(t_j)|, \quad j = 0, \ldots, n, \tag{3.11}$$

where $t_0 := -1$, $t_n := 1$ and t_j, $j = 1, \ldots, n-1$, are the zeros of q'? This is a very interesting question. The next theorem from [26] provides conditions that secure the property (3.9) of q.

Theorem 3.11. Let $x_1 < \cdots < x_n$ be fixed points in $(-1, 1)$ and $q(x) := (x - x_1) \cdots (x - x_n)$. Suppose that the zeros $\alpha_1 < \cdots < \alpha_{n-1}$ of the polynomial $((1+x)q'(x))'$ and the zeros $\beta_1 < \cdots < \beta_{n-1}$ of the polynomial $((1 - x)q'(x))'$ interlace with $x_1, \ldots x_n$. Then,

$$|f'(x_j)| \le |q'(x_j)|, \quad j = 1, \ldots, n,$$

for every polynomial $f \in \pi_n$ satisfying conditions (3.11).

Proof: Since the zeros $-1, t_1, \ldots, t_{n-1}$ of $(1 + x)q'(x)$ interlace with the zeros $t_1, \ldots t_{n-1}, 1$ of $(1 - x)q'(x)$, it follows from Markov's interlacing property that $\alpha_i < \beta_i$ for $i = 1, \ldots, n - 1$. By the same argument the zeros $\delta_{j1} < \cdots < \delta_{j,n-1}$ of $\ell_{nj}'(x)$, where

$$\ell_{nj}(x) := \prod_{i=0, i \ne j}^{n} \frac{x - t_i}{t_j - t_i}, \quad j = 0, \ldots, n,$$

are the fundamental polynomials for Lagrange interpolation at $t_0, \ldots, t_n$, interlace with $\{\alpha_i\}_1^{n-1}$ and $\{\beta_i\}_1^{n-1}$. Thus, for each $j = 1, \ldots, n-1$, we have

$$\alpha_i = \delta_{ni} < \delta_{ji} < \delta_{0i} = \beta_i, \quad i = 1, \ldots, n-1.$$

The interlacing of $\{x_i\}$ with $\{\alpha_i\}$ and $\{\beta_i\}$ implies that no zero x_i of $q(x)$ lies in the subintervals $[\alpha_j, \beta_j]$, $j = 1, \ldots, n-1$.

Now following the idea of V. Markov, note that for every x such that

$$\text{sign } \ell'_{nj}(x) = \epsilon(-1)^j, \quad j = 0, \ldots, n, \tag{3.12}$$

with some $\epsilon = 1$ or $\epsilon = -1$, we have

$$|f'(x)| = \left| \sum_{j=0}^n f(t_j)\ell'_{nj}(x) \right| \le \sum_{j=0}^n |f(t_j)| \, |\ell'_{nj}(x)|$$

$$= \left| \sum_{j=0}^n q(t_j)\ell'_{nj}(x) \right| = |q'(x)|.$$

Observe that the numbers in the sequence $\ell_{n0}(x), \ldots, \ell_{nn}(x)$ have alternating sign for $x > 1$. Then taking into account the ordering of the zeros of $\ell'_{nj}(x)$ and the fact that $\ell'_{nj}(x)$ changes sign only at the zeros δ_{ji}, $i = 1, \ldots, n-1$, we see that the relation (3.12) holds for every x outside the subintervals $[\alpha_i, \beta_i]$, $i = 1, \ldots n-1$. In particular, in view of our observation above, it holds for $x = x_j$ and therefore $|f'(x_j)| \le |q'(x_j)|$, which was to be shown. Note that the equality is attained if and only if $f = \pm q$. The proof is complete. $\square$

Note that, according to Lemma 2 of [26], for $-\frac{1}{2} \le \alpha = \beta \le 0$ the zeros of the ultraspherical polynomials $y(x) = P_n^{(\alpha,\beta)}(x)$ interlace with the zeros of $((1+x)y')'$ and $((1-x)y')'$, respectively.

The question of extending the Duffin-Schaeffer inequality for arbitrary polynomials q with real zeros was studied by Shadrin [108] who proved that under the conditions (3.11), we have

$$|f^{(k)}(x)| \le \max\left\{ |q^{(k)}(x)|, \, \left| \frac{1}{k}(x^2 - 1)q^{(k+1)}(x) + xq^{(k)}(x) \right| \right\}. \tag{3.13}$$

Vladimir Markov's inequality (and the Duffin-Schaeffer extension for real x) follows immediately from Shadrin's estimate since

$$\max\left\{ |T_n^{(k)}(x)|, \, \left| \frac{1}{k}(x^2 - 1)T_n^{(k+1)}(x) + xT_n^{(k)}(x) \right| \right\} \le |T_n^{(k)}(1)|.$$

Using (3.13), the following was proved in [32]: Let $-1 = t_0 < t_1 < \cdots < t_n = 1$ be the extremal points of the ultraspherical polynomial $P_n^{(\alpha,\alpha)}$ and $\alpha \geq -\frac{1}{2}$. Suppose that $f \in \pi_n$ and

$$|f(t_j)| \leq |P_n^{(\alpha,\alpha)}(t_j)|, \quad j = 0, \ldots, n.$$

Then

$$|f^{(k)}(x)| \leq \|(P_n^{(\alpha,\alpha)})^{(k)}\| = |(P_n^{(\alpha,\alpha)})^{(k)}(1)|, \quad -1 \leq x \leq 1,$$

for all $k \in \{1, \ldots, n\}$.

To extend the last inequality to the complex plane, like in the Duffin -Schaeffer Theorem, we need to know that

$$|P_n^{(\alpha,\alpha)}(x + iy)| \leq |P_n^{(\alpha,\alpha)}(1 + iy)|$$

for $-1 \leq x \leq 1$, $-\infty \leq y \leq \infty$, and $\alpha \geq -\frac{1}{2}$. Unfortunately, the latter is still an open question.

The next remarkable inequality of Duffin-Schaeffer type was established recently by Nikolov [92].

Theorem 3.12. *Let $\{t_\nu\}_{\nu=0}^n$ satisfy $1 \geq t_0 > \xi_1 > t_1 > \cdots > \xi_n > t_n \geq -1$, where $\{\xi_\nu\}_{\nu=1}^n$ are the zeros of T_n. If $P \in \pi_n$ and $|P(t_\nu)| \leq |T_n(t_\nu)|$ for $\nu = 0, \ldots, n$, then*

$$\|P'\| \leq \|T_n'\| = T_n'(1).$$

Moreover, equality is possible if and only if $P = cT_n$ with $|c| = 1$.

Verifying (3.9) for the Tchebycheff polynomials, we proved above that for every polynomial $f \in \pi_n$ such that $|f(\eta_j)| \leq 1$, $j = 0, \ldots, n$, we have

$$|f'(\xi_k)| \leq |T_n'(\xi_k)|, \quad k = 1, \ldots, n, \tag{3.14}$$

with equality only for $f = \pm T_n$. The next exact bound for the weighted L_2-norm of the derivative follows easily from this observation (see [26]):

For each polynomial $f \in \pi_n$ satisfying the conditions $|f(\eta_j)| \leq 1$, $j = 0, \ldots, n$, we have

$$\int_{-1}^1 \frac{f'^2(x)}{\sqrt{1 - x^2}} \, dx \leq \pi n^3.$$

The equality is attained only for $f = \pm T_n$.

The proof is short: We shall make use of the Gaussian quadrature formula in $[-1, 1]$ with a weight $1/\sqrt{1 - x^2}$. Since the Tchebycheff polynomial T_n

is orthogonal with this weight to each polynomial of degree $n - 1$, the nodes of the Gaussian quadrature lie at the zeros $\xi_1, \ldots, \xi_n$ of T_n. As is well-known, the quadrature is

$$\int_{-1}^{1} \frac{f(x)}{\sqrt{1 - x^2}} \, dx \approx \frac{\pi}{n} \sum_{k=1}^{n} f(\xi_k)$$

and is exact for all polynomials of degree $2n - 1$.

Let $f \in \pi_n$. Then clearly $[f'(x)]^2$ is a polynomial of degree $2n - 2$, and therefore

$$\int_{-1}^{1} \frac{f'^2(x)}{\sqrt{1 - x^2}} \, dx = \frac{\pi}{n} \sum_{k=1}^{n} f'^2(\xi_k).$$

Assume now that

$$\max_{x \in [-1,1]} |f(x)| \leq 1.$$

Then, by (3.14),

$$|f'(\xi_k)| \leq |T_n'(\xi_k)|,$$

and hence

$$\int_{-1}^{1} \frac{f'^2(x)}{\sqrt{1 - x^2}} \, dx = \frac{\pi}{n} \sum_{k=1}^{n} f'^2(\xi_k) \leq \frac{\pi}{n} \sum_{k=1}^{n} T_n'^2(\xi_k) = \int_{-1}^{1} \frac{T_n'^2(x)}{\sqrt{1 - x^2}} \, dx.$$

The last integral can be easily computed. Performing the change $x = \cos\theta$, we obtain

$$C := \int_{-1}^{1} \frac{T_n'^2(x)}{\sqrt{1 - x^2}} \, dx = n^2 \int_{0}^{\pi} \left(\frac{\sin n\theta}{\sin \theta} \right)^2 d\theta.$$

But the integrand is a symmetric function with respect to $\pi/2$. Thus

$$C = 2n^2 \int_{0}^{\pi/2} \left(\frac{\sin n\theta}{\sin \theta} \right)^2 d\theta = n^2 \int_{-\pi/2}^{\pi/2} \left(\frac{\sin n\theta}{\sin \theta} \right)^2 d\theta$$

$$= \frac{1}{2} n^2 \int_{-\pi}^{\pi} \left(\frac{\sin n\theta/2}{\sin \theta/2} \right)^2 d\theta.$$

Now using the property of the Fejer kernel:

$$\frac{1}{2\pi n} \int_{-\pi}^{\pi} \left(\frac{\sin n\theta/2}{\sin \theta/2} \right)^2 d\theta = 1,$$

we find $C = \pi n^3$. The inequality is proved. $\square$

Exact inequalities of this type for higher derivatives and other weights are proved in [26,44,59,64,123]. Note that Bernstein's inequality on the unit disk was also extended to Duffin-Schaeffer's form. More precisely, the following was proved in [56].

Theorem 3.13. *If P is a real algebraic polynomial of degree n, then*

$$\max_{|z|=1} |P'(z)| \le n \max_{1 \le m \le 2n} |P(e^{\frac{im\pi}{n}})|$$

Inequalities of this type with restrictions on less than $2n$ points cannot be true.

§4. Oscillating Polynomials

Although there are several proofs of the Markov inequality that are based on completely different ideas, it is not clear yet what is the mechanism that makes the Tchebycheff polynomials the only extremals in Markov's problem. The intuition suggests that the alteration property of T_n is the main reason for its extremality: The bigger the variation of the graph of a polynomial P in a bounded strip, the bigger should be the derivative of P. Following this belief, we showed in collaboration with others that the oscillation of a polynomial (as well as of other special functions as trigonometric polynomials and splines) is closely related to the growth of its derivative. The relation is explicitly revealed in the case of oscillating polynomials and splines.

We say that a polynomial P is **oscillating** in $[a, b]$ if P has all its zeros in $[a, b]$. As we shall see below, such a polynomial is uniquely determined by the values of its local extrema (that is, the values of P at the zeros of P' and at the end-points). The purpose of this section is to show that every functional of a quite wide class, including the L_p-norms of derivatives, is an increasing function of the local extrema of P. This implies immediately that the Tchebycheff polynomial is extremal in all problems dealing with the maximization of the L_p-norms of a derivative in the class of oscillating polynomials bounded by 1. This is a new approach to extremal problems for polynomials with real zeros. The main tool is an *inheritance theorem* which says that a certain relation between two polynomials P and Q is inherited by their derivatives P' and Q'.

4.1 Interpolation at extremal points

The starting point of the subject we are going to present in this section is an interesting problem raised by Chandler Davis [43], from Columbia University, in the section Advanced Problems of the American Mathematical Monthly in 1956. Let us reproduce the question in its original form.

Problem 4.1. *For a real polynomial $P(x)$, denote the zeros of $P'(x)$ (multiplicities counted) by ξ_i, $i = 1, \ldots, n - 1$. Assume all ξ_i real, and let $\xi_1 \le \xi_2 \le \cdots \le \xi_{n-1}$. Now to what extent are the numbers $P(\xi_i)$ arbitrary? More precisely, give necessary and sufficient conditions on an*

$(n - 1)$-tuple of real numbers $\eta_1, \ldots, \eta_{n-1}$, in order that there exists a polynomial P such that

$$P'(\xi_i) = 0, \quad P(\xi_i) = \eta_i, \quad i = 1, \ldots, n - 1.$$

The solution, given by Davis, was published a year later in the *Monthly* [43]. He proved the following.

Theorem 4.2. *Let $\{e_i\}_0^n$ be a sequence of numbers satisfying $(-1)^i(e_i - e_{i-1}) < 0$, $i = 1, \ldots, n$. Then there exists a unique polynomial $P(x)$ of degree n and a unique system of points $-1 =: t_0 < t_1 < \cdots < t_{n-1} < t_n := 1$ such that*

$$P(t_i) = e_i, \quad i = 0, \ldots, n,$$
$$P'(t_i) = 0, \quad i = 1, \ldots, n - 1.$$

Note that the Polish mathematicians Mycielski and Paszkowski [89] arrived at a similar problem independently and solved it using a topological method dealing with covering mappings. In their result the numbers $\{e_i\}_0^n$ are supposed to satisfy the more restrictive condition $(-1)^i e_i > 0$, $i = 0, \ldots, n$.

It is worth mentioning the fact that the analogous question for the complex domain was also raised at that time. In an *Editorial Note*, following the proof of Davis [43], it is remarked that A. W. Goodman has asked:

Can the branch points of (the image domain of) a polynomial be prescribed in advance? That is, given $n - 1$ complex numbers $\eta_1, \ldots, \eta_{n-1}$, are there a polynomial $P(x)$ of degree n, and $n - 1$ complex numbers $w_1, \ldots, w_{n-1}$ such that

$$P'(w_i) = 0, \quad P(w_i) = \eta_i, \quad i = 1, \ldots, n - 1.$$

One may assume the η_k are all distinct, for simplicity.

The same problem has been announced also by Andrushkiw [2] in 1956 and solved 9 years later by Thom [115]. Another shorter proof was given by Mycielski [88].

Harald Kuhn, obviously being unaware of the Davis note and the subsequent publications on this topic, considers in [74] an equivalent problem and proves the following:

For any given n and positive numbers $F_1, \ldots, F_n$ there exists a unique polynomial Q of the form $Q(x) = (x - x_0) \cdots (x - x_n)$ with $0 =: x_0 < x_1 < \cdots < x_n$ such that

$$\int_{x_{j-1}}^{x_j} Q(x) \, dx = (-1)^{n-j-1} F_j, \quad j = 1, \ldots, n.$$

Clearly, the derivative of the solution P of Davis's problem gives the solution to a corresponding Kuhn's problem and vice versa.

Another approach to interpolation at extremal points was presented by Fitzgerald and Schumaker [55]. It is based on a study of topological properties of a mapping defined by a system of differential equations. The authors proved the existence and uniqueness of the interpolant in case of interpolation by trigonometric polynomials and T-systems (periodic and nonperiodic), splines with fixed knots, monosplines. The uniqueness part of this method requires the existence of a unique solution for certain particular data. As such, a datum corresponding to the alternating solution is usually taken.

It is just amazing that a question raised in the Problem section of *Monthly* provoked such an interesting development and produced considerable literature in Approximation Theory. The reason is that this is a natural question, related to various investigations in Analysis, like Tchebycheff approximations, error estimation of interpolation schemes, and function inequalities. Curiously, the same problem arose recently again in another branch of mathematics as *The Conjecture of Arnold*. In his paper [4] in *Uspehi Mat. Nauk* of 1996, Vladimir Arnold formulated the following conjecture:

Conjecture 4.3. *The number of connectedness components of the space* $\tilde{T}g_n$ *of trigonometric polynomials of degree* n, *having* $2n$ *different real critical values, coincides with the number of connectedness components of the space* $\mathcal{F}_{2n}$ *of all smooth functions on the unit circle with the same property (i.e., having exactly* $2n$ *distinct critical values).*

The problem is reduced to the following: Given any set of numbers $\{e_j\}_0^{2n}$ satisfying the condition $(-1)^j(e_j - e_{j-1}) < 0$, $i = 1, \ldots, 2n$, and $e_{2n} < e_0$, prove that there exists a unique trigonometric polynomial $s(t)$ of degree n and a unique set of points $0 = t_0 < t_1 < \cdots < t_{2n} < 2\pi$ such that

$$s(t_j) = e_j, \quad j = 0, \ldots, 2n,$$

$$s'(t_j) = 0, \quad j = 1, \ldots, 2n.$$

As we mentioned already, the latter was proved by Fitzgerald and Schumaker [55] in 1969 and thus the conjecture was a known fact in Approximation Theory.

A year after Arnold's Conjecture was published, it was proved by Shapiro [111]. An independent proof was given also by Arnold [5].

In all the papers mentioned above, about interpolation at extremal points, the proofs are different but all are based on the study of topological properties of certain nonlinear mappings and use facts from topology or differential equations. In [19], we showed by a direct inductive proof that the problem in its more general setting, including multiple critical points, has the following solution.

Theorem 4.4. *Let $(\nu_k)_1^n$ be arbitrary fixed natural numbers and let $\mu(x)$ be a continuous function defined and nonnegative on $[x_0, \infty)$, having a finite number of zeros in any finite subinterval $[x_0, x]$. Given positive numbers $(e_k)_{k=1}^n$, there exists a unique system of points $\bar{x} = (x_k)_1^n$, $x_0 < x_1 < \cdots < x_n$, such that*

$$g_k(\bar{x}) := \left| \int_{x_{k-1}}^{x_k} \mu(x) \prod_{i=1}^{n} (x - x_i)^{\nu_i} \, dx \right| = e_k, \quad k = 1, \ldots, n.$$

Proof: First we show that the Jacobian $J(\bar{x})$ of the nonlinear system $g_k(\bar{x}) - e_k = 0$, $k = 1, \ldots, n$, with respect to $x_1, \ldots, x_n$ is nonzero at each $x_0 < x_1 < \cdots < x_n$. Indeed, using the notation

$$W(x) := \prod_{i=1}^{n} (x - x_i)^{\nu_i - 1}, \quad \omega_j(x) := \prod_{i=1, i \neq j}^{n} (x - x_i),$$

we see that the Jacobian matrix is (up to a nonzero factor)

$$J(\bar{x}) = \left\{ \int_{x_{k-1}}^{x_k} \mu(x) W(x) \omega_j(x) \, dx \right\}_{k=1, j=1}^{n, n}.$$

Assume that $\det J(\bar{x}) = 0$ for some $x_0 < x_1 < \cdots < x_n$. Then there is a linear dependence between the columns of the matrix, that is, there exist numbers $b_1, \ldots, b_n$, at least one distinct from zero, such that

$$\int_{x_{k-1}}^{x_k} \mu(x) W(x) (b_1 \omega_1(x) + \cdots + b_n \omega_n(x)) \, dx = 0, \quad \text{for} \quad k = 1, \ldots, n.$$

This means that the polynomial $Q(x) := b_1 \omega_1(x) + \cdots + b_n \omega_n(x)$, which is plainly of degree $n - 1$, must vanish at some point from (x_{k-1}, x_k) for each $k = 1, \ldots, n$. Thus $Q \equiv 0$, a contradiction.

Next we prove the theorem by induction on n. For $n = 1$ we have only one equation which obviously has a unique solution. Suppose that the theorem holds for every choice of the weight function $\mu(x)$ and the multiplicities in case of $n - 1$ points. Then, in particular, the problem

$$\left(\begin{array}{c} e_1, \ldots, e_{n-2}, e_{n-1} \\ \nu_1, \ldots, \nu_{n-2}, \nu_{n-1} + \nu_n \end{array} \middle| \mu(x) \right)$$

(i.e., the $(n-1)$-points problem with the corresponding parameters) has a unique solution $(t_k)_1^{n-1}$ in (x_0, ∞). Now we introduce a parameter $\xi \geq x_0$ and consider the unique solution $\{x_k(\xi)\}_{k=1}^{n-1}$ of the problem

$$\left(\begin{array}{c} e_1, \ldots, e_{n-1} \\ \nu_1, \ldots, \nu_{n-2}, \nu_{n-1} \end{array} \middle| \mu_\xi(x) \right)$$

with a weight $\mu_\xi(x) = \mu(x)|x - \xi|^{\nu_n}$. By the Implicit Function Theorem, $x_k(\xi)$ depends continuously on ξ. For $\xi = t_{n-1}$ we have $x_k(\xi) = t_k$, $k = 1, \ldots, n-1$, and in particular, $x_{n-1}(\xi) = \xi$. Because of the uniqueness of the solution in the case of $n - 1$ points (by the induction hypothesis), there is no other value of ξ for which $x_{n-1}(\xi) = \xi$. Therefore $x_{n-1}(\xi) < \xi$ for $\xi > t_{n-1}$. Since the first $n - 1$ equations of the system are satisfied by $x_1(\xi), \ldots, x_{n-1}(\xi)$ for any ξ, the theorem will be proved if we show that the last equation

$$g(\xi) := \left| \int_{x_{n-1}(\xi)}^{\xi} \mu(x)(\xi - x)^{\nu_n} \prod_{i=1}^{n-1} (x - x_i(\xi))^{\nu_i}\, dx \right| = e_n$$

has a unique solution with respect to $\xi > t_{n-1}$. Assume the contrary. Then there exists a point $\eta > t_{n-1}$ such that $g'(\eta) = 0$. On the other hand, a careful computation of the derivative shows that

$$g'(\eta) = \frac{J(x_1(\eta), \ldots, x_{n-1}(\eta), \eta; \mu(x))}{J(x_1(\eta), \ldots, x_{n-1}(\eta); \mu_\eta(x))}.$$

As we have proven at the beginning, any of the Jacobians above are nonzero. Therefore $g'(\eta) \neq 0$, a contradiction. $\square$

Choosing $\mu(x) \equiv 1$, by an appropriate integration and compression of the solution in the last problem, one can get (see [19]) the solution of the following general problem about interpolation at extremal points.

Corollary 4.5. *Let $(\nu_k)_1^n$ be a fixed system of arbitrary natural numbers. Let the real numbers $(y_k)_0^{n+1}$ satisfy the requirements $y_k \neq y_{k-1}$, $k = 1, \ldots, n + 1$, and*

$$|y_{k-1} - y_k| + |y_k - y_{k+1}| = |y_{k-1} - y_{k+1}| \quad \text{if } \nu_k \text{ is even,}$$

$$|y_{k-1} - y_k| + |y_k - y_{k+1}| > |y_{k-1} - y_{k+1}| \quad \text{if } \nu_k \text{ is odd,}$$

$k = 1, \ldots, n$. Given an interval $[a, b]$, there exists a unique polynomial $P \in \pi_N$, $N := \nu_1 + \cdots + \nu_n + 1$, and a unique system of points $(x_k)_1^n$, $a = x_0 < x_1 < \cdots < x_n < x_{n+1} = b$, such that

$$P(x_k) = y_k, \quad k = 0, \ldots, n + 1,$$

$$P^{(\lambda)}(x_k) = 0, \quad k = 1, \ldots, n, \; \lambda = 1, \ldots, \nu_k.$$

Note that the case $\nu_1 = \cdots = \nu_n = 1$ corresponds to the problem of Davis. An algorithm for the construction of the oscillating polynomial (that solves Davis's problem) was proposed by Ranko Bojanic [34]. A very nice proof of the convergence of this algorithm was given by Naidenov [90].

We learned recently that a similar algorithm was studied much earlier by Kammerer [65].

The observation (see Lemma 1 in [19]) that the leading coefficient of the solution $P(x) = a_0 x^N + \cdots$ in Corollary 4.5 is an increasing function of $|y_k - y_{k-1}|$ is crucial for our study of oscillating polynomials. For the sake of simplicity we formulate this property in a particular case.

Corollary 4.6. *Let $(\nu_k)_1^n$ be a fixed system of arbitrary natural numbers and let $(h_k)_0^{n+1}$ be positive. Then, there exists a unique system of points*

$$-1 =: t_0 < x_1 < t_1 < x_2 < \cdots < x_{n-1} < t_{n-1} < x_n < t_n := 1$$

such that the polynomial $P(x) = c \prod_{k=1}^n (x - x_k)^{\nu_k}$ satisfies the conditions

$$P(t_k) = (-1)^{\sigma_k} h_k, \quad k = 0, \ldots, n,$$

$$P'(t_k) = 0, \quad k = 1, \ldots, n-1,$$

where $\sigma_k = \nu_{k+1} + \cdots + \nu_n$, $k = 1, \ldots, n-1$, $\sigma_n := 0$. Moreover, the leading coefficient c is an increasing function of $h_0, h_1, \ldots, h_n$ in the domain $h_k > 0$, $k = 0, \ldots, n$.

As an immediate consequence of Corollary 4.6, we obtain the characterization of the solution in the next extremal problem.

For given $\bar{\nu} = (\nu_k)_1^n$ and $\bar{h} := \{h_k\}_0^{n+1}$, denote by $\tau(\bar{\nu}, \bar{h}; x)$ the polynomial from Corollary 4.6. In case $h_0 = \cdots = h_{n+1} = 1$ we shall abbreviate the notation to $\tau(\bar{\nu}; x)$.

Theorem 4.7. *Let $\bar{\nu} = (\nu_k)_1^n$ be a fixed system of arbitrary natural numbers, and let C be the leading coefficient of $\tau(\bar{\nu}; x)$. Then*

$$\|\tau(\bar{\nu}; \cdot)\| = \inf\left\{ \left\| C \prod_{k=1}^n (x - x_k)^{\nu_k} \right\| : -1 < x_1 < \cdots < x_n < 1 \right\}.$$

Proof: Assume the contrary. Then there is an oscillating polynomial $P(x)$ of the type considered which has a uniform norm on $[-1, 1]$ smaller than 1. But P is uniquely determined by its extremal values $h_k = |P(t_k)|$ in $[-1, 1]$. Since $h_k < 1$ for all k, we conclude on the basis of Corollary 4.6 that the leading coefficient of P is smaller than C. The contradiction ends the proof. □

The polynomials $\tau(\bar{\nu}; x)$ can be considered as generalizations of the Tchebycheff polynomials. They preserve some of the characteristic extremal property of the classical Tchebycheff polynomials. Here is one example from [33].

Theorem 4.8. Let $\nu_1, \ldots, \nu_n$ be given natural numbers. Then, for every fixed point z lying outside the closed unit disk, $|\tau(\bar{\nu}, \bar{h}; z)|$ is a strictly increasing function of h_k $(0 \leq k \leq n)$ in the domain $h_0 > 0, \ldots, h_n > 0$.

As a corollary we get $|\tau(\bar{\nu}, \bar{h}; z)| \leq |\tau(\bar{\nu}; z)|$ for each $h_k \leq 1$, $k = 0, \ldots, n$, which is a known property of the Tchebycheff polynomials in case $\nu_1 = \cdots = \nu_n = 1$, discussed already in Section 3.3.

4.2 The inheritance theorem

The Markov interlacing property given in Lemma 3.1 is a typical example of a relation between two polynomials that is inherited by their derivatives. Another interesting example is Theorem 3.7, due to Bernstein and de Bruijn.

In [33] we proved with Q. I. Rahman a property of the oscillating polynomials which can be viewed as another inheritance theorem. For the sake of simplicity, we shall formulate it here in the case of simple zeros.

For any given algebraic polynomial $P \in \pi_n$ with n simple zeros in $(-1, 1)$, we denote by $h_j(P)$, $j = 0, \ldots, n$, the absolute values of the local extrema of P (including the values at the end-points -1 and 1).

Theorem 4.9 (Inheritance Theorem). *If P and Q are polynomials from π_n with n simple zeros in $(-1, 1)$ and*

$$h_j(P) \leq h_j(Q) \quad \text{for } j = 0, \ldots, n,$$

then

$$h_j(P') \leq h_j(Q') \quad \text{for } j = 0, \ldots, n-1.$$

Proof: This is a simplified version of the proof presented in [33]. Denote by $\tau(\bar{h}; t)$ the oscillating polynomial in $[-1, 1]$, determined as in Corollary 4.6 by the parameters $\bar{h} = (h_0, \ldots, h_n)$. Let $t_1 < \cdots < t_{n-1}$ be the zeros of $\tau'(\bar{h}; t)$ and $t_0 = -1$, $t_n = 1$. Since the Jacobian of the system

$$\tau(\bar{h}; t_k) = (-1)^{n-k} h_k, \quad k = 0, \ldots, n,$$
$$\tau'(\bar{h}; t_k) = 0, \quad k = 1, \ldots, n-1,$$

is nonzero, it follows by the Implicit Function Theorem that the coefficients of $\tau(\bar{h}; t)$ are differentiable functions of h_k. We shall show that for every point x,

$$\frac{\partial \tau'(\bar{h}, x)}{\partial h_k} = \frac{g_k'(x)}{|g_k(t_k)|}, \tag{4.1}$$

where $g_k(t) := (t - t_0) \cdots (t - t_{k-1})(t - t_{k+1}) \cdots (t - t_n)$. Notice that

$$G_k(\cdot) := \frac{\partial \tau(\bar{h}; \cdot)}{\partial h_k} \in \pi_n.$$

Therefore

$$G'_k(x) = \frac{\partial \tau'(\bar{h}, x)}{\partial h_k}.$$

Thus, our claim (4.1) will be proved if we show that

$$G_k(x) = (-1)^{n-k} \frac{g_k(x)}{g_k(t_k)}.$$

We have

$$(-1)^{n-j}\delta_{jk} = (-1)^{n-j}\frac{\partial h_j}{\partial h_k} =: \frac{\partial \tau(\bar{h}, t_j(\bar{h}))}{\partial h_k}$$

$$= \frac{\partial \tau(\bar{h}; x)}{\partial h_k}\bigg|_{x=t_j} + \frac{\partial \tau(\bar{h}; x)}{\partial x}\bigg|_{x=t_j} \cdot \frac{\partial t_j(\bar{h})}{\partial h_k}$$

$$= \frac{\partial \tau(\bar{h}; x)}{\partial h_k}\bigg|_{x=t_j} = G_k(t_j).$$

Note that the second term in the sum vanishes because either

$$\frac{\partial \tau(\bar{h}; x)}{\partial x}\bigg|_{x=t_j} = \tau'(\bar{h}; t_j) = 0, \quad j = 1, \ldots, n-1,$$

or $\partial t_j(\bar{h})/\partial h_k = 0$, $j = 0, n$, since $t_0(\bar{h}) = -1$, $t_n(\bar{h}) = 1$. Thus, we proved that the polynomial G_k from π_n satisfies the conditions

$$G_k(t_j) = (-1)^{n-j}\delta_{jk}, \quad j = 0, \ldots, n.$$

This shows that $(-1)^{n-k}G_k$ is the Lagrange fundamental polynomial, that is, $G_k(x) = (-1)^{n-k}g_k(x)/g_k(t_k)$ and the claim (4.1) is proved.

Let $\xi(\bar{h})$ be a peak-point of $\tau'(\bar{h}; t)$ of fixed index. In other words, $\xi(\bar{h})$ is a zero of $\tau'(\bar{h}; t)$ or an end-point of the interval $[-1, 1]$. We shall show next that $|\tau'(\bar{h}; \xi(\bar{h}))|$ is a strictly increasing function of h_k, $0 \le k \le n$, in $H := \{\bar{h} : h_0 > 0, \ldots, h_n > 0\}$.

Assume first that ξ is an interior point of $[-1, 1]$. Then

$$\tau''(\bar{h}; \xi(\bar{h})) = 0,$$

and consequently,

$$\frac{\partial \tau'(\bar{h}; \xi(\bar{h}))}{\partial h_k} = \frac{\partial \tau'(\bar{h}; t)}{\partial h_k}\bigg|_{t=\xi} + \tau''(\xi)\frac{\partial \xi}{\partial h_k} = \frac{\partial \tau'(\bar{h}; t)}{\partial h_k}\bigg|_{t=\xi}.$$

The last equality holds also if ξ is an end-point since then ξ is a constant and $\partial \xi/\partial h_k = 0$.

Further, using the expression for the derivative we derived above, we obtain

$$\frac{\partial \tau'(\bar{h}; \xi(\bar{h}))}{\partial h_k} = (-1)^{n-k} \frac{g_k'(\xi)}{g_k(t_k)}.$$

By the definition of g_k, $g_k(t) > 0$ for $t > t_n$, and thus sign $g_k(t_k) = (-1)^{n-k}$. Let $\xi_1 < \cdots < \xi_{n-2}$ be the zeros of $\tau''(\bar{h}; t)$. Note that g_k has a maximal number of zeros, namely $\{t_0, \ldots, t_n\} \backslash t_k$ and they interlace with the zeros $\{t_1, \ldots, t_{n-1}\}$ of $\tau'(\bar{h}; t)$. Then by Lemma 3.1 (Markov interlacing property), the zeros $\xi_1 < \cdots < \xi_{n-2}$ of $\tau''(\bar{h}; t)$ interlace strictly with the zeros $\{\eta_1, \ldots, \eta_{n-1}\}$ of $g_k'(t)$. But $g_k'(t)$ changes sign at the zeros $\{\eta_j\}$. Then

$$\text{sign } g_k'(\xi_i) = (-1)^{i+1+n}$$

and, consequently,

$$\text{sign } \frac{\partial \tau'(\bar{h}; \xi_i(\bar{h}))}{\partial h_k} = (-1)^{i+1+n} = \text{sign } \tau'(\bar{h}; \xi_i(\bar{h})).$$

This shows that $|\tau'(\bar{h}; \xi(\bar{h}))|$ is a strictly increasing function of h_k. Now the assertion of the theorem follows immediately. □

The inheritance theorem implies in an elegant way V. Markov's inequality for oscillating polynomials. Indeed, let P be any oscillating polynomial of degree n on $[-1, 1]$. Assume that $|P(x)| \le 1$ on $[-1, 1]$. Then

$$h_j(P) \le 1 = h_j(T_n), \quad j = 0, \ldots, n.$$

Now, by the Inheritance Theorem,

$$h_j(P') \le h_j(T_m'), \quad j = 0, \ldots, n-1.$$

The latter implies the corresponding inequalities for the extremal values of the second derivative of P and T_n, and so on, until we get

$$h_j(P^{(k)}) \le h_j(T_n^{(k)}), \quad j = 0, \ldots, n-k.$$

Finally, it remains to note that

$$\|P^{(k)}\| = \max_{0 \le j \le n-k} h_j(P^{(k)}) = h_{j_0}(P^{(k)}) \le h_{j_0}(T_n^{(k)}) \le \|T_n^{(k)}\|,$$

which is the Markov inequality.

As we mentioned, the Inheritance Theorem was proved in [33] in the general case, treating polynomials with multiple zeros. In the same way as above, we then can derive the following inequality of Markov-type.

For given $\bar{\nu} = (\nu_1, \ldots, \nu_n)$, denote by $\pi(\bar{\nu})$ the set of all polynomials of the form

$$c \prod_{i=1}^{n} (x - x_i)^{\nu_i}, \quad -1 \le x_1 < \cdots < x_n \le 1.$$

Recall that $\tau(\bar{\nu}; x)$ is the polynomial from $\pi(\bar{\nu})$ for which all nonzero local extrema and end-point values are equal to 1 in absolute value.

Corollary 4.10. *Let $\bar{\nu} = (\nu_1, \ldots, \nu_n)$ be fixed. Then for every polynomial $P \in \pi(\bar{\nu})$ with $\|P\| = 1$ and $0 < k < |\bar{\nu}| := \nu_1 + \cdots + \nu_n$, we have*

$$\|P^{(k)}\| \le \|\tau^{(k)}(\bar{\nu}; \cdot)\|.$$

The equality is attained only for $P = \pm \tau(\bar{\nu}; \cdot)$.

4.3 Inequalities in L_p

Let us denote

$$\pi_n^\circ := \{P \in \pi_n : |P(x)| \le 1 \ \text{ on } \ [-1, 1]\}.$$

According to A. Markov's inequality,

$$\sup\{\|P'\| : P \in \pi_n^\circ\} = \|T_n'\|.$$

It is also evident that the Tchebycheff polynomial T_n has the biggest variation on $[-1, 1]$ among all polynomials from π_n°. This means that

$$\sup\{\|P'\|_1 : P \in \pi_n^\circ\} = \|T_n'\|_1.$$

Then the question of characterizing the extremal polynomial in the corresponding L_p problem, for $1 < p < \infty$, comes naturally. We proved in [21] that the Tchebycheff polynomial is again the only extremal for each $1 < p < \infty$. Namely, the following inequality holds:

$$\|P'\|_p \le \|T_n'\|_p \, \|P\|_\infty, \qquad \forall P \in \pi_n.$$

Let us denote by Φ the class of all continuously differentiable, strictly increasing convex functions on $[0, \infty)$. The following slightly more general result [23] is true.

Theorem 4.11. *Let $\phi \in \Phi$ and $M > 0$. Then, for every $P \in \pi_n$, $\|P\| \le M$, we have*

$$\int_{-1}^1 \phi(|P'(x)|) \, dx \le \int_{-1}^1 \phi(M \, |T_n'(x)|) \, dx.$$

Since the inequality above obviously holds for linear functions $\phi(x) = A + Bx$ with $B \ge 0$, without loss of generality we may assume in what follows that $\phi(0) = \phi'(0) = 0$. Also, our proof does not depend on M, so we put for definiteness $M = 1$.

The main steps in the proof of Theorem 4.11 are:

Step 1: We show that every extremal polynomial P has only maximal local extrema in $[-1, 1]$. This means that $|P(t)| = 1$ for $t = -1$, $t = 1$, and all $t \in (-1, 1)$ for which $P'(t) = 0$.

Step 2: Assume that all local extrema of P in $[-1, 1]$ are maximal. Let $-1 = t_0 < \cdots < t_m = 1$ be the points at which $|P(t)| = 1$, and let $t_i < 0 \leq t_{i+1}$. Consider the partition of $[-1, 1]$ into subintervals

$$[t_0, t_1], \ldots, [t_i, 0], [0, t_{i+1}], \ldots, [t_{m-1}, t_m],$$

on which P is monotone. We denote these subintervals by $I_0, \ldots, I_m$ and associate to them the subintervals $I_0^*, \ldots, I_m^*$, defined similarly by the extremal points $\{\eta_j\}$ of T_n as

$$[\eta_0, \eta_1], \ldots, [\eta_i, \tau_1], [\tau_2, \eta_{i+n-m+1}], \ldots, [\eta_{n-1}, \eta_n],$$

where the points τ_1, τ_2 are chosen from $[\eta_i, \eta_{i+1}]$, $[\eta_{i+n-m}, \eta_{i+n-m+1}]$, respectively, so that $T_n(\tau_1) = T_n(\tau_2) = P(0)$.

In Step 2 we prove that

$$\int_I \phi(|P'(x)|) \, dx \leq \int_{I^*} \phi(|T_n'(x)|) \, dx,$$

for each pair $(I, I^*) = (I_j, I_j^*)$, $j = 0, \ldots, m$, of associated subintervals.

To verify these steps we need the following auxiliary lemmas.

Lemma 4.12. *Assume that $\phi \in \Phi$ and τ is any fixed point in $(-1, 1)$. Then for every polynomial $f \in \pi_n^\circ$ such that $f'(\tau) = 0$, we have*

$$\int_{-1}^1 \phi'(|f'(x)|) \left(\frac{x^2 - 1}{x - \tau} f'(x) \right)' \text{ sign } f'(x) \, dx > 0.$$

Proof: Note first that the integrand $F(x)$ is a piecewise continuous function. For every sufficiently small $\delta > 0$ we denote by $I(\delta)$ the integral of F over the subintervals $E(\delta) := [-1, \tau - \delta] \cup [\tau + \delta, 1]$. Clearly, $\lim_{\delta \to 0} I(\delta) = I(0)$, the last being the integral under consideration. Next we compute $I(\delta)$ by a careful integration by parts and then pass to the limit as $\delta \to 0$ to show that $I(0) > 0$. We have

$$I(\delta) = \int_{E(\delta)} F(x) \, dx$$

$$= \int_{E(\delta)} \phi'(|f'(x)|) \text{ sign } f'(x)$$

$$\times \left\{ \frac{x^2 - 1}{x - \tau} f''(x) + \left(1 + \frac{1 - \tau^2}{(x - \tau)^2} \right) f'(x) \right\} \, dx > 0$$

$$= \int_{E(\delta)} \frac{x^2 - 1}{x - \tau} \, d\phi(|f'(x)|)$$

$$+ \int_{E(\delta)} \phi'(|f'(x)|)|f'(x)| \left(1 + \frac{1 - \tau^2}{(x - \tau)^2} \right) \, dx > 0.$$

An integration by parts in the first integral yields

$$I(\delta) = A(\delta) + \int_{E(\delta)} \left\{ \phi'(|f'(x)|)|f'(x)| - \phi(|f'(x)|) \right\} \left(1 + \frac{1 - \tau^2}{(x - \tau)^2} \right) dx,$$

where we have used the notation

$$A(\delta) := -\left. \frac{x^2 - 1}{x - \tau} \phi(|f'(x)|) \right|_{t-\delta}^{\tau+\delta}.$$

Now it follows from the properties of ϕ that $A(\delta) \to 0$ as $\delta \to 0$. Since $\phi'(x) \geq \phi(x)/x$ for each $x \geq 0$ (with equality if and only if ϕ is a linear function on $(0, x)$), we see that the second term in $I(\delta)$ tends to the integral over $[-1, 1]$ of a nonnegative function, which is not identically zero. Thus $I(0) > 0$ and the proof is complete. $\square$

Lemma 4.13. *Let $\varphi \in C^1[a, b]$ be a strictly increasing function on $[a, b]$ and $f \in C^1[a_1, b_1]$. Assume that f is monotone on $[a_1, b_1]$, $f(a_1) = \varphi(a)$, $f(b_1) = \varphi(b)$, and the equality $f(t) = \varphi(y)$ for some $t \in [a_1, b_1]$ and $y \in [a, b]$ implies*

$$|f'(t)| \leq |\varphi'(y)|.$$

Then

$$\int_{a_1}^{b_1} \phi(|f'(x)|) \, dx \leq \int_a^b \phi(|\varphi'(x)|) \, dx.$$

Proof: We shall establish the inequality by a direct comparison of the corresponding integrals for f and φ. Set

$$m_1 := f(a_1), \qquad m_2 := f(b_1).$$

Consider the inverse function $x = f^{-1}(y)$ on $[m_1, m_2]$. Note that x is continuous on $[m_1, m_2]$ and is even differentiable on the open interval (m_1, m_2). Performing the change of variables $y = f(x)$, we obtain

$$\int_{a_1}^{b_1} \phi(f'(x)) \, dx = \int_{a_1}^{b_1} \frac{\phi(f'(x))}{f'(x)} \, df(x) = \int_{m_1}^{m_2} \frac{\phi(f'(x(y)))}{f'(x(y))} \, dy.$$

Note that in case $f'(a_1) = 0$ or $f'(b_1) = 0$ we still have integrability in the proper sense since the limit $\lim_{x \to 0+} \phi(x)/x = \phi'(0)$ exists by our assumption about ϕ.

Because of the convexity of ϕ, the function $\phi(t)/t$ is nondecreasing on $[0, \infty)$. Then, making use of the majorization property of φ, we get

$$\int_{a_1}^{b_1} \phi(f'(t)) \, dt = \int_{m_1}^{m_2} \frac{\phi(f'(t(y)))}{f'(t(y))} \, dy$$

$$\leq \int_{m_1}^{m_2} \frac{\phi(\varphi'(x(y)))}{\varphi'(x(y))} \, dy = \int_a^b \phi(\varphi'(x)) \, dx.$$

Clearly, equality is attained only if $b_1 - a_1 = b - a$ and f is a translation of φ. □

The proof of Theorem 4.11 now follows easily: Assume that f is an extremal polynomial and $|f(\tau)| < 1$ for some interior critical point τ of f. Then the polynomial

$$g(x) := f(x) + \epsilon \frac{x^2 - 1}{x - \tau} f'(x)$$

will satisfy $\|g\| \leq \|f\| + o(\epsilon)$ while, by Lemma 4.12,

$$\int_{-1}^{1} \varphi(|g'(x)|)\, dx = \int_{-1}^{1} \varphi(|f'(x)|)\, dx + C\epsilon,$$

with some positive absolute constant C. This means that an appropriate normalization of g will produce an admissible polynomial for which the integral is bigger. This contradicts the optimality of f. Therefore, all interior local extrema of f must be equal to 1. Similarly one can show that the values of f at the end-points are equal to 1 in absolute value. Step 1 is completed. Next using the property of the Tchebycheff polynomials described in Lemma 2.2, together with Lemma 4.13, show that the integral of $\varphi(|f'(x)|)$ over every interval of monotonicity of f is majorized by the integral of $\varphi(|T_n'(x)|)$ over the corresponding interval of monotonicity of $T_n(x)$. This finishes the proof.

In 1939 (see [49]), P. Erdős proved that $\cos n\theta$ has the biggest arc-length on $[0, 2\pi]$ among all trigonometric polynomials of degree n, bounded by 1, and made the conjecture that T_n has the biggest arc-length on $[-1, 1]$ among the algebraic polynomials of degree n bounded by 1. In the particular case $\phi(x) = \sqrt{1 + x^2}$ we obtain from Theorem 4.11 that the Tchebycheff polynomial $MT_n(x)$ is the longest one (i.e., it has the biggest arc-length) in the set $\{f \in \pi_n : \|f\| \leq M\}$, thus proving the Erdős conjecture even for any bound M of $\|f\|$. This was proved directly in [20]. Later, Bakan [12] gave an independent proof of Erdős conjecture about the longest polynomial, using essentially that $\|f\| \leq 1$. Note that, although intuitively clear, the claim that "if f is extremal for $M = 1$, then $Mf(x)$ is extremal for any positive M" is not obvious. Even more, the claim, if true, implies easily Erdős conjecture letting M tend to infinity (see the comments in [20]).

The proof of Theorem 4.11 sketched above was presented in [21] in the case $\phi(x) = x^p$, $1 \leq p < \infty$ and reproduced to cover the general case of $\phi \in \Phi$, in [23] (a publication of our lecture given in the international conference of Constructive Function Theory in Varna, May, 1981). The trigonometric variant of Theorem 4.11 was established for the first time by Calderon and Klein [39], and rediscovered later by Taikov [114] and Kristiansen [71]. The part of [71] concerning algebraic polynomials, although repaired subsequently in [72] and [73], is not clear.

The extension of Theorem 4.11 to higher derivatives is still an open problem. The case of the second derivative was studied by Avvakumova [8]. In [21] we formulated the following

Conjecture 4.14. *For every* $P \in \pi_n$ *and* $0 < k < n$,

$$\|P^{(k)}\|_p \leq \|T_n^{(k)}\|_p \|P\|_\infty.$$

The above inequality was proved in [33] for polynomials $P \in \pi_n$ that have n zeros in $[-1, 1]$. Actually, it was derived as a consequence from the following more general result.

Theorem 4.15. ([33]) *Let* $\phi \in \Phi$ *and* $\bar{\nu} = (\nu_1, \ldots, \nu_n)$ *be any fixed set of multiplicities,* $N := \nu_1 + \cdots + \nu_n$. *Then, for each* k, $0 < k \leq N$, *the integral*

$$\int_{-1}^1 \phi(|\tau^{(k)}(\bar{\nu}, \bar{h}; x)|) \, dx$$

is a strictly increasing function of $h_0, \ldots, h_n$ *in the domain* $H := \{\bar{h} : h_0 \geq 0, \ldots, h_n \geq 0\}$.

Choosing $\nu_1 = \cdots = \nu_n = 1$, as an immediate consequence of Theorem 4.15 we get the inequality: For every polynomial $P \in \pi_n^\circ$ which has all its zeros in $[-1, 1]$, and $0 < k < n$,

$$\int_{-1}^1 \phi(|P^{(k)}(x)|) \, dx \leq \int_{-1}^1 \phi(|T_n^{(k)}(x)|) \, dx.$$

The equality is attained only for $P = \pm T_n$.

A Schur-type version of the last inequality also follows from Theorem 4.15 for derivatives of any order $k \leq N$. Moreover, for $k = 1$, the following L_p extension of the classical Schur inequality was obtained in [25]:

$$\|P'\|_p \leq \|\tilde{T}_n'\|_p \|P\|, \quad 1 \leq p < \infty,$$

for every $P \in \pi_n$ such that $P(-1) = P(1) = 0$.

Theorem 4.15 provides a tool for solving Turan-type extremal problems. In 1939 Paul Turan [118] initiated a study of inequalities of the form

$$\|P'\| \geq C(n)\|P\|$$

for polynomials $P \in \pi_n$ which have all their zeros in $[-1, 1]$ (This subset of π_n will be denoted in what follows by $\mathcal{P}_n$). He proved that

$$\|P'\| \geq \frac{1}{6}\sqrt{n}\|P\|$$

for each $P \in \mathcal{P}_n$.

Eröd [54] investigated this problem again, and showed that

$$\|P'\| \ge A_n \|P\| \quad \text{for } P \in \mathcal{P}_n,$$

where $A_2 = 1$, $A_3 = \frac{3}{2}$ and

$$A_{2k} = \frac{2k}{\sqrt{2k-1}} \left(1 - \frac{1}{2k-1}\right)^{k-1},$$

$$A_{2k+1} = \frac{(2k+1)^2}{2k\sqrt{2k+2}} \left(1 - \frac{\sqrt{2k+2}}{2k}\right)^{k-1} \left(1 + \frac{1}{\sqrt{2k+2}}\right)^{k}$$

for $k = 2, 3 \ldots$. The equality is attained for $P(t) = (1-t)^n$ if $n = 1, 2, 3$ and for $P(t) = (1-t)^{n-\lceil n/2 \rceil}(1+t)^{\lceil n/2 \rceil}$, if $n \ge 4$.

The long technical paper of Eröd treating the first derivative might have been a signal that an eventual study of higher derivatives would be really hard. In contrast to these expectations Babenko and Pichugov [9] gave a very short and elegant proof of the corresponding inequality concerning the second derivative. They showed that

$$\|P''\| \ge B_n \|P\|, \quad P \in \mathcal{P}_n$$

with $B_n = \min\{n, n(n-1)/4\}$. For $n = 2, 3, 4, 5$, $B_n = n(n-1)/4$ and the equality is attained only for polynomials P of the form $P(x) = a(1 \pm x)^n$. For $n \ge 6$, $B_n = n$ and in case $n = 2m$ the equality is attained if and only if $P(x) = a(1 - x^2)^m$. In case of odd $n \ge 7$ the exact estimation and the extremal polynomial are not known. The conjecture is that $(1 - x)^m(1 + x)^{m+1}$ is an extremal polynomial if $n = 2m + 1$.

An application of Theorem 4.15 and the technique based on the inheritance theorem (Theorem 4.9) allow us to distinguish a small group of polynomials that are the only candidates for an extremal function in Turan-type inequalities for any L_p norm (see [25]).

Consider the polynomials

$$p_{nm}(x) := (-1)^{n-m} \frac{n^n}{2^n m^m (n-m)^{n-m}} (x+1)^m (x-1)^{n-m}$$

for any natural number n and $m = 0, 1, \ldots, n$. The normalization factor is chosen so that $\|p_{nm}\| = 1$.

The maximum of $p_{nm}(x)$ in $[-1, 1]$ is attained at the point $(2m-n)/n$. Note that all local extrema $\{h_j(p_{nm}\}_{j=0}^n$ of p_{nm} are equal to zero except h_m which is 1.

Theorem 4.16. *Let $\phi \in \Phi$. Then, for every natural n and $k \in \{1, \ldots, n\}$,*

$$\int_{-1}^{1} \phi(|P^{(k)}(x)|)\,dx \ \geq \ A_{nk}\,\|P\| \quad \text{for every } \ P \in \mathcal{P}_n$$

with

$$A_{nk} := \min\left\{ \int_{-1}^{1} \phi(|p_{nm}^{(k)}(x)|)\,dx \ : \ m = 0, \ldots, n \right\}$$

and this is the exact constant. The equality can be attained only for a polynomial from the set $\{p_{n0}, p_{n1}, \ldots, p_{nn}\}$.

Proof: We have to show that the integral

$$I(P) := \int_{-1}^{1} \phi(|P^{(k)}(x)|)\,dx$$

attains its minimal value in $\{P \in \mathcal{P}_n \ : \ \|P\| = 1\}$ for one of the polynomials $\{p_{n0}, p_{n1}, \ldots, p_{nn}\}$. To do this, note that if $P \in \mathcal{P}_n$ and $\|P\| = 1$, then $h_m(P) = 1$ for some $m \in \{0, \ldots, n\}$. But, according to Theorem 4.15, $I(P)$ is a strictly increasing function of $h_0(P), \ldots, h_n(P)$. Therefore, if $P \neq p_{nm}$, then

$$I(P) > I(p_{nm}) \geq A_{nk}$$

and the proof is complete. $\square$

Similarly, using the inheritance theorem we obtain the next Turán-type inequality for the kth derivative in the uniform norm (see [25]).

Theorem 4.17. *For any given n and k, and every $P \in \mathcal{P}_n$,*

$$\|P^{(k)}\| \geq c_{nk}\,\|P\|,$$

with $c_{nk} := \min\{\|p_{nm}^{(k)}\| \ : \ m = 0, \ldots, n\}$.

It would be of interest to find a simpler way of proving that $c_{n1} = \|p'_{nm}\|$ for $m = [n/2]$. Eröd's proof is rather elaborate.

Similar problems under additional conditions of the form $P(1) = 1$ and/or $P(-1) = 1$ are solved in [25]. One example:

Let $P \in \mathcal{P}_n$ and $P(1) = 1$. Then, for each $k = 1, \ldots, n$ and $1 \leq q < \infty$,

$$\|P^{(k)}\|_q \ \geq \ \frac{n!}{(n-k)!}\frac{1}{2^k}\left(\frac{2}{(n-k)q+1}\right)^{1/q}.$$

The equality is attained only for $P = p_{nn}(x) = \frac{1}{2^n}(x+1)^n$.

The particular case corresponding to $q = 2$ and $k - 1$ was obtained earlier by Varma [122] in a different way.

The trigonometric versions of Theorem 4.15 and the inheritance theorem were proved in [24]. They imply, as in the algebraic case, Turan-type inequalities. Because of the periodicity, the candidates for extremal polynomials (i.e., the trigonometric analog of $\{p_{nm}\}$) reduce to a single polynomial and this fact leads to the complete characterization of the extremal polynomial.

Let us denote by $\mathcal{T}_n$ the subset of trigonometric polynomials s_n of degree n which have $2n$ zeros in $[0, 2\pi)$ and $\|s_n\| := \max_{\theta \in [0,2\pi]} |s_n(\theta)| = 1$. The following was proved in [24].

Theorem 4.18. *For any $\phi \in \Phi$ and every $s_n \in \mathcal{T}_n$,*

$$\min_{s_n \in \mathcal{T}_n} \int_0^{2\pi} \phi(|s_n^{(k)}(\theta)|) \, d\theta = \int_0^{2\pi} \phi \left(\left| \left[\left(\cos \frac{\theta}{2} \right)^{2n} \right]^{(k)} \right| \right) d\theta,$$

for every natural k. The polynomial $(\cos \frac{\theta}{2})^{2n}$ is the unique extremal function (up to translation).

The case $k = 1$ was considered earlier by Tyrygin [120]. We remark that Turan [118] also established an exact estimate from below of the derivative of a polynomial P on the unit disk D. He proved the following: If $P(z)$ is a polynomial of degree n having all its zeros in the closed unit disk $\{z : |z| \leq 1\}$, then

$$\max_{|z|=1} |P'(z)| \geq \frac{n}{2} \max_{|z|=1} |P(z)|.$$

The result is best possible and equality holds for $P(z) = az^n + b$ where $|a| = |b|$.

There are further extensions of this inequality, the most striking one due to Malik [80]: If $P(z)$ is a polynomial of degree n having all its zeros in $|z| \leq 1$, then for each $q > 0$

$$n \left(\int_0^{2\pi} |P(e^{i\theta})|^q \, d\theta \right)^{1/q} \leq (A_q)^{\frac{1}{q}} \max_{|z|=1} |P'(z)|$$

where

$$A_q = 2^{q+1} \sqrt{\pi} \Gamma(\frac{1}{2}q + \frac{1}{2}) / \Gamma(\frac{1}{2}q + 1).$$

The result is best possible and the equality holds for $P(z) = az^n + b$ where $|a| = |b|$. In the more restricted situation when P is supposed to have all its zeros on the circle $|z| = 1$, the same result was proved in [100].

§5. Landau-Kolmogorov Inequality

The intensive research over the years on the size of a derivative of a function from a given class in terms of its norm and the norm of a higher derivative can be looked upon as a further development of the topic initiated by the brothers Markov. The literature on this subject is vast. We shall concentrate only on those results that give exact inequalities and complete characterization of the extremal function. Kolmogorov's inequality distinguishes among these results as a completed study of an important natural question in its general form. He proved in [67] that

$$\|f^{(k)}\|_{L_\infty(\mathbf{R})} \le C_{rk} \|f\|_{L_\infty(\mathbf{R})}^{(r-k)/r} \|f^{(r)}\|_{L_\infty(\mathbf{R})}^{k/r}$$

for every function $f : \mathbb{R} \to \mathbb{R}$ from the class

$$W_\infty^r(\mathbb{R}) := \{f : f^{(r-1)} \text{ locally abs. cont. with } \|f^{(r)}\|_{L_\infty(\mathbf{R})} < \infty\}.$$

The exact constant is

$$C_{rk} = \frac{K_{r-k}}{K_r^{1-k/r}},$$

where K_r is the famous Favard-Akhiezer-Krein constant,

$$K_r = \frac{4}{\pi} \sum_{\nu=0}^\infty \frac{(-1)^{\nu(r+1)}}{(2\nu+1)^{r+1}}.$$

The only extremal (up to translation and multiplication by a non-zero constant) to this inequality is the Euler perfect spline

$$\varphi_r(x) := \frac{4}{\pi} \sum_{m=0}^\infty \frac{\sin((2m+1)x - \frac{\pi}{2}r)}{(2m+1)^{r+1}}.$$

Let us recall that (see, e.g., [68], §2.3)

$$\varphi_r \text{ is a } 2\pi\text{-periodic function from } W_\infty^r(\mathbb{R});$$

$$\varphi_r^{(r)}(x) = \text{sign } \sin x,$$

and $\|\varphi_r\|_{C(\mathbf{R})} = K_r$.

The main ingredient of the Kolmogorov proof is the following wonderful extremal property of the Euler spline $\varphi_r(x)$ which resembles the property of $\cos n\theta$ proved by de La Vallée Poussin and given in Section 2 as Theorem 2.1. Kolmogorov calls a function having this property a comparison function. Let us formulate the property.

For each $f \in W_\infty^r(\mathbb{R})$ we set

$$M_0(f) := \|f\|_{C(\mathbb{R})}, \quad M_r(f) := \|f^{(r)}\|_{L_\infty(\mathbb{R})}.$$

Then, on the basis of φ_r, define the function

$$\varphi(f; x) := c\varphi_r(\lambda x)$$

where c and λ are chosen so that

$$\|\varphi(f; \cdot)\|_{C(\mathbb{R})} = M_0(f), \quad \|\varphi^{(r)}(f; \cdot)\|_{L_\infty(\mathbb{R})} = M_r(f).$$

It is easy to verify that

$$c = \frac{M_0(f)}{K_r} \quad \text{and} \quad \lambda^r = \frac{M_r(f)K_r}{M_0(f)}.$$

The following was established by Kolmogorov [67].

Theorem 5.1. *Assume that $f \in W_\infty^r(\mathbb{R})$. Then $\varphi(f; \cdot)$ is a comparison function for f, that is,*

$$|f'(x_0)| \leq |\varphi'(f; x_1)|$$

provided $f(x_0) = \varphi(f; x_1)$.

Kolmogorov's inequality follows from Theorem 5.1. As was mentioned in [37] and [67], Shilov has proved earlier the inequality for $r \leq 4$, $k \leq r$, and $r = 5$, $k = 2$. He also conjectured the explicit form of the extremal function for all r.

Note that the problem of estimating the first derivative of f on a given interval I on the basis of the uniform norm of f and f'' on I comes from E. Landau. He found in [76] the exact upper bound of $\|f'\|_{C(I)}$ in the case I is the half-line $\mathbb{R}_+$. That is why the problems of estimating intermediate derivatives are often referred to as Landau-Kolmogorov problems. Hadamard [60] solved the same problem on the whole line $\mathbb{R}$. Schoenberg and Cavaretta [105] described the extremal function to the problem on the half-line $\mathbb{R}_+$ for each r. Motorin [87] did it earlier for $r = 3$ and $k = 1, 2$. There are only a few cases in which exact inequalities of the type considered, involving other norms, are known. Hardy, Littlewood and Pólya [61] (see also [62]) established such an inequality for the L_2-norm on $\mathbb{R}$ and Stein [112] did this for L_1. A detailed account on this subject is given in [70,75,109,116]. See also [69] and the comments by Tikhomirov in [67] on the Kolmogorov result and other sharp inequalities on infinite intervals.

One of the oldest open problems in constructive function theory is the characterization of the extremal function for Kolmogorov's inequality in the case of a finite interval. We shall present in this section some recent results from [29,30,31] which are related to this open problem.

The Kolmogorov inequality is equivalently reduced to the extremal problem

$$\|f^{(k)}\|_{C(\mathbf{R})} \to \sup$$

over all functions $f \in W_\infty^r(\mathbb{R})$ such that $\|f^{(r)}\| \le M_r$, $|f(x)| \le M_0$ on $(-\infty, \infty)$.

For given $\phi \in \Phi$, positive numbers M_0 and M_r, and a finite interval $[a, b] \subset I$, we define the **Landau-Kolmogorov problem** as:

$$J_\phi(f^{(k)}; [a, b]) := \int_a^b \phi(|f^{(k)}(x)|)\, dx \qquad \to \qquad \sup$$

over the set of all functions from $W_\infty^r(I)$ satisfying the conditions

$$\|f\|_{C(I)} \le M_0, \qquad \|f^{(r)}\|_{L_\infty(I)} \le M_r.$$

The solution of this problem was given in [29] in case $I = \mathbb{R}$ and arbitrary bounded interval $[a, b]$. In order to formulate the result we need some notation.

Let $[a, b]$ be a fixed subinterval of $\mathbb{R}$. For a given $f \in W_\infty^r(\mathbb{R})$ with

$$M_0(f) := \|f\|_{C(I)} < \infty \quad \text{and} \quad M_r(f) := \|f^{(r)}\|_{L_\infty(I)} < \infty,$$

we define by appropriate compression and normalization of φ_r the spline $\varphi_r([a, b], f; x)$ with the properties:

$$\|\varphi_r([a, b], f; \cdot)\|_{C(\mathbf{R})} = M_0(f)$$

$$\|\varphi_r^{(r)}([a, b], f; \cdot)\|_{L_\infty(\mathbf{R})} = M_r(f).$$

Let ω be the half-period of $\varphi_r([a, b], f; x)$ and $b - a = N\omega + 2\theta$, $2\theta < \omega$ (N is an integer). We assume in addition that $\varphi_r([a, b], f; x)$ is shifted so that for a given k (which is clear from the context),

$$\varphi_r^{(k-1)}([a, b], f; a + \theta) = 0, \quad \varphi_r^{(k)}([a, b], f; a + \theta) > 0.$$

The following Landau-Kolmogorov type inequality was proved in [29].

Theorem 5.2. *Assume that $\phi \in \Phi$ and $[a, b]$ is any finite subinterval of $\mathbb{R}$. Then for every function $f \in W_\infty^r(\mathbb{R})$ and $k = 1, \ldots, r - 1$,*

$$\int_a^b \phi(|f^{(k)}(x)|)\, dx \le \int_a^b \phi(|\varphi_r^{(k)}([a, b], f; x)|)\, dx.$$

Moreover, the equality holds only for $f(x) = \pm\varphi_r([a, b], f; x)$, if $b - a \ne m\omega$ for every integer m, and only for the translations of $\varphi_r([a, b], f; x)$, if $b - a = m\omega$ for some integer m.

Choosing $\phi(x) = x^p$ we get an estimate of the L_p-norm of the derivative.

Corollary 5.3. *Let* $1 \leq p < \infty$ *and* $[a, b]$ *be any finite subinterval of* $\mathbb{R}$ *with* $b - a = N\frac{\pi}{\lambda} + 2\theta$, $2\theta < \frac{\pi}{\lambda}$. *Then for every function* $f \in W^r_\infty(\mathbb{R})$ *and* $k = 1, \ldots, r - 1$,

$$\|f^{(k)}\|_{L_p[a,b]} \leq A_{rkp\lambda}([a, b]) M_0(f)^{1 - \frac{k}{r}} M_r(f)^{\frac{k}{r}},$$

where

$$A_{rkp\lambda}([a, b]) = \frac{1}{K_r^{1 - \frac{k}{r}}} \|\varphi_{r-k}(\lambda(\cdot - a - \theta) + (r - k + 1)\frac{\pi}{2}\|_{L_p[a,b]}.$$

Let us recall that the quantity λ was defined in the text just before Theorem 5.1.

Under the additional restriction that the functions f are 2π-periodic, Ligun [78] obtained the above inequality in the particular case when $[a, b] = [0, 2\pi]$.

The method of proof of Theorem 5.2 uses essentially the fact that the Euler perfect spline φ_r is a comparison function in the class $W^r_\infty[a, b]$. According to Theorem 2.1, $\cos n\theta$ plays the same role in the space τ_n of trigonometric polynomials of degree n. This allowed the authors of [29] to modify their method and apply it to solve an extremal problem for trigonometric polynomials raised by Erdős a long time ago (see [48, 51]). We now formulate the result.

As in the previous case we write $b - a$ in the form $b - a = N\frac{\pi}{n} + 2\theta$ with $2\theta < \frac{\pi}{n}$, and then set

$$sc_k(x) := \begin{cases} \cos n(x - a - \theta) & \text{if } k \text{ is even} \\ \sin n(x - a - \theta) & \text{if } k \text{ is odd.} \end{cases}$$

Theorem 5.4 ([29]). *Assume that* $\phi \in \Phi$ *and that* M *is an arbitrary fixed positive number. Let* $[a, b]$ *be any real finite interval. Then*

$$\int_a^b \phi(|f^{(k)}(x)|) \, dx \leq \int_a^b \phi(M|sc_k^{(k)}(x)|) \, dx$$

for every trigonometric polynomial $f \in \tau_n$ *with* $\|f\|_{C(\mathbb{R})} \leq M$. *Moreover, equality holds only for* $f(x) = \pm M \, sc_k(x)$, *if* $2\theta \neq 0$, *and only for the translates of* $M \sin nx$, *if* $b - a = N\frac{\pi}{n}$.

Paul Erdős proved in [49] that the trigonometric polynomial $\cos nx$ has the longest arc-length on the period $[0, 2\pi]$ among all trigonometric polynomials of degree n which are bounded by 1. He posed the same question for the arc-length of trigonometric polynomials on any fixed subinterval $[a, b]$ (see [48,51]). Taking $\phi(x) = \sqrt{1 + x^2}$ and $k = 1$ in Theorem 5.4, we get the answer.

Corollary 5.5. *Assume that M is an arbitrary fixed positive number. Let $[a,b]$ be any real finite interval, $b - a = N\frac{\pi}{n} + 2\theta$. Then the arc-length over $[a,b]$ of any trigonometric polynomial $f \in \tau_n$ bounded by M on $\mathbb{R}$ is less than the arc-length of $s(x) := M\sin n(x - a - \theta)$ (unless f coincides with s).*

Note that Theorem 5.4 contains as a particular case (when $[a,b] = [0, 2\pi]$) the result of Calderon and Klein [39].

The similar question (see Erdős [51]) about the polynomial of longest arc-length on a subinterval $[a,b]$ of $[-1,1]$, in the class of all algebraic polynomials of degree n that are bounded by 1 on $[-1,1]$, is still open.

Kolmogorov's inequality has been extended in various directions – with respect to L_p-norms (see [69,70,75,116]), to nonsymmetric norms (see, for example, [63]), and to the half-line ([11]). However, despite the effort of many mathematicians, the problem is still unsettled in the most interesting case of functions given on a finite interval $[a,b]$. Chui and Smith [40] found the solution for $r = 2$, and the case $r = 3$ was studied in [101] and [126]. A thorough study of the problem for each r was done by Pinkus [94]. He showed that the extremal function is in the set of perfect splines of degree r which have almost maximal number of zeros in $[a,b]$. This was our, with N. Naidenov, motivation to consider in [30] the problem of maximizing the functional $J_\phi(f^{(k)})$ in the set of oscillating perfect splines (i.e., perfect splines that have maximal number of zeros in $[a,b]$). Extending the method we used for oscillating polynomials, we gave a complete characterization of the extremal spline.

Recall that a **perfect spline** of degree r with knots $\xi_1 < \cdots < \xi_{n-r}$ is any expression of the form

$$s(t) = \sum_{j=1}^{r} \alpha_j t^{j-1} + c\left(t^r + 2\sum_{i=1}^{n-r}(-1)^i(t - \xi_i)_+^r\right) \tag{5.1}$$

with real coefficients $c, \alpha_1, \ldots, \alpha_r$. In other words, s is a perfect spline if $s^{(r)}(t)$ is a piecewise constant function which changes sign only at the knots and $|s^{(r)}(t)| = \text{const} = r!|c|$ for each t (except the knots, since the r-th derivative does not exist at the knots). The perfect splines are natural generalizations of the algebraic polynomials and preserve some of their properties. For example, the fundamental theorem of algebra holds for perfect splines in the following form:

Theorem 5.6. *For any given set of points $x_1 \leq \cdots \leq x_n$, with $x_j < x_{j+r}$ $\forall j$, there exists a unique (up to a constant factor) perfect spline $s(t)$ of degree r with at most $n - r$ knots which vanishes at $x_1, \ldots, x_n$. The multiple points are counted as multiple zeros.*

The proof can be seen, for example, in [28]. Beautiful extremal properties of the perfect splines are discovered in [35,66,83,95].

It follows from Rolle's theorem that if s vanishes at n points (counting multiplicities up to r), then s must have at least $n - r$ knots. Therefore, any perfect spline of degree r and $n - r$ knots may have at most n zeros. Let us denote by $\mathcal{S}_{nr}$ the set of all perfect splines (5.1) which have exactly n (thus, the maximal number) simple zeros in $[-1, 1]$.

It was shown recently in [27] that the perfect splines from $\mathcal{S}_{nr}$ possess the Markov interlacing property. In particular, the following is true.

Let us denote by $p(\mathbf{x}; t)$ the spline from $\mathcal{S}_{nr}$ that vanishes at the points $\mathbf{x} = (x_1, \ldots, x_n)$.

Theorem 5.7. *Let* $\mathbf{x}$ *and* $\mathbf{y}$ *interlace. Then the zeros of* $p'(\mathbf{x}; t)$ *and* $p'(\mathbf{y}; t)$ *also interlace. If the points* $\mathbf{x} = (x_1, \ldots, x_n)$ *and* $\mathbf{y} = (y_1, \ldots, y_n)$ *satisfy the strict interlacing conditions*

$$x_1 < y_1 < x_2 < \cdots < x_n < y_n,$$

then the zeros of $p'(\mathbf{x}; t)$ *and* $p'(\mathbf{y}; t)$ *strictly interlace.*

Other properties and applications of perfect splines are given in [28]. Next we return to Markov-type problems for perfect splines. For each $s \in \mathcal{S}_{nr}$ there exist exactly $n - 1$ points $t_1 < \cdots < t_{n-1}$ in $(-1, 1)$ such that

$$s'(t_j) = 0, \quad j = 1, \ldots, n - 1.$$

As in the polynomial case, we denote by $\{h_j(s)\}_{j=0}^n$ the absolute values of the local extrema of $s(t)$ in $[-1, 1]$ including the values at the end-points, namely, $h_j(s) := |s(t_j)|$, $j = 1, \ldots, n - 1$, $h_0(s) := |s(-1)|$, $h_n(s) := |s(1)|$. The following is a particular case of a result from [16] (see also [28], Theorem 6.10).

Theorem 5.8. *Let* $[a, b]$ *be a fixed bounded interval. Then for every set of numbers* $h_0 > 0, \ldots, h_n > 0$, *there exists a unique system of points*

$$a =: t_0 < t_1 < \cdots < t_{n-1} < t_n := b$$

and a unique perfect spline $s(t) \in \mathcal{S}_{nr}$ *of degree* r *with* $n - r$ *knots such that*

$$s(t_k) = (-1)^k h_k, \quad k = 0, \ldots, n,$$
$$s'(t_k) = 0, \quad k = 1, \ldots, n - 1.$$

Moreover, $\|s^{(r)}\|_{L_\infty[a,b]}$ *is a strictly increasing function of* $h_0, h_1, \ldots, h_n$ *in the domain* $h_k > 0$, $k = 0, \ldots, n$.

Therefore, every spline $s \in \mathcal{S}_{nr}$ is completely determined by its local extrema $\{h_k(s)\}_0^n$. It turned out that the inheritance theorem concerning $\{h_k\}$ is true also for perfect splines. This was shown in [30].

Theorem 5.9. *Let s and f be any two perfect splines from the set S_{nr}. Assume that*

$$h_j(s) \le h_j(f) \quad \text{for } j = 0, \dots, n.$$

Then

$$h_j(s') \le h_j(f') \quad \text{for } j = 0, \dots, n-1.$$

As an immediate consequence of this theorem we get as in the polynomial case the following Markov-type inequality. Let us denote by $T_{rn}(t)$ the Tchebycheff perfect spline. This is the unique (up to a constant factor) perfect spline from S_{nr} that takes it maximal value on $[-1, 1]$ with alternating sign at $n + 1$ points in $[-1, 1]$. We shall normalize T_{rn} by the condition

$$\max_{-1 \le x \le 1} |T_{rn}(x)| = 1.$$

Corollary 5.10 (Markov inequality). *For each $s \in S_{nr}$ and $1 \le k \le r$, we have*

$$\|s^{(k)}\| \le \|T_{rn}^{(k)}\| \, \|s\|.$$

Using the inheritance theorem, the Markov interlacing property, and following the method used in the polynomial case ([33]), a Markov-type inequality with respect to a general integral norm was proved in [30].

Theorem 5.11 ([30]). *Let $\phi \in \Phi$. The integral*

$$\int_{-1}^{1} \phi(|s^{(k)}(\bar{h}; x)|) \, dx$$

is a strictly increasing function of $h_0, \dots, h_n$ in the domain $H := \{\bar{h} : h_0 > 0, \dots h_n > 0\}$.

Corollary 5.12. *Let $\phi \in \Phi$. Then*

$$\int_{-1}^{1} \phi(|s^{(k)}(x)|) \, dx \le \int_{-1}^{1} \phi(|T_{rn}^{(k)}(x)|) \, dx, \quad 1 \le k \le r,$$

for every perfect spline s of degree r with $n-r$ knots and n zeros in $[-1, 1]$ such that $\|s\| = 1$. The equality holds only for $s = \pm T_{rn}$.

Taking $\phi(t) = t^p$, $1 \le p < \infty$, we get a Markov-type inequality in L_p for oscillating perfect splines:

$$\|s^{(k)}\|_{L_p[-1,1]} \le \|T_{rn}^{(k)}\|_{L_p[-1,1]} \, \|s\|, \quad 1 \le k \le r.$$

Another interesting example is $\phi(t) = \sqrt{1 + t^2}$, $k = 1$, in which one gets that the Tchebycheff perfect spline T_{nr} has a longest arc-length among all perfect splines s of degree r with $n - r$ knots in $[-1, 1]$ and n zeros, counting the multiplicities, provided $\|s\| \le 1$.

All inequalities presented here for perfect splines hold also in the space $S_r(\xi_1, \dots, \xi_{n-r})$ of splines of a given degree r with fixed knots $\xi_1 < \cdots < \xi_{n-r}$ (see [30]).

§6. Open Problems

Some investigations of the Kolmogorov inequality for a bounded interval in the simplest case $r = 2$ indicate that the characterization of the extremal function for an arbitrary interval $[a, b]$ is a very difficult problem. On the other hand, the results from [91] and [31] show that the Tchebycheff perfect spline T_{rn} is a strong candidate to be extremal in important particular cases.

Conjecture 1. *Let* $f \in W_\infty^r[-1, 1]$, $\|f\| \leq 1$ *and* $\|f^{(r)}\| \leq \|T_{rn}^{(r)}\|$. *Then, for each* $1 \leq k \leq r$, *we have*

$$\|f^{(k)}\|_p \leq \|T_{rn}^{(k)}\|_p, \quad 1 \leq p \leq \infty.$$

The conjecture was proved for $r = 2$ and for $r = 3$ in [31]. The more general problem with an arbitrary bound for $\|f^{(r)}\|$ was completely solved earlier by Naidenov [91] in the case $r = 2$. For $p = \infty$ one arrives at Kolmogorov's inequality for a bounded interval.

A particular case of the above problem concerning the Tchebycheff polynomials T_r is of special interest. Let us formulate it.

Conjecture 2. *Let* $f \in W_\infty^r[-1, 1]$, $\|f\| \leq 1$ *and* $\|f^{(r)}\| \leq T_r^{(r)}(1)$. *Then, for each* $1 \leq k \leq r$, *we have*

$$\|f^{(k)}\|_p \leq \|T_r^{(k)}\|_p, \quad 1 \leq p \leq \infty.$$

The case $p = \infty$ was proved by Eriksson [53]. A further simplification of Conjecture 2 concerning a purely polynomial problem was announced earlier in [21]. We recall it.

Conjecture 3. *Let* P *be any polynomial of degree* r *such that* $\|P\| \leq 1$ *on* $[-1, 1]$. *Then*

$$\|P^{(k)}\|_p \leq \|T_r^{(k)}\|_p, \quad 1 \leq p < \infty.$$

This would be a beautiful extension of the Vladimir Markov inequality. The conjecture is proved for $k = 1$ (see [21]). The case $k = 2$ was studied by Avvakumova [8]. For arbitrary k, the conjecture is proved for every polynomial that has all its zeros in $[-1, 1]$ (see [33]). Estimates for the quantities $\|T_r'\|_p$, $1 \leq p < \infty$, are given by Ciesielski [41].

An eventual way to prove Conjecture 3 is to show that the extremal polynomial belongs to the subset of polynomials which have all their zeros in $[-1, 1]$. This would follow, for example, if one can prove the following.

Conjecture 4. *Let G be a bounded convex domain containing $[-1, 1]$. Let $\hat{\pi}_n$ be the set of all polynomials of degree n which have no zeros in G. For a fixed q, $1 \leq q \leq \infty$, let P_* be the extremal polynomial for the problem*

$$\frac{\|P^{(k)}\|_q}{\|P\|} \quad \to \quad \sup \quad over \quad P \in \hat{\pi}_n.$$

Prove that all zeros of P_* lie on the boundary of G.

The above problem was announced by the author for the first time in the conference "Open Problems in Approximation Theory", Voneshta voda, June 18–24, 1993. A result of this kind concerning the L_2-norm was obtained recently by Akopyan [1].

In connection with the extension of the Duffin-Schaeffer inequality in the complex plane, with respect to other polynomials $Q(x)$ (in the role of T_n), it would be helpful to describe a wide class of polynomials $Q \in \pi_n$ with all their zeros in $[-1, 1]$, having the property

$$|Q(x + iy)| \leq |Q(1 + iy)|, \quad -1 \leq x \leq 1, \quad -\infty < y < \infty. \tag{6.1}$$

Question 5. *Let $Q(x) = (x - x_1) \cdots (x - x_n)$ with $-1 \leq x_1 < \cdots < x_n \leq 1$, and $x_j = -x_{n-j}$ for all j. Suppose that Q satisfies one of the following two conditions:*

a) $\quad |Q^{(k)}(x)| \leq Q^{(k)}(1)$ *for* $k = 0, \ldots, n$, $x \in (-1, 1)$;

b) *The local extrema of $|Q(x)|$ in $[0, 1]$, including the values $|Q(0)|$ and $|Q(1)|$, form an increasing sequence.*

Is it true that Q satisfies (6.1)?

Geno Nikolov and Alexei Shadrin conjectured (in a private communication) that if a polynomial Q has only real zeros and they lie in $[-1, 1]$, then the condition $\max_{-1 \leq x \leq 1} |Q(x)| = |Q(1)|$ would imply that $|Q'(x)|$ attains its maximal value in $[-1, 1]$ at -1 or 1. In view of this, Question 5 can be formulated in the following stronger version: Let Q be as above, and let $|Q(x)| \leq |Q(1)|$ on $[-1, 1]$. Is it true that Q satisfies (6.1)?

There is hope that a more detailed study of the monotone dependence of the integral $\int_{-1}^{1} \phi(|P'(x)|) \, dx$ and the leading coefficient of P on the local extrema $\{h_j(P)\}$ of P will result in the characterization of the extremal polynomial in the next problem.

Problem 6. *Find the monic polynomial $P(x) = (x - x_1) \cdots (x - x_n)$, $-1 \leq x_1 \leq \cdots \leq x_n \leq 1$, of minimal (maximal) arc-length in $[-1, 1]$.*

A considerable effort was expanded to find the exact error estimate in the class $W_\infty^n[-1, 1]$ of the differentiation formula

$$f^{(k)}(x) \approx \sum_{j=1}^{n} f(x_j)\ell_{nj}^{(k)}(x), \quad 1 \leq k \leq n - 1,$$

based on Lagrange interpolation at any fixed system of nodes $\{x_j\}$ in $[-1, 1]$. This reduces to the extremal problem

$$\| f^{(k)} \| \quad \rightarrow \quad \sup$$

over all functions $f \in W^n_\infty[-1, 1]$ such that $\| f^{(n)} \|_\infty \leq 1$ and $f(x_1) = \cdots = f(x_n) = 0$. It was shown recently by Shadrin [110] that the supremum is attained for the polynomial $\omega_n(x) := (x - x_1) \cdots (x - x_n)/n!$. This gives rise to the following.

Question 7. *Is it true that for each $1 \leq p < \infty$ the supremum of $\| f^{(k)} \|_p$ over the same subset of $W^n_\infty[-1, 1]$ as above is attained again at ω_n?*

Acknowledgments. The author is grateful to his colleagues Geno Nikolov and Nikola Naidenov for a number of constructive suggestions. The research was supported by the University of Sofia Science Foundation under Contract No 328, 2000/2001.

References

1. Akopyan, R. R., Bernstein–Jackson inequality for algebraic polynomials with restrictions on their zeros, East J. Approx. **7** (2001), 351–370.

2. Andrushkiw, J. W., Polynomials with prescribed values at critical points, Bull. Amer. Math. Soc. **62** (1956), 243.

3. Arestov, V. V., On integral inequalities for trigonometric polynomials and their derivatives, Izv. Akad. Nauk SSSR (Ser. Mat.) **45** (1981), 3–22 (in Russian); English transl.: Math. USSR Izvestiya **18** (1982), 1–17.

4. Arnold, V. I., Topological problems of the theory of wave diffusion, Uspehi Mat. Nauk **51**, 1 (1996), 1–50.

5. Arnold, V. I., Topological classification of real trigonometric polynomials and cyclic serpents polyhedron, *Arnold-Gelfand Math. Sem.* Basel, Birkhäuser, 1997, 101 – 106.

6. Arsenault, M., and Q. I. Rahman, On two polynomial inequalities of Erdős related to those of brothers Markov, J. Approx. Theory **84** (1996), 197–235.

7. Atkinson, K., and A. Sharma, A partial characterization of poised Hermite-Birkhoff interpolation problems, SIAM J. Numer. Anal. **6** (1969), 230–235.

8. Avvakumova, L. S., An extension of V. A. Markov's inequality for second derivative to a wider polynomial class, East J. Approx. **6**, 4 (2000), 493–519.

9. Babenko, V. F., and S. A. Pichugov, On inequalities for derivatives of polynomials with real zeros, Ukrainian Math. J. **38**, 4 (1986), 411–416.

10. Babenko, V. F., V. A. Kofanov, and S. A. Pichugov, Inequalities for norms of intermediate derivatives of periodic functions and their applications, East J. Approx. **3**, 3 (1997), 351–376.

11. Babenko, V. F., V. A. Kofanov, and S. A. Pichugov, Landau-Kolmogorov-Hörmander inequalities on the semiaxis, Mat. Zametki **65**, 2 (1999), 175–185; English transl.: Math. Notes **65**, 2 (1999), 144–152.

12. Bakan, A., On a certain extremal property of the Tchebycheff polynomials, in *Mathematics Today*, Visha shkola, Kiev, 1983, pp. 167–173 (in Russian).

13. Bernstein, S. N., On the best approximation of continuous functions by polynomials of a given degree, Soobsht. Khark. mat. obshchestva. Ser. 2, v.**13** (1912), 49–194 (in Russian).

14. Bernstein, S. N., Sur une propriété des polinômes, Comm. Soc. Math. Kharkow Sér. 2, **14**, 1-2 (1913), 1–6.

15. Bernstein, S. N., Sur la limitation des dérivées des polynômes, C. R. Acad. Sci. Paris **190** (1930), 338–341.

16. Bernstein, S. N., On V. A. Markov's theorem, Trudy Leningr. industr. instituta, No 5, Razdel fiz.-mat. nauk 1 (1938), 8–13; Collected works, Volume 2, 281–286 (in Russian).

17. Bernstein, S. N., Collected Works, AN USSR, Moscow, 1952–1954, Vol. 1–4.

18. Bojanov, B., Perfect splines of least uniform deviation, Anal. Math. **6** (1980), 185–193.

19. Bojanov, B., A generalization of Chebyshev polynomials. J. Approx. Theory **26**, 4 (1979), 293–300.

20. Bojanov, B., Proof of a conjecture of Erdös about the longest polynomial. Proc. Amer. Math. Soc. **84**, 1 (1982), 99–103.

21. Bojanov, B., An extension of the Markov inequality. J. Approx. Theory **35**, 2 (1982), 181–190.

22. Bojanov, B., A generalization of the Markov inequality, Dokl. Akad. Nauk USSR **262**, 1(1982); English transl.: Soviet Math. Dokl. **25** 1 (1982), 4–5.

23. Bojanov, B., An extremal problem for polynomials, in *Constructive Function Theory'81"*, BAN, Sofia, 1983, 229–233.

24. Bojanov, B., Turan's inequalities for trigonometric polynomials, J. London Math. Soc. (2), **53** (1996), 539–550.

25. Bojanov, B., Polynomial inequalities, in B. Bojanov (ed.), *Open Problems in Approximation Theory* , 25–42, CST Publishing, Singapore, 1994.

26. Bojanov, B., An inequality of Duffin and Schaeffer type, East J. Approx. **1**, 1 (1995), 37–46.

27. Bojanov, B., Markov interlacing property for perfect splines, J. Approx. Theory **100** (1999), 183–201.

28. Bojanov, B., H. Hakopian, and A. Sahakian, *Spline Functions and Multivariate Interpolations*, Kluwer Academic Publishers, Dordrecht, 1993.

29. Bojanov, B., and N. Naidenov, An extension of the Landau-Kolmogorov inequality. Solution of a problem of Erdős, J. Analyse Math. **78** (1999), 263–280.

30. Bojanov, B., and N. Naidenov, Exact Markov-type inequalities for oscillating perfect splines, Constr. Approx. **18** (2002), 37–59.

31. Bojanov, B., and N. Naidenov, Examples of Landau-Kolmogorov inequality in integral norm on a finite interval, J. Approx. Theory, to appear.

32. Bojanov, B., and G. Nikolov, Duffin and Schaeffer type inequality for ultraspherical polynomials, J. Approx. Theory **84**, 2 (1996), 129–138.

33. Bojanov, B., and Q. I. Rahman, On certain extremal problems for polynomials, J. Math. Anal. Appl. **189** (1995), 781–800.

34. Bojanic, R., Private communication.

35. de Boor, C., A remark concerning perfect splines, Bull. Amer. Math. Soc. **80** (1974), 723–727.

36. Borwein, P., and T. Erdely, *Polynomials and Polynomial Inequalities*, Springer, Berlin-Heidelberg-New York, 1995.

37. Bosse, Yu. G. (G. E. Shilov), On inequalities between derivatives, Sb. Rabot Stud. Nauch. Kruzhkov MGU **1** (1937), 17–27 (in Russian).

38. de Bruijn, N. G., Inequalities concerning polynomials in the complex domain, Nederl. Akad. Wetensch. Proc. 50 (1947), 1265–1272 = Indag. Math. **IX** (1947), 591–598.

39. Calderon, A. P., and G. Klein, On an extremum problem concerning trigonometric polynomials, Studia Math. **12** (1951), 166–169.

40. Chui, C. K., and P. W. Smith, A note on Landau's problems for bounded intervals, Amer. Math. Monthly **82** (1975), 927–929.

41. Ciesielski, Z., On the A. A. Markov inequality for polynomials in the L^p case, in *Approximation Theory* (Memphis, TN, 1991), 257–262, Lecture Notes in Pure and Appl. Math., **138**, Dekker, New York, 1992.

42. van der Corput, J. G., and G. Schaake, Ungleichungen für Polinome und trigonometrische Polinome, Compositio Math. **2** (1935), 321–361; Correction ibid., **3** (1936), 128.

43. Davis, Chandler, Problem 4714, Amer. Math. Monthly **63** (1956), 729; Solution, Amer. Math. Monthly **64** (1957), 679–680.

44. Dimitrov, D. K., Markov inequalities for weight functions of Chebyshev type, J. Approx. Theory **83** (1995), 175–181.

45. Dimitrov, D. K., On a conjecture concerning monotonicity of zeros of ultraspherical polynomials, J. Approx. Theory **85** (1996), 88–97.

46. Dryanov, D., and Q. I. Rahman, On certain mean values of polynomials on the unit interval, J. Approx. Theory **101** (1999), 92–120.

47. Duffin, R. J., and A. C. Schaeffer, A refinement of an inequality of the brothers Markoff, Trans. Amer. Math. Soc. **50** (1941), 517–528.

48. Erdélyi, T., Paul Erdös and polynomials, in *In Memoriam: Paul Erdös (1913–1996)*, J. Approx. Theory **94**, 1 (1998), 1–41.

49. Erdős, P., An extremum-problem concerning trigonometric polynomials, Acta Sci. Math. (Szeged) **9** (1939), 113–115.

50. Erdős, P., Some remarks on polynomials, Bull. Amer. Math. Soc. **53** (1947), 1169–1176.

51. Erdős, P., Open problems, in B. Bojanov (ed.), *Open Problems in Approximation Theory* , 238–242, SCT Publishing, Singapore, 1994.

52. Erdős, P., and G. Szegő, On a problem of I. Schur, Annals Math. **43**, 3 (1942), 451–470.

53. Eriksson, B. O., Some best constants in the Landau inequality on a finite interval, J. Approx. Theory **94**, 3 (1998), 420–454.

54. Eröd, J., Bizonyos polinomok maximumának, Mat. Fiz. Lapok **46** (1939), 58–82.

55. Fitzgerald, C. H., and L. L. Schumaker, A differential equation approach to interpolation at extremal points, J. Analyse Math. **22** (1969), 117–134.

56. Frappier, C., Q. I. Rahman, and St. Ruscheweyh, New inequalities for polynomials, Trans. Amer. Math. Soc. **288** (1985), 69–99.

57. Golitschek, M. v., and G. G. Lorentz, Bernstein inequalities in L_p for $0 \le p \le +\infty$, Rocky Mountain Math. J. **19** (1989), 145–156.

58. Goluzin, G. M., *Geometric Theory of Functions of Complex Variable*, Nauka, Moscow, 1966 (in Russian); English transl.: Transl. Math. Monographs, **26**, Amer. Math. Soc., Providence, R.I., 1969.

59. Guessab, A., and Q.I. Rahman, Quadrature formulae and polynomial inequalities, J. Approx. Theory **90** (1997), 255–282.

60. Hadamard, J., Sur le module maximum d'une fonction et de ses dérivées, C. R. de Séances de l'année 1914, Soc. Math. de France (1914), 66 – 72.

61. Hardy, G. H., and J. E. Littlewood, Contribution to the arithmetic theory of series, Proc. London Math. Soc. **11**, 2 (1912), 411–478.

62. Hardy, G. H., J. E. Littlewood, and G. Pólya, *Inequalities*, 2nd edn., Cambridge University Press, 1964.

63. Hörmander, L., New proof and generalization of inequality of Bohr, Math. Scand. **2** (1954), 33–45.

64. Hunter, D. B., and G. Nikolov, Gegenbauer weight functions admitting L_2 Duffin and Schaeffer type inequalities, ISNM, Vol. 131, Birkhäuser Verlag Basel, 1999, 121–131.

65. Kammerer, W. J., Polynomial approximations to finitely oscillating functions, Math. Comp. **15** (1961), 115–119.

66. Karlin, S., Interpolation properties of generalized perfect splines and the solution of certain extremal problems, I, Trans. Amer. Math. Soc. **206** (1975), 25–66.

67. Kolmogorov, A. N., On inequalities for suprema of consecutive derivatives of an arbitrary function on an infinite interval, Uchen. Zap. MGU **30**, 3 (1939), 3–16 (in Russian); English transl.: V. M. Tikhomirov (ed.), *Selected Works of A. N. Kolmogorov, Vol. I: Mathematics and Mechanics*, Kluwer Academic Publishers, Dordrecht, 1985, pp. 277–290.

68. Korneichuk, N. P., *Splines in Approximation Theory*, Mir, Moscow, 1984 (in Russian).

69. Korneichuk, N. P., *Exact Constants in Approximation Theory*, Nauka, Moscow, 1987; English transl.: Cambridge University Press, 1991.

70. Korneichuk, N. P., V. F. Babenko, and A. A. Ligun, *Extremal Properties of Polynomials and Splines*, Naukova Dumka, Kiev, 1992 (in Russian).

71. Kristiansen, G. K., Some inequalities for algebraic and trigonometric polynomials, J. London Math. Soc. (2), **20** (1979), 300–314.

72. Kristiansen, G. K., Corrigendum: Some inequalities for algebraic and trigonometric polynomials, J. London Math. Soc. (2), **22** (1980), 575–576.

73. Kristiansen, G. K., Some inequalities for algebraic and trigonometric polynomials II, J. London Math. Soc. **28** (1983), 83–92.

74. Kuhn, H., Interpolation vorgeschriebener Extremwerte, J. reine angew. Math. **238** (1969), 24–31.

75. Kwong, M. K., and A. Zettl, *Norm Inequalities for Derivatives and Differences*, Lecture Notes in Math. **1536**, Springer-Verlag, Berlin, 1992.

76. Landau, E., Einige Ungleichungen für zweimal differenzierbare Funktionen, Proc. London Math. Soc. **13** (1913), 43–49.

77. Lax, P., Proof of a conjecture of Erdős on the derivative of a polynomial, Bull. Amer. Math. Soc. **50** (1944), 509–513.

78. Ligun, A. A., On inequalities between the norms of derivatives of periodic functions, Mat. Zametki **33** (1983), 385–391 (in Russian).

79. Malik, M. A., On the derivative of a polynomial, J. London Math. Soc. (2), **1** (1969), 57–60.

80. Malik, M. A., On integral mean estimate for polynomials, Proc. Amer. Math. Soc. **91** (1984), 281–284.

81. Markov, A. A., On a question of D. I. Mendeleev, Zapiski Peterb. Akademii Nauk **62** (1890), 1–24 (in Russian).

82. Markov, V. A., On the functions of least deviation from zero in a given interval, St. Petersburg, 1892 (in Russian); German translation with condensation: W. Markoff: Über Polynome die in einem gegebenen Intervalle möglichst wenig von Null abweichen, Math. Ann. **77** (1916), 213–258.

83. Micchelli, C. A., T. J. Rivlin, and S. Winograd, Optimal recovery of smooth functions, Numer. Math. **26** (1976), 191–200.

84. Milev, L., and G. Nikolov, On the inequality of Schur, J. Math. Anal. Appl. **216** (1997), 421–427.

85. Milovanović, G. M., Various extremal problems of Markov's type for algebraic polynomials, Facta Univ. Ser. Math. Inform. **2** (1987), 7–28.

86. Milovanović, G. M., D. S. Mitrinović, and Th. M. Rassias, *Topics in Polynomials: Extremal Problems, Inequalities, Zeros*, World Scientific, Singapore, 1994.

87. Motorin, A. P., On inequalities between the maximum absolute values of a function and its derivatives on the half-line, Ukrain. Mat. Zh. **7** (1955), 262–266 (in Russian).

88. Mycielski, J., Polynomials with preassigned values at their branching points, Amer. Math. Monthly **77**, (1970), 853–855.

89. Mycielski, J., and S. Paszkowski, A generalization of Chebyshev polynomials, Bull. Acad. Polonaise Sci. Série math. astr. et phys. **8**, 7 (1960), 433–438.

90. Naidenov, N., Algorithm for the construction of the smoothest interpolant, East J. Approx. **1**, 1 (1995), 83–97.

91. Naidenov, N., Landau-type extremal problem for the triple $\|f\|_\infty$, $\|f'\|_p$, $\|f''\|_\infty$ on a finite interval, submitted.

92. Nikolov, G., Inequalities of Duffin-Schaeffer type, SIAM J. Math. Anal. **33** (2001), 686–698.

93. Pierre, R., and Q. I. Rahman, On a problem of Turán about polynomials II, Canad. J. Math. **33**, (1981), 701–733.

94. Pinkus, A., Some extremal properties of perfect splines and the pointwise Landau problem on the finite interval, J. Approx. Theory **78** (1978), 37–64.

95. Pinkus, A., On smoothest interpolation, SIAM J. Math. Anal. **19** (1988), 1431–1441.

96. Rahman, Q. I., On a problem of Turán about polynomials with curved majorants, Trans. Amer. Math. Soc. **163** (1972), 447–455.

97. Rahman, Q. I., and G. Schmeisser, Markov-Duffin-Schaeffer inequality for polynomials with a circular majorant, Trans. Amer. Math. Soc. **310** (1988), 693–702.

98. Rahman, Q. I., and G. Schmeisser, *Les inégalités de Markoff et de Bernstein*, Les Presses de Univ. Montréal, Montréal, 1983.

99. Rivlin, T. J., *Chebyshev Polynomials, Second Edition*, John Wiley & Sons, Inc. New York, 1990.

100. Saff, E. B., and T. Sheil-Small, Coefficient and integral mean estimates for algebraic and trigonometric polynomials with restricted zeros, J. London Math. Soc. (2), **9** (1974), 16–22.

101. Sato, M., The Landau inequality for bounded intervals with $\|f^{(3)}\|$ finite, J. Approx.Theory **34** (1982), 159–166.

102. Schaeffer, A. C., Inequalities of A. Markoff and S. Bernstein for polynomials and related functions, Bull. Amer. Math. Soc. **47** (1941), 565–579.

103. Schaeffer, A. C., and R. J. Duffin, On some inequalities of S. Bernstein and W. Markoff for derivatives of polynomials, Bull. Amer. Math. Soc. **44** (1938), 289–297.

104. Schmidt, E., Über die nebst ihren Ableitungen orthogonalen Polynomensysteme und das zugehörige Extremum, Math. Ann. **119** (1944), 165–204.

105. Schoenberg, I. J., and A. Cavaretta, Solution of Landau's problem, concerning higher derivatives on halfline, MRC Report No. 1104, Univ. of Wisconsin, 1970; Also in Proc. of the Intern. Conf. on Constructive Function Theory, Golden Sands (Varna) May 19–25, 1970, Publ. House Bulgarian Acad. Sci., Sofia, (1972) 297–308.

106. Schönhage, A., *Approximationstheorie*, de Gruyter, Berlin, 1971.

107. Schur, I., Über das Maximum des absoluten Betrages eines Polynoms in einem gegebenen Intervall, Math. Z. **4** (1919), 217–287.

108. Shadrin, A. Yu., Interpolation with Lagrange polynomials: A simple proof of Markov inequality and some of its generalizations, Approx. Theory Appl. **8** 3(1992), 51–61.

109. Shadrin, A. Yu., To the Landau-Kolmogorov problem on a finite interval, in: *Open Problems in Approximation Theory* (B. Bojanov, Ed.), pp. 192–204, SCT Publishing, Singapore, 1994.

110. Shadrin, A. Yu., Error bounds for the Lagrange interpolation, J. Approx. Theory **80** 1995, 25–49.

111. Shapiro, B., On the number of components of the space of trigonometric polynomials of degree n with $2n$ distinct critical values, Mat. Zametki **62**, 4 (1997), 635–640.

112. Stein, E. M., Functions of exponential type, Ann. Math. **65**, 3 (1957), 582–592.

113. Szegö, G., Über einen Satz des Herrn Serge Bernstein, Schriften der Königsberger Gelehrten Gesellschaft **5** (1928), 59–70.

114. Taikov, L. V., A generalization of S. N. Bernstein's inequality, Trudy Mat. Inst. Steklov LXXVIII, Nauka, Moscow, 1965, pp. 43–47 (in Russian).

115. Thom, R., L'équivalence d'une fonction différenciable et d'un polinome, Topology **3**, Suppl. 2 (1965), 297–307.

116. Tikhomirov, V. M., *Certain Questions in Approximation Theory*, Moscow State University, Moscow, 1976.

117. Tikhomirov, V. M., Private communication.

118. Turan, P., Über die Ableitung von Polynomen, Compositio Math. **7** (1939), 85–95.

119. Turan, P., Remark on a theorem of Erhard Schmidt, Mathematica **2** (25)(1960), 373–378.

120. Tyrygin, I. Y., On Turan's type inequalities in certain integral metrics, Ukrainian Math. J. **40**, 2 (1988), 256–260.

121. de la Vallée-Poussin, Ch. J., Sur le maximum du module de la dérivée d'une expression trigonométrique d'ordre et de module bornés, C. R. Acad. Sci. Paris **166** (1918), 843–846.

122. Varma, A. K., Some inequalities of algebraic polynomials having all zeros inside $[-1, 1]$, Proc. Amer. Math. Soc. **88**, 2 (1983), 227–233.

123. Varma, A. K., On some extremal properties of algebraic polynomials, J. Approx. Theory **69** (1992), 48–54.

124. Videnskii, V. S., A generalization of an inequality of V. A. Markov, Dokl. Akad. Nauk SSR **120**, 3 (1958), 447–449.

125. Voronovskaya, E.V., *The Functional Method and its Applications*, LEIS, Leningrad, 1963; English transl.: AMS, Providence, Rhode Island, 1970.

126. Zviagintsev, A. I., and A. J. Lepin, On the Kolmogorov inequalities between the upper bounds of the derivatives of functions for $n = 3$, Latv. Mat. Ezhegodnik **26** (1982), 176–181 (in Russian).

127. Zygmund, A., A remark on conjugate series, Proc. London Math. Soc. (2), **34** (1932), 392–400.

Borislav Bojanov
Dept. of Mathematics
University of Sofia
Blvd. James Boucher 5
1164 Sofia, Bulgaria
boris@fmi.uni-sofia.bg

Nonlinear Approximation by Refinable Functions and Piecewise Polynomials

Yu. Brudnyi

Abstract. The paper introduces a newly–developed algorithm for nonlinear approximation and describes its several applications to approximation by refinable functions and piecewise polynomials

§0. Introduction

The initial version of the algorithm considered in this paper was introduced in 1989 in a joint project with I. Irodova. We had two goals in that project. First we aimed to apply the algorithm to computerized quality control of microchips. In the devastating atmosphere of the late "perestroika", this goal could not be achieved because of technical and financial problems. Our second goal, application of the algorithm to nonlinear L_q–approximation ($0 < q \leq \infty$) of functions in Besov spaces by multivariate B-splines was partially realized in [6]. This problem was studied independently in [14] for $q < \infty$ and in [15] for $q = \infty$. The results of these latter two papers are essentially more general than those of [6], since they exploit a wide class of compactly supported refinable functions including B-splines. The paper [14] also presents remarkably clear and elegant proofs of the basic facts of this field for nonlinear L_q-approximation ($q < \infty$) by *wavelets of noncompact support*. The basic key of the approach in [14] is a generalization of the Littlewood–Paley theory developed in [16], and the main tool is the classical *greedy algorithm*. Since the approach of the paper [6] seems to be of importance for nonlinear approximation and also for signal and image processing, I decided to return to this project.

In [7] I. Kozlov and I developed a new algorithm based on some ideas in [6]. My coauthor then tested it in a series of computer experiments.

Approximation Theory X: Abstract and Classical Analysis
Charles K. Chui, Larry L. Schumaker, and Joachim Stöckler (eds.), pp. 91–112.
Copyright ⓒ 2002 by Vanderbilt University Press, Nashville, TN.
ISBN 0-8265-1415-4.

Her extensive and devoted efforts produced results which clearly indicate the effectiveness of this algorithm in applications. In this survey I present a new abstract version of the algorithm designed for use with vector functions on *tree-like graphs* for nonlinear N-term approximation. The applications mainly aim to convince the reader of the usefulness of the algorithm for this rapidly developing field. We include a number of previously known results, but there are also some new ones, in particular, Theorem 12 whose corollaries are a classical approximation theorem by M. Birman and M. Solomyak [10], and a broad generalization of the recent important result of [12] on nonlinear approximation of BV-functions.

Since I was one of several mathematicians who already recognized the importance of this field at the early stages of its development, I have taken the liberty here of presenting several of my results from those times which have since been used for nonlinear approximation. The reader can find many other interesting facts about nonlinear approximation in the excellent survey [13]. Here I have only referred to papers which are not mentioned in [13].

§1. Basic Notions

The algorithm under consideration is designed to vector functions defined on tree-like ordered graphs with chain structure. We now examine in some detail several important terms of this phrase.

1.1. Ordered Graphs

Let $G = (\mathcal{V}, \mathcal{E})$ be an ordered graph with the sets of vertices $\mathcal{V} = \{v\}$ and (ordered) edges $\mathcal{E} = \{e\}$. If edge e goes from vertex v' to vertex v'', we will write $e = [v', v'']$ or $v' \to v''$ and call v' a son of v'' and v'' a parent of v'. We let $\sigma(v)$ and $\pi(v)$ the sets of sons and parents of v, respectively.

The subset $A := \{v_1, \ldots, v_n\}$ is said to be an array, if $v_1 \to v_2 \to \ldots \to v_n$. The extreme points of this array, v_1, and v_n are named respectively, the bottom and top endpoints of A. For these we use the notations

$$v_1 := v_A^-, \quad v_n := v_A^+. \tag{1}$$

We say that v' it is majorized by v'' and write

$$v' \prec v'', \tag{2}$$

if there is an array A such that

$$v' = v_A^- \quad \text{and} \quad v'' = v_A^+.$$

In this case we also say that A connects v' with v'' (in this order!). Lastly, we introduce a bundle $\beta(v', v'')$, the set of all arrays connecting v' with v''.

1.2. Definition

$G = (\mathcal{V}, \mathcal{E})$ is said to be a **tree-like graph**, if there are the (unique) vertex v_0, the **root** of G, and a constant $C = C_G \geq 1$ such that for every $v \neq v_0$

$$1 \leq \operatorname{card} \beta(v, v_0) \leq C, \tag{3}$$

and, moreover, all arrays from $\beta(v, v_0)$ have the same cardinality.

For such a graph one introduces a $\mathbb{Z}_+$-**graded structure** as follows. Define the **height** of $v \in \mathcal{V}$ letting

$$h(v) := \operatorname{card} A, \tag{4}$$

where A is (any) array connecting v with v_0; specially, $h(v_0) := 0$. Then we set for $j \in \mathbb{Z}_+$

$$\mathcal{V}_j := \{v; h(v) = j\}. \tag{5}$$

In particular, $\mathcal{V}_0 = \{v_0\}$, $\mathcal{V}_1 = \sigma(v_0)$, $\mathcal{V}_2 = \bigcup_{v \in \mathcal{V}_1} \sigma(v)$ and so forth.

It is clear that

$$\mathcal{V} = \bigcup_{j \in \mathbb{Z}_+} \mathcal{V}_j \tag{6}$$

and $\mathcal{V}_j \bigcap \mathcal{V}_{j'} = \emptyset$ provided $j \neq j'$. Therefore for each v there is the unique $j \in \mathbb{Z}_+$ such that $v \in \mathcal{V}_j$. This number we denote by $j_v (= h(v))$.

1.3. Basic Assumption

There is a constant $D = D_G \geq 1$ such that for every $v \in \mathcal{V}$

$$1 \leq \operatorname{card} \sigma(v) \leq D. \tag{7}$$

Hereafter $G = (\mathcal{V}, \mathcal{E})$ is a tree-like graph.

1.4. Examples

(a) A tree is clearly a tree-like graph with $C = 1$. In this case each $v \neq v_0$ has a *unique* parent, and each array is *uniquely determined by its endpoints*. In this (and only in this) case we use the notation

$$A = [v_1, v_n] = \left[v_A^-, v_A^+\right] \tag{8}$$

for $A = \{v_1 \to v_2 \to \ldots \to v_n\}$. We will also write

$$[v_1, v_n) := A \backslash \{v_n\} \tag{9}$$

for a "half open from the top" array.

(b) Let $\Sigma \subset 2^{\mathbb{R}^d}$ be a countable family of measurable subsets from $\mathbb{R}^d$. We equip Σ with an ordered graph structure regarding its subsets as vertices and connecting S', S'' from Σ by the edge directed from S' to S'', if $S' \subset S''$ and there is no different subset of Σ situated between them. This graph will be denoted by $G_\Sigma = (\mathcal{V}_\Sigma, \mathcal{E}_\Sigma)$; hence $\mathcal{V}_\Sigma = \Sigma$. In this case, majoration $S' \prec S''$, see (2), is equivalent to embedding $S' \subset S''$.

(c) If, in particular, $\Sigma = \mathcal{D}(Q_0)$ is the family of dyadic cubes situated in the unit cube $Q_0 := [0, 1]^d$, the $G_{\mathcal{D}(Q_0)}$ is a tree satisfying the basic assumption, see (7), with $D = 2^d$. In fact, here $\sigma(v) = 2^d$ for each $v \in \mathcal{D}(Q_0)$. Note that a "generic" element of $\mathcal{D}(Q_0)$, a dyadic cube, will be denoted by Q.

(d) We introduce an anisotropic version of the last example. Let $\bar{\lambda} = (\lambda_1, \ldots, \lambda_d) \in \mathbb{R}_+^d$ and

$$\langle \bar{\lambda} \rangle := \left(\frac{1}{d} \sum_{i=1}^d \frac{1}{\lambda_i} \right)^{-1}. \tag{10}$$

Set

$$\mu_i := \frac{\langle \bar{\lambda} \rangle}{\lambda_i}, \quad 1 \leq i \leq d. \tag{11}$$

We introduce the dyadic $\bar{\lambda}$-cube of the j-th level by

$$Q_{j,k}^{\bar{\lambda}} := \prod_{s=1}^d \{ 2^{-j_s} [0, 1] + k_s \}, \quad j \in \mathbb{Z}, \quad k \in \mathbb{Z}^d, \tag{12}$$

where j_s is the integer part of $\mu_s j$.

Denote the family of these $\bar{\lambda}$-cubes by $\mathcal{D}_j^{\bar{\lambda}}$ and set

$$\mathcal{D}^{\bar{\lambda}} := \bigcup_{j \in \mathbb{Z}} \mathcal{D}_j^{\bar{\lambda}}.$$

Lastly, set

$$\mathcal{D}^{\bar{\lambda}}(Q_0) := \{ Q^{\bar{\lambda}} \in \mathcal{D}^{\bar{\lambda}} ; Q^{\bar{\lambda}} \subset Q_0 \}. \tag{13}$$

Each $\bar{\lambda}$-cube of the latter family is a result of consequent subdivisions of Q_0 by hyperplanes orthogonal to the faces of Q_0. It is easily seen that G_Σ with $\Sigma := \mathcal{D}^{\bar{\lambda}}(Q_0)$ is a tree, and (7) holds with $D \leq 2^{2d}$.

(e) We now introduce self–similar coverings. Let M be $d \times d$ matrix of integer entries, and $\det(M) > 1$. *Assume that there is a subset $S_0 \subset \mathbb{R}^d$ of nontrivial finite measure and a finite collection $\mathcal{K} \subset \mathbb{Z}^d$ such that*

$$S_0 = \bigcup_{k \in \mathcal{K}} S_{1,k}. \tag{14}$$

Here we set

$$S_{j,k} := M^{-j}(S_0 + k), \quad j \in \mathbb{Z}_+, \quad k \in \mathbb{Z}^d.$$

We introduce now the collection of subsets $\mathcal{D}^M(S_0) \subset 2^{\mathbb{R}^d}$ as follows. For $S \subset \mathbb{R}^d$, let

$$\Delta(S) := \{M^{-1}(S + k); \ k \in \mathcal{K}\}.$$

Then we introduce the desired collection letting

$$\mathcal{D}^M(S_0) := \bigcup_{j \in \mathbb{Z}_+} \mathcal{D}_j^M(S_0), \tag{15}$$

where

$$\mathcal{D}_0^M(S_0) := \{S_0\} \quad \text{and} \quad \mathcal{D}_j^M(S_0) := \bigcup_{S \in \mathcal{D}_{j-1}^M(S_0)} \Delta(S).$$

It is easily checked that G_Σ with $\Sigma := \mathcal{D}^M(S_0)$ is a tree-like graph satisfying (7) with $D = \text{card}(\mathcal{K})$. Besides, the constant C_G in (3) in this case does not exceed the multiplicity of the covering $\Delta(S_0)$.

1.5. Chain structure

Let $G = (\mathcal{V}, \mathcal{E})$ be an *arbitrary* ordered graph, and $B_1(X)$ the unit ball of linear bounded operators acting in a given Banach space X. A chain of G is an operator-valued function

$$C : \mathcal{E} \to B_1(X)$$

defined on the set $\mathcal{E}$ of edges. If $e = [v', v'']$ we will write

$$C(e) := C(v', v'').$$

We extend this function to the set of pairs (v', v'') satisfying $v' \prec v''$ as follows.

For an array $A := \{v_1, \ldots, v_n\}$ we let

$$C(A) := \prod_{s=1}^{n-1} C(v_s, v_{s+1}), \tag{16}$$

and then set for $v' \prec v''$

$$C(v', v'') := \sum_{A \in \beta(v', v'')} C(A). \tag{17}$$

Note that for tree-like graphs the right hand sum is finite. Besides, $C(A)$ belongs to $B_1(X)$. Hence in this case

$$\|C(v', v'')\|_{X \to X} \leq C_G, \tag{18}$$

see (3).

1.6. Translations

Let $l_0(\mathcal{V}, X)$ be the linear space of vector functions $f : \mathcal{V} \to X$, and let $v \in \mathcal{V}$ be given. We introduce a translation

$$T_v : l_0(\mathcal{V}, X) \to l_0(\mathcal{V}, X)$$

by letting

$$T_v f(v') := \begin{cases} 0, & \text{if } v' = v \\ f(v') + C(v', v) f(v), & \text{if } v' \in \sigma(v) \\ f(v'), & \text{otherwise.} \end{cases} \tag{19}$$

Note that T_v and $T_{v'}$ commute if v, v' do not connect by an edge, and do not commute otherwise. Now for $\Omega \subset \mathcal{V}_j$, we set

$$T_\Omega := \prod_{v \in \Omega} T_v,$$

(the order is unimportant). For the general case of $\Omega = \bigcup_{j' \leq j \leq j''} \Omega_j$, where $\Omega_j \subset \mathcal{V}_j$, we then set

$$T_\Omega := T_{\Omega_{j''}} \ldots T_{\Omega_{j'}} \tag{20}$$

(the order is important!).

§2. The Up and Down Algorithm

Let $G = (\mathcal{V}, \mathcal{E})$ be a tree-like graph with chain $C : \mathcal{E} \to B_1(X)$ satisfying the basic assumption 1.3. The *Input of the algorithm* is an integer $N \geq 1$ and a function $f \in l_p^w(X)$, $0 < p < \infty$. Here $w : \mathcal{V} \to \mathbb{R}_+$ is a weight and

$$\|f\|_{l_p^w(X)} := \left\{ \sum_{v \in \mathcal{V}} (w(v) \|f(v)\|_X)^p \right\}^{\frac{1}{p}}.$$

Without loss of generality we assume

$$\|f\|_{l_p^w(X)} = 1. \tag{21}$$

In the sequel we use the notation

$$\mathcal{I}(\Omega) := \left\| f \cdot \mathbf{1}_\Omega \right\|_{\ell_p^w(X)}^p, \quad \Omega \subset \mathcal{V}. \tag{22}$$

In the case

$$\Omega = \mathcal{V}(v) := \{v' \in \mathcal{V}; \, v' \prec v\} \tag{23}$$

we simplify this notation setting

$$\mathcal{I}(v) := \mathcal{I}(\mathcal{V}(v)). \tag{24}$$

Note that $\mathcal{I}(\{v\}) \neq \mathcal{I}(v)$.

The algorithm begins with the set

$$\mathcal{V}_N := \{v \in \mathcal{V}; \ \mathcal{I}(v) \geq N^{-1}\} \tag{25}$$

and its subset of minimal elements

$$\mathcal{M}_N := \{v_j^m\}.$$

Case 1. $\mathcal{V}_N$ *is a tree.* In this case we use the notations of Example 1.4(a), see (8) and (9). Then we introduce the set

$$\mathcal{A}_N := \{A_j\}$$

of arrows from $\mathcal{V}_N$ by letting

$$A_0 := \{v_0\} \quad \text{and} \quad A_j := [v_j^m, v_0] \Big\backslash \Big(\bigcup_{s<j} A_s\Big). \tag{26}$$

Each A_j with $j \geq 1$ has a form

$$A_j = [v_j^m, v_j^c) \quad \text{where} \quad v_j^c \in \bigcup_{s<j} A_s.$$

Using the set

$$\mathcal{C}_N := \{v_j^c\}$$

of the **contact vertices**, we divide each A_j with $j \geq 1$ into a collection of subarrays $[v', v'')$ where $v' \in \mathcal{M}_N \bigcup \mathcal{C}_N$ and $v'' \in \mathcal{C}_N$. Denote the set of these subarrays by $\mathcal{R}_N$. We proceed with the subdivision procedure as follows. Let $R \in \mathcal{R}_N$, and $\{v_R^l; \ 1 \leq l \leq l_R\} \subset R$ be a collection of vertices defined by induction on l starting from $v_R^1 := v_R^-$, see (1) for the notation. If v_R^l has already been defined, we introduce the next one by the conditions

$$\mathcal{I}\big([v_R^l, v_R^{l+1}]\big) \geq N^{-1} \quad \text{while} \quad \mathcal{I}\big([v_R^l, v_R^{l+1})\big) < N^{-1}.$$

Such a vertex may not exist in the following two cases:

(a) $v_R^l = v_R^+$ or $\mathcal{I}\big([v_R^l, v_R^+]\big) < N^{-1}$. We then define v_R^{l+1} as the (unique) parent of v_R^+ and terminate the procedure.

(b) $v_R^l \neq v_R^+$ and $\mathcal{I}(\{v_R^l\}) \geq N^{-1}$. Here we define v_R^{l+1} as the parent of v_R^l (it belongs to R in this case).

Set

$$\mathcal{B}_N := \{[v_R^l, v_R^{l+1}); \ 1 \leq l \leq l_R, \ R \in \mathcal{R}_N\}.$$

Statements.

(a) $\mathcal{V}_N = \bigcup_{A \in \mathcal{B}_N} A$;

(b) For $A \in \mathcal{B}_N$, $\mathcal{I}\big([v_A^-, v_A^+)\big) < N^{-1}$;

(c) $\mathrm{card}\,(\mathcal{B}_N) \leq 4N$.

This completes the upward part of the algorithm. As a result, we obtain the collection of bottom endpoints

$$\mathcal{T}_N := \{v_A^-; \ A \in \mathcal{B}_N\}, \tag{27}$$

which we will name the **traps**.

In the downpart of the algorithm, we first define a function $\tilde{f} : \mathcal{V}_N \to X$ by the translation T_{Ω_N} with

$$\Omega_N := \mathcal{V}_N \backslash \mathcal{T}_N,$$

that is,

$$\tilde{f} := T_{\Omega_N} f, \tag{28}$$

see (20). Note that $\tilde{f}(v) = 0$ for $v \notin \mathcal{T}_N$, and

$$\tilde{f}(v) = \sum_{v \in A} C(v, v_A^+) f(v) \tag{29}$$

for $v = v_A^-$.

Hence this part of the algorithm gives the collection of elements

$$\{\tilde{f}(v_A^-); \ A \in \mathcal{B}_N\} \subset X. \tag{30}$$

Output of the algorithm consists of the collection of the traps (27) and the elements (30).

Case 2. $\mathcal{V}_N$ *is a tree-like graph.* As in the previous case, we introduce the sets $\mathcal{V}_N$ and $\mathcal{M}_N$ and consider the set of bundles

$$\{\beta(v_j^m, v_0); \ v_j^m \in \mathcal{M}_N\},$$

see 1.1 for the definition. Without loss of generality, we assume that

$$\mathrm{card}\,(\beta(v_1^m, v_0)) \geq \mathrm{card}\,(\beta(v_2^m, v_0)) \geq \cdots$$

Then we index the arrays of each $\beta(v_j^m, v_0)$ in some order. At last, define inductively the set of vertices of subtree T^j and function $f_j : \mathcal{V}_N \to X$ supported by T^j. To introduce T^1 and f_1, we choose the first array A_s^1 from each bundle $\beta(v_s^m, v_0)$, and let

$$T^1 := \bigcup_s A_s^1, \quad f_1 := f \cdot \mathbf{1}_{T^1}.$$

We also define $f^1 : \mathcal{V}_N \to X$ by

$$f^1 = f \cdot \mathbf{1}_{\mathcal{V}_N} - f_1.$$

If we have already defined T^j, f_j and f^j, we then set

$$T^{j+1} = \bigcup_s A_s^{j+1},$$

where A_s^{j+1} is the $(j+1)$th array of bundle $\beta(v_s^n, v_0)$. Then we let

$$f_{j+1} := f^j \cdot \mathbf{1}_{T^{j+1}}, \quad f^{j+1} := f^j - f_{j+1}.$$

Statement II.

(a) Each T^j is a tree with the root v_0, and the number of these is at most C_G, see (3).

(b) Arrays from different T^j do not coincide.

(c) $\sum_j f_j = f \cdot \mathbf{1}_{\mathcal{V}_N}$ and $\bigcup_j T^j = \mathcal{V}_N$.

(d) $\mathcal{I}(f_j; \Omega) \leq \mathcal{I}(f; \Omega)$.

We now apply the algorithm from Case 1 to each tree T^j and function f^j. This gives us the collection of the traps $\mathcal{T}_N^j$, see (27). Then we set

$$\mathcal{T}_N := \bigcup_j \mathcal{T}_N^j \tag{31}$$

and, moreover,

$$\tilde{f} := T_{\Omega_N} f,$$

where $\Omega_N := \mathcal{V}_N \backslash \mathcal{T}_N$.

Now the *output of the algorithm* is the set of traps (31) and the set of elements

$$\{\tilde{f}(v); \ v \in \mathcal{T}_N\}. \tag{32}$$

Note that in this case a formula for $\tilde{f}(v)$ has a more complicated form in comparison to (29).

Remark 1. The choice of trees may be organized in a special way regarding the size of norms in the chain $C : \mathcal{E} \to B_1(X)$. Under this choice, the number of the trees T^j can be essentially reduced.

§3. An Approximation Theorem

In order to formulate the main result we first recall the definition of a vector valued *refinable function*. Let μ be a locally bounded Borel measure on $\mathbb{R}^d$, and $\varphi : \mathbb{R}^d \to X$ a strongly μ-measurable function assigning its value in a Banach space X. Then φ is a refinable function if it is a solution to the refinement equation

$$\varphi(x) = \sum_{k \in \mathbb{Z}^d} c(k)\, \varphi\,(Mx - k), \qquad x \in \mathbb{R}^d. \tag{33}$$

Here M is a $d \times d$ matrix of integral entries and $\det(M) > 1$, and the mask $c := \{c(k)\}$ is an operator valued function on $\mathbb{Z}^d$ taking its values in the space $L(X)$ of linear bounded operators acting on X.

We assume that φ satisfies the following conditions:

$$\varphi \in L_\infty(\mu; X). \tag{34}$$

$$\operatorname{card}(\operatorname{supp} c) < \infty. \tag{35}$$

$$\sup_k \|c(k)\|_{X \to X} \leq 1. \tag{36}$$

Here and for the rest of the paper, $L_p(\mu; X)$ is the linear space of strongly μ-measurable $f : \mathbb{R}^d \to X$ with the finite (quasi) norm

$$\|f\|_{L_p(\mu;X)} := \left\{ \int_{\mathbf{R}^d} \|f\|_X^p d\mu \right\}^{\frac{1}{p}}.$$

In the case of $X = \mathbb{R}$ we write simply $L_p(\mu)$. In particular, (35) implies φ is *compactly supported*, while (36) says that $c(k)$ belongs to the unit ball $B_1(X)$ of $L(X)$.

We now set $\mathcal{K} := \operatorname{supp} c$ and consider the family of subsets of $\mathbb{R}^d$

$$\Sigma_\varphi := \{M^{-j}(\operatorname{supp}\varphi + k);\ j \in \mathbb{Z}_+,\ k \in \mathcal{K}\}. \tag{37}$$

Note that $M^{-j}(\operatorname{supp}\varphi + k)$ is the support of $\varphi(M^j x - k)$ and there is a one-to-one correspondence between $\sum_\varphi$ and the set of dyadic cubes of $\mathcal{D}(\mathbb{R}^d)$

$$\mathcal{D}_\varphi := \{Q_{j,k} := 2^{-j}(Q_0 + k);\ j \in \mathbb{Z}_+,\ k \in \mathcal{K}\}. \tag{38}$$

Therefore the notions

$$\hat{Q} := M^{-j}(\text{supp}\,\varphi + k) = \text{supp}\,\varphi(M^j \cdot -k), \tag{39}$$

$$\varphi_Q := \varphi(M^j \cdot -k) \tag{40}$$

where $Q := Q_{j,k}$ lead to no ambiguity. We also set for $Q \subset \mathcal{D}(\mathbb{R}^d)$ and $j \in \mathbb{Z}_+$

$$\mathcal{D}_\varphi^j(Q) := \{Q_{j,k} \in \mathcal{D}_\varphi;\ Q_{j,k} \subset Q\}.$$

Since Σ_φ satisfies the condition of Example 1.4(e), it determines the tree-like graph G_{Σ_φ}. We introduce a chain structure for this graph, see 1.5, as follows. By scaling we rewrite the refinable equation (33) as

$$\varphi_Q = \sum_{\hat{Q}' \in \sigma(\hat{Q})} c(\hat{Q}', \hat{Q})\,\varphi_{Q'}, \tag{41}$$

where $\sigma(\hat{Q})$ stands for the set of sons for $\hat{Q}$ in G_{Σ_φ}, see 1.1, and $c(\hat{Q}', \hat{Q}) = c(k)$ for a suitable $k \in \mathcal{K}$. By (36) the function C defined on an edge $e = [\hat{Q}', \hat{Q}]$ by

$$C(e) := c(\hat{Q}', \hat{Q})$$

takes its values in $B_1(X)$. Hence C determines a chain structure for G_{Σ_φ}.

We now introduce the **approximation space** $\mathcal{A}_p^{w,\theta}(\varphi)$, $0 < p, \theta \leq \infty$, of *scalar valued* functions $f : \mathbb{R}^d \to \mathbb{R}$ defined by finiteness of the (quasi) norm

$$\|f\|_{\mathcal{A}_p^{w,\theta}(\varphi)} := \inf \left\{ \sum_{j \in \mathbb{Z}_+} \left[\sum_{Q \in \mathcal{D}_\varphi^j} (w(Q)\|f_Q\|_{X^*})^p \right]^{\frac{\theta}{p}} \right\}^{\frac{1}{\theta}}. \tag{42}$$

Hence $w : \mathcal{D}_\varphi \to \mathbb{R}_+$ is a weight, and the infimum is taken over all decompositions

$$f = \sum_{Q \in \mathcal{D}_\varphi} \langle f_Q, \varphi_Q \rangle \quad (\text{convergence in } L_p(\mathbb{R}^d)) \tag{43}$$

with suitable linear bounded functional $f_Q \in X^*$. Hence $\langle f_Q, \varphi_Q(x) \rangle$ denotes the value of f_Q at $\varphi_Q(x) \in X$. We will write $\mathcal{A}_p^w(\varphi)$, if $\theta = p$, and $\mathcal{A}_p^{\lambda,\theta}(\varphi)$, if $w(Q) := (\text{vol}\,(Q))^\lambda$, $\lambda > 0$.

Assume now that for $0 < p < q \leq \infty$ the continuous embedding

$$\mathcal{A}_p^w(\varphi) \subset L_q(\mu) \tag{44}$$

holds. Under this condition the following is true.

Theorem 1. *Let $f \in \mathcal{A}_p^w(\varphi)$ and $N \geq 1$. Then there is an N-term linear combination*

$$T_N(f) := \sum_Q \left\langle \tilde{f}_A, \varphi_Q \right\rangle$$

such that

$$\|f - T_N(f)\|_{L_q(\mu)} \leq CN^{\frac{1}{q} - \frac{1}{p}} \|f\|_{\mathcal{A}_p^w}.$$

Here the constant C depends only on φ and the embedding constant in (44).

In order to introduce the desired approximation aggregate T_N we apply the up and down algorithm to a function $F : \Sigma_\varphi \to X^*$ defined on the tree-like graph G_{Σ_φ} (with chain structure) by

$$F(\hat{Q}) := f_Q, \qquad \hat{Q} \in \Sigma_\varphi.$$

Here f_Q is taken from an almost optimal decomposition (43).

The algorithm gets the collection of traps $\mathcal{T}_N \subset \Sigma_\varphi$ and elements $\{\tilde{f}_Q; \hat{Q} \in \mathcal{T}_N\}$. Then we simply set

$$T_N(f) := \sum_{\hat{Q} \in \mathcal{T}_N} \left\langle \tilde{f}_Q, \varphi_Q \right\rangle$$

to get the required linear combination.

§4. Applications: Besov Spaces

Let X be a finite dimensional space, say $\mathbb{R}^l$, and suppose $\varphi : \mathbb{R}^d \to X$ satisfies the regular refinement equation

$$\varphi(x) = \sum_{k \in \mathcal{K}} c(k)\, \varphi(2x - k), \qquad x \in \mathbb{R}^d \tag{45}$$

with mask $c = \{c(k)\}$ compactly supported by $\mathcal{K}$. It is worth noting that a collection of compactly supported wavelets $\{\Psi_\varepsilon;\ \varepsilon \in \{0,1\}^d\}$ form a refinable vector-valued function of this type with $\dim X = 2^d$.

We will say that the φ **belongs to the class** $B_+^\lambda(\mathbb{R}^d)$ if for all $0 < p, \theta \leq \infty$

$$B_p^{\lambda, \theta}(\mathbb{R}^d) \subset \mathcal{A}_p^{\lambda, \theta}(\varphi). \tag{46}$$

Reversing this embedding we define the class $\mathcal{B}_-^\lambda(\mathbb{R}^d)$ and then set

$$B^\lambda(\mathbb{R}^d) = (B_-^\lambda \cap B_+^\lambda)(\mathbb{R}^d). \tag{47}$$

A wide class of scalar refinable functions satisfying (45) which belongs to $B^\lambda(\mathbb{R}^d)$ was described in [14: §4]. We will consider a general situation related to the refinable equation (33) elsewhere.

Now let $\varphi \in B_+^\lambda$ and $0 < p < q \leq \infty$. Then with notations of Section 3 the following holds:

Theorem 2. *Assume that* $f \in B_p^\lambda(Q_0)$ *and*

$$\frac{\lambda}{d} = \frac{1}{p} - \frac{1}{q}. \tag{48}$$

Then for each $N \geq 1$ there is an N-term linear combination

$$T_N(f) := \sum_{Q \in \mathcal{D}_\varphi} \left\langle \tilde{f}_Q, \varphi_Q \right\rangle$$

such that

$$\|f - T_N(f)\|_{L_q(Q_0)} \leq C N^{-\frac{\lambda}{d}} \|f\|_{B_p^\lambda(Q_0)}. \tag{49}$$

In the case of $q = \infty$ the result holds either for $p \leq 1$ or for $1 < p < \infty$ and $L_q(Q_0)$ replaced by $BMO(Q_0)$. The constant depends only on φ and λ.

With the exception of a small distinction in the assumptions, this result is due to R. DeVore and his collaborators, see Theorem 6.2 in [14] and Theorem 3.1 in [15]. The proof of the next result is based on the latter theorem in combination with the reiteration theorem for approximation spaces from [8].

Theorem 3. *Let $\varphi \in B_+^\lambda$ and $0 < p < q \leq \infty$ satisfy*

$$\frac{\lambda}{d} > \frac{1}{p} - \frac{1}{q}. \tag{50}$$

Then for $f \in B_p^{\lambda,\infty}(Q_0)$ and each $N \geq 1$ there is an N-term linear combination

$$T_N(f) := \sum_{Q \in \Omega_N} \left\langle \tilde{f}_Q, \varphi_Q \right\rangle$$

where $\Omega_N \subset \mathcal{D}_\varphi$ is now a collection of nonoverlapping dyadic cubes such that

$$\|f - T_N(f)\|_{L_q(Q_0)} \leq C N^{-\frac{\lambda}{d}} \|f\|_{B_p^{\lambda,\infty}(Q_0)} \tag{51}$$

holds with C depending only on φ and $\varepsilon := \frac{\lambda}{d} - \frac{1}{p} + \frac{1}{q}$.

Note that $C \to \infty$ as $\varepsilon \to 0$. Note also that the collection of supports $\{\hat{Q}; Q \in \Omega_N\}$ covers $\hat{Q}_0 := \operatorname{supp} \varphi$ with multiplicity at most $C(\varphi, d)$.

We now briefly discuss an anisotropic case. Let φ_s be a univariate scalar refinable function of class $B_+^{\lambda_s}(\mathbb{R})$, $1 \leq s \leq d$. We introduce a refinable function $\varphi : \mathbb{R}^d \to \mathbb{R}$ by

$$\varphi := \bigotimes_{s=1}^d \varphi_s.$$

For the dyadic $\bar{\lambda}$-cube $Q = Q_{j,k}^{\bar{\lambda}}$, $j \in \mathbb{Z}_+$, $k \in \mathbb{Z}^d$, $\bar{\lambda} = (\lambda_1, \ldots, \lambda_s)$, see (12), we set

$$\varphi_Q(x) := \varphi\left(2^{j_1} x_1 - k_1, \ldots, 2^{j_d} x_d - k_d\right).$$

Recall that $j_s := [\mu_s j]$ and μ_s is defined in (11).

Under these conditions, the results of this section hold for the aniso-tropic case with λ replaced by $\langle \bar{\lambda} \rangle$, see (10), and $B_p^{\lambda,\theta}(Q_0)$ replaced by anisotropic Besov space $B_p^{\bar{\lambda},\theta}(Q_0)$, $\theta = p$ or ∞, respectively. We recall that these are defined by finiteness of the (quasi) seminorm

$$\sup_{1 \leq s \leq d} \left\{ \int_{\mathbf{R}_+} \left(\frac{\omega_{r_s}^{(s)}(f;t)_{L_p}}{t^{\lambda_s}} \right)^\theta \frac{dt}{t} \right\}^{\frac{1}{\theta}},$$

where $r_s > \lambda_s$ and $\omega_r^{(s)}(f; \cdot)_{L_p}$ is the partial r-modulus of continuity with respect to x_s, $1 \leq s \leq d$; see [7] for this analog of Theorem 2.

Using now the anisotropic analog of Theorem 3 and the method firstly introduced in [10] we immediately get the following result:

Corollary 4. *Let $\bar{\lambda}$, $0 < p < q \leq \infty$, satisfy*

$$\frac{\langle \bar{\lambda} \rangle}{d} > \frac{1}{p} - \frac{1}{q}.$$

Then the ε-entropy of the unit ball of $B_p^{\bar{\lambda},\infty}(Q_0)$ in the space $L_q(Q_0)$ has sharp order $\varepsilon^{-\frac{d}{\langle \bar{\lambda} \rangle}}$, $\varepsilon \to 0$.

In the case $p = q = \infty$, this asymptotic was firstly established in [5]; for an isotropic case the result was proved in [11].

§5. A Digression: Local Polynomial Approximation Theory

In this section we present some aspects of the theory mentioned in the title which were already developed by the author in the sixties. The main results were proved in the author's Doctor of Science dissertation of 1965 and announced in the doctoral thesis. Because of the situation in Soviet mathematics of that time (see, e.g., [1] for its description) these results were published only partially. Fortunately, the basic result described in Theorem 5 did get published, see [2] for for the case L_p, $1 \leq p \leq \infty$, and [3] for that of translation invariant Banach lattices. In order to formulate this result which plays an important role for the results of the next section, we introduce the set $\mathcal{K}(Q_0)$ of cubes in $Q_0 := [0,1]^d$ homothetic to Q_0, and denote by Π_t, $0 < t \leq 1$, the family of disjoint collections of cubes from $\mathcal{K}(Q_0)$ of volume t^d. For $f \in L_p(Q_0)$ we define its local approximation of order k in L_p as a function of $Q \in \mathcal{K}(Q_0)$ determined by

$$E_k(f;Q)_{L_p} := \inf_{m \in \mathcal{P}_k} \|f - m\|_{L_p(Q)}. \tag{52}$$

Here $\mathcal{P}_k \subset \mathbb{R}[x_1, \ldots, x_d]$ is the space of polynomials of degree $k - 1$. Specially, $\mathcal{P}_1 = \{\text{const}\}$, $\mathcal{P}_0 = \{0\}$.

Theorem 5. *For the k-modulus of continuity of $f \in L_p(Q_0)$, $1 \le p \le \infty$, the two-sided inequality*

$$\omega_k(f;t)_{L_p(Q_0)} \approx \sup_{\pi \in \Pi_t} \left\{ \sum_{Q \in \pi} E_k(f;Q)_{L_p}^p \right\}^{\frac{1}{p}} \tag{53}$$

holds uniformly in f, t and p.

Remark 2. This result can be easily reformulated in another form. Actually, let

$$\mathcal{E}_k(f;x,t)_{L_p} := \frac{E_k(f;Q)_{L_p}}{(vol\, Q)^{\frac{1}{p}}}, \tag{54}$$

where x and t^d are the center and volume of Q, respectively. Then (53) is equivalent to

$$\omega_k(f;t)_{L_p(Q_0)} \approx \left\{ \int_{Q_0} \mathcal{E}_k(f;x,t)_{L_p}^p \, dx \right\}^{\frac{1}{p}}. \tag{55}$$

The following consequence of (53) is of importance.

Corollary 6. *For $f \in L_p(Q_0)$ and $N \ge 1$ there is a collection of polynomials $\{m_Q; Q \in \pi_N\} \subset \mathcal{P}_k$ related to the subdivision π_N of Q_0 into N^d congruent cubes such that*

$$\left\| f - \sum_{Q \in \pi_N} m_Q \cdot 1_Q \right\|_{L_p(Q_o)} \le C\omega_k(f; N^{-1})_{L_p(Q_0)}. \tag{56}$$

Here C is independent of f and $1 \le p \le \infty$.

This result was then extended to $0 < p < 1$ in the paper [19]. For a further generalization including the case of rearrangement invariant Banach spaces and anisotropic polynomial approximation, see [4].

An important goal of the theory under consideration is the study of "classical" function spaces. In order to achieve this goal one first has to represent every such object as a local approximation space, that is, a space determined by the behaviour of local polynomial approximation of its elements. Actually, all "classical" function spaces, including those of Sobolev, Besov, Triebel–Lizorkin, BMO and H_p can be represented in this fashion. At the second stage one can study these spaces by tools of local approximation space theory.

As an example of this approach, we present several results for a "new" (well forgotten) family of approximation spaces introduced in [2]. In order to introduce this family, we first define the function space v_p^λ, $0 < p \le \infty$, $\lambda \le 0$, of functions $F : \mathcal{K}(Q_0) \to \mathbb{R}$ satisfying

$$\|F\|_{v_p^\lambda} := \sup_\pi \|F\|_{l_p^\lambda(\pi)} < \infty. \tag{57}$$

Here π runs over all *disjoint* collections $\pi \subset \mathcal{K}(Q_0)$, and

$$\|F\|_{l_p^\lambda(\pi)} := \left\{ \sum_{Q \in \pi} \left[\frac{|F(Q)|}{(vol\, Q)^\lambda} \right]^p \right\}^{\frac{1}{p}} \tag{58}$$

is a weighted l_p-norm.

Definition 7. The approximation space $V_p^{k,\lambda}(L_q)$ consists of functions $f \in L_q(Q_0)$ satisfying

$$|f|_{V_p^{k,\lambda}(L_q)} := \|E_k(f;\cdot)_{L_q}\|_{v_p^\lambda} < \infty. \tag{59}$$

In the case $\lambda = 0$, which is the main case for the next section, we will omit λ and write simply $V_p^k(L_q)$. We will also write $V_p^{k,\lambda}(X)$, if L_q is replaced by a functional space X.

The important characteristics of the space introduced are its **variation type** p and **smoothness** s defined by

$$\frac{s}{d} := \lambda + \frac{1}{p} - \frac{1}{q}. \tag{60}$$

Having these, we also define the **limiting exponent** of $V_p^{k,\lambda}(L_q)$ letting

$$\hat{p} := \left[\left(\frac{1}{p} - \frac{s}{d} \right)_+ \right]^{-1}. \tag{61}$$

Specially, $\hat{p} = \infty$, if $\frac{s}{d} \ge \frac{1}{p}$. The next two results demonstrate the role of these characteristics.

Theorem 8. Let $k \geq 1$ and $q < \hat{p}$. Then

$$V_p^{k,\lambda}(L_q) = V_p^k(L_{\hat{p}}), \tag{62}$$

and the weak $L_{\hat{p}}$ should be replaced by BMO, if $\hat{p} = \infty$ and $\frac{s}{d} = \frac{1}{p}$. In the remaining case $\frac{s}{d} > \frac{1}{p}$, we get

$$V_p^{k,\lambda}(L_q) = V_p^{k,\hat{\lambda}}(C(Q_0)), \tag{63}$$

where $\hat{\lambda} := \frac{s}{d} - \frac{1}{p}$.

This result is a consequence of the inequality for the rearrangement f^* of $f \in V_p^{k,\lambda}(L_q)$ proved in [2]. In particular, two V-spaces coincide if they have the same variation type p and smoothness s, and the L_q spaces related to them contain $L_{\hat{p}}$.

Theorem 9.
(a) If $0 < s < k$, then

$$B_q^s(Q_0) \subset V_p^{k,\lambda}(L_q) \subset B_p^{s,\infty}(Q_0). \tag{64}$$

(b) If $s = k$ and $1 < p \leq \infty$, then

$$V_p^{k,\hat{\lambda}}(L_{\hat{p}}) = W_p^k(Q_0). \tag{65}$$

In this case $\hat{\lambda} := \left(\frac{k}{d} - \frac{1}{p}\right)$, if $\frac{k}{d} > \frac{1}{p}$ and $\hat{\lambda} := 0$, otherwise.

(c) If $s = k$ and $p = 1$ the previous holds with Sobolev space replaced by

$$BV^k(Q_0) := \{f \in W_1^{k-1}(Q_0); \ \nabla^k f \in BV(Q_o)\}. \tag{66}$$

Remark 3. In the case $s = 0$, the space $V_p^{k,\lambda}(L_q)$ is situated between L_p and $L_{p\infty}$ if $q < p$, and coincides with L_p if $q = p$. In the case $s < 0$ it is contained in the Morrey space of (negative) smoothness s.

Our last result concerns the Peetre K-functional of the couple

$$\bar{V} := \left(V_{p_0}^{k,\lambda_0}(L_q), V_{p_1}^{k,\lambda_1}(L_q)\right). \tag{67}$$

For its formulation, we introduce invariants $\alpha(\bar{V})$ and $\lambda(\bar{V})$ of this couple letting for $p_0 \neq p_1$

$$\alpha(\bar{V}) := (\lambda_0 - \lambda_1) \left/ \left(\frac{1}{p_1} - \frac{1}{p_0}\right)\right.,$$

and for $p_0 = p_1$

$$\alpha(\bar{V}) := \pm\infty,$$

where the sign equals $\mathrm{sgn}(\lambda_0 - \lambda_1)$. Then we set

$$\lambda(\bar{V}) := \lambda_i + \frac{\alpha(\bar{V})}{p_i}$$

with $i = 0$ (or 1) for the first case, and

$$\lambda(\bar{V}) := \alpha(\bar{V})$$

for the second case.

Theorem 10 [9]. *Assume that $\alpha(\bar{V}) \notin [0, 1 - \frac{1}{d}]$, and suppose $\operatorname{sgn} \lambda(\bar{V}) = \operatorname{sgn} \alpha(\bar{V})$. Then uniformly in f and $0 < t \leq 1$,*

$$K(t; f; \bar{V}) \approx \sup_{\pi} K\big(t; E_k(f; \cdot)_{L_q}; \bar{l}(\pi)\big), \qquad (68)$$

where, see (58),

$$\bar{l}(\pi) := \big(l_{p_0}^{\lambda_0}(\pi), l_{p_1}^{\lambda_1}(\pi)\big).$$

Using this lifting result and well-known results on K-functionals of weighted l_p-spaces, one can obtain several important interpolation theorems for couples of "classical" functional spaces. As an example, we demonstrate the result for the couple $\big(BMO(Q_0), W_\infty^k(Q_0)\big)$.

Corollary 11. *Uniformly in f and $0 < t \leq 1$,*

$$K(t, f; BMO, W_\infty^k) \approx \sup_{Q \in \mathcal{K}(Q_0)} \left\{ \min\left(1, \frac{t}{(\operatorname{vol} Q)^{\frac{k}{d}}}\right) \right\} \mathcal{E}_k(f; Q)_{L_1},$$

see (54) for the definition of $\mathcal{E}_k$. In particular,

$$(BMO, W_\infty^k)_{\theta\infty} = B_\infty^{k\theta,\infty}, \quad 0 < \theta < 1.$$

§6. Applications: BV_p-functions

We now apply the algorithm to a function from the space $\overset{\circ}{V}_p^k(L_q)$, $1 \leq p < q \leq \infty$, where $\overset{\circ}{V}_p^k(L_q)$ is the closure in $V_p^k(L_q)$ of the set $C^k(Q_0)$. In this case we are dealing with the tree $G_{\mathcal{D}(Q_0)}$, generated by the collection of dyadic cubes $\mathcal{D}(Q_0)$, see Example 1.3(c), and with a decomposition

$$f = \sum_{Q \in \mathcal{D}(Q_0)} P_Q \quad (\text{convergence in } L_q),$$

where we set

$$P_Q := m_Q \cdot \mathbf{1}_Q - \sum_{Q' \in \sigma(Q)} m_{Q'} \cdot \mathbf{1}_{Q'}$$

with m_Q being an optimal polynomial of degree $k - 1$ for $E_k(f; Q)_{L_q}$ (best approximant).

Lastly, we define $\mathcal{I} : 2^{\mathcal{D}(Q_0)} \to \mathbb{R}$, cf. (2.2), by

$$\mathcal{I}(\Omega) := \sup_{\pi} \left\{ \sum_{Q \in \pi} E_k(f; Q)^p_{L_q} \right\}, \quad \Omega \subset \mathcal{D}(Q_0),$$

where π runs over all disjoint families of cubes from $\mathcal{K}(Q_0)$ contained in $\Omega_{\max} \backslash \Omega_{\min}$. Here $\Omega_{\max}(\Omega_{\min})$ is the union of all maximal (minimal) cubes of Ω. In these settings, the algorithm leads to the following result.

Theorem 12. *Let f belong to the space $\overset{\circ}{V}{}^k_p(L_q)$ of smoothness $s \in (0, k]$, and $d \geq 2$. Then for each $N \geq 1$ there is an N-term linear combination*

$$T_N(f) := \sum_{Q \in \Omega_N} f_Q \cdot \mathbf{1}_Q \tag{69}$$

with $\Omega_N \subset \mathcal{D}(Q_0)$ and $\{f_Q\} \subset \mathcal{P}_k$ such that

$$\|f - T_N(f)\|_{L_q(Q_0)} \leq C N^{-\frac{s}{d}} \|f\|_{V^k_p(L_q)}. \tag{70}$$

Here C depends only on k and d.

According to Section 5, the family of spaces in question contains classical spaces of bounded variation ($BV(Q_0)$, $V_p(0,1)$ and so on) and Sobolev spaces. Hence the above theorem implies similar consequences for these spaces. In particular, we have the following two results, the first of which is a generalization of the main theorem from [12] (case $k = 1$), and the second one is an extension of the main theorem from [10] (the case of Sobolev spaces).

Corollary 13. *If $f \in BV^k(Q_0)$, see (65), and $k < d$, then the assertion of Theorem 12 holds with $s = k$ and $q = \frac{d}{d-k}$.*

Corollary 14. *If $f \in B^{s,\infty}_p(Q_0)$ and*

$$\frac{s}{d} > \frac{1}{p} - \frac{1}{q} > 0,$$

then the assertion of Theorem 12 holds with Ω_N in (69) being a subdivision of Q_0.

We leave to the reader the formulation of an analog of Corollary 6.2 for Sobolov spaces $W^k_p(Q_0)$; see also the next section for wavelet approximation of functions from the latter spaces.

Finally, we formulate a consequence of a different type (cf. the corresponding result in [2]) concerning a Luzin type theorem for BV_p-functions.

Theorem 15. *Let $f \in V^k_p(L_q)$ and $0 < s \leq k$. Then for every $\varepsilon > 0$ there is f_ε from $\mathrm{Lip}(s; Q_0)$ such that*

$$mes_d\{x \in Q_0; \, f(x) \neq f_\varepsilon(x)\} < \varepsilon.$$

Here $\mathrm{Lip}(s; Q_0)$ is the Lipschitz–Besov space $B^{s,\infty}_\infty(Q_0)$, if $s < k$, and is the space $C^k(Q_0)$ of k-times continuously differentiable functions if $s = k$.

§7. Applications: Sobolev Spaces

We apply the algorithm to nonlinear approximation of functions from $W_p^k(Q_0)$ by compactly supported wavelets. In this case $\varphi : \mathbb{R}^d \to \mathbb{R}^l$, $l :=$ 2^d, is a refinable vector function with the components Ψ_ε, $\varepsilon \in \{0,1\}^d$, satisfying (45). Here we set

$$\Psi_\varepsilon := \bigotimes_{s=1}^{d} \Psi_{\varepsilon_s},$$

where $\Psi_0, \Psi_1 : \mathbb{R} \to \mathbb{R}$ are compactly supported wavelet generators, see, e.g., [13] for more details.

Using the settings of Sections 3 and 4, and applying the algorithm to the tree-like graph $G_{\mathcal{D}_\varphi}$, but with the function J in (22) replaced by

$$\mathcal{I}(\Omega) := \left\| |\nabla^k f| \right\|_{L_p(\Omega_{\max} \setminus \Omega_{\min})}, \qquad \Omega \subset \mathcal{D}_\varphi,$$

where $\Omega_{\max}$ and $\Omega_{\min}$ are, respectively, the union of maximal and minimal subsets from Ω, we obtain

Theorem 16. *Let $f \in W_p^k(Q_0)$ and $1 \le p < q < \infty$ satisfy*

$$\frac{k}{d} = \frac{1}{p} - \frac{1}{q}.$$

Then for every $N \ge 1$ there is an N-term linear combination

$$T_N(f) := \sum_Q \langle \tilde{f}_Q, \varphi_Q \rangle \tag{71}$$

with suitable $\hat{Q} \in \mathcal{D}_\varphi$ and $\tilde{f}_Q \in \mathbb{R}^l$, $l := 2^d$, such that

$$\| f - T_N(f) \|_{L_q(Q_o)} \le C N^{-\frac{k}{d}} \| f \|_{W_p^k(Q_0)}. \tag{72}$$

Here C depends only on k, d and q.

Remark 4. (a) The constant in (72) tends to infinity together with q.
(b) The result does not hold for $q = \infty$.
(c) While we use in this case the wavelet decomposition

$$f = \sum_{Q \in \mathcal{D}_\varphi} \langle f_Q, \varphi_Q \rangle$$

with the corresponding wavelet coefficients, $T_N(f)$ contains only terms generated by the "father" wavelet $\Psi_{0,\dots,0}$.

As a consequence of this theorem and the Bernstein inequality from [14], we obtain the following interpolation result.

Theorem 17. *Let integers* $0 \leq k_0 < k$ *and real numbers* $1 \leq p_0 < p_1 < \infty$ *satisfy*

$$\frac{k_1 - k_0}{d} \geq \frac{1}{p_0} - \frac{1}{p_1}.$$

Then for $0 < \theta < 1$

$$\left(W_{p_0}^{k_0}(Q_0),\ W_{p_1}^{k_1}(Q_1) \right)_{\theta p_\theta} = B_{p_\theta}^{k_\theta}(Q_0).$$

Here $\frac{1}{p_\theta} := \frac{1-\theta}{p_0} + \frac{\theta}{p_1}$ *and* $k_\theta := (1 - \theta)k_0 + \theta k_1$.

In the case $k_0 = 0$, that is, with L_{p_0} replacing $W_{p_0}^{k_0}$, and $\frac{k_1}{d} > \frac{1}{p_0} - \frac{1}{p_1}$ this result was established in [9] as a consequence of the main theorem, see Theorem 10. A simplified proof of this result was presented in [18]. In the case $k_0 = 0$, $k_1 = 1$, $p_0 = 1$ and $p_1 = \frac{d}{d-1}$, the result was initially proved in [12].

Acknowledgments. The author gratefully acknowledges support by the Fund for the Promotion of Research at the Technion.

References

1. Artin, M., et al., Letter to Editor, Notices of AMS, November 1978.

2. Brudnyi, Yu., Spaces defined by local polynomial approximation, Trudy Mosk. Mat. Ob. **24** (1971), 69–132; English translation in Trans. Moscow Math. Soc. **24** (1973).

3. Brudnyi, Yu., A multidimensional analog of Whitney's theorem, Mat. Sb. **82 (124)** (1970), 169-191; English translation in Math. USSR – Sb **11** (1970).

4. Brudnyi, Yu., Local polynomial approximation of multivariate functions, in *Approximation Theory*, a volume dedicated to B. Sendov (2002), 1–19.

5. Brudnyi, Yu., and B. Cotlyar, The order of growth of ε-entropy, in *Studies Contemporary Problems of Constructive Function Theory*, Baku, 1965.

6. Brudnyi, Yu., and I. Irodova, Nonlinear approximation and the B-spaces, Algebra and Analysis **4**, No. 6 (1992), 45–79.

7. Brudnyi, Yu., and I. Kozlov, An algorithm of nonlinear approximation, submitted.

8. Brudnyi, Yu., and N. Krugljak, A family of approximation spaces, in *Studies in the Theory of Functions of Several Variables*, Yaroslavl State Univ., Yaroslavl, 1978, 15–42.

9. Brudnyi, Yu., and N. Krugljak, The real interpolation for couples of spaces of smooth functions, Doklady RAN **349**, No. 6 (1966), 721–723.

10. Birman, M., and M. Solomyak, Piecewise polynomial approximation of functions of the classes W_p, Mat. Sb. **73** (1967), 295–317.

11. Cohen, A., W. Dahmen, J. Daubechies, and R. DeVore, Tree approximation and optimal encoding, Preprint, 1999, 1–27.

12. Cohen, A., R. DeVore, P. Petrushev, and H. Xu, Nonlinear approximation and the space $BV(\mathbb{R}^2)$, Amer. J. Math. **121** (1999), 587–628.

13. DeVore, R., *Nonlinear Approximation*, Acta Numerica, Birkhäuser, 1998.

14. DeVore, R., B. Jawerth, and V. Popov, Compression of wavelet decompositions, Amer. J. Math. **114** (1992), 737–785.

15. DeVore, R., P. Petrushev and X. M. Yu, Nonlinear wavelet approximation in the space $C(\mathbb{R}^d)$, in *Progress in Approximation Theory*, Springer, 1992, 261–283.

16. Frazier, M., and B. Jawerth, A discrete transform of distribution spaces, Func. Analysis **93** (1990), 34–170.

17. Irodova, I., Dyadic Besov spaces, Algebra and Analysis **12**, 3 (2000), 40–80.

18. Krugljak, N., Calderòn–Zygmund type decompositions, K-functional and quantitative covering theorems, Algebra and Analysis **8** (1996), 110–160.

19. Oswald, P., and E. Storojenko, The Jackson theorem in the space $L^p(\mathbb{R}^k)$, $0 < p < 1$, Sib. Mat. Journal **19**, No. 4 (1978), 888–901.

20. Temlyakov, V. N., Nonlinear methods of approximation, preprint.

Prof. Dr. Yu. Brudnyi
Department of Mathematics
Technion – Israel Institute of Technology
32000 Haifa, Israel
ybrudnyi@math.technion.ac.il

On Spaces Admitting Minimal Projections which are Orthogonal

Bruce L. Chalmers and Boris Shekhtman

Abstract. Let ν be a symmetric measure on $[-1, 1]$. In this paper we show that any 2-dimensional unconditional subspace V of $L^{4/3}(\nu)$, admitting a minimal projection (from $L^{4/3}$) onto itself which is also orthogonal, must have the form $[\cos\theta(t), \sin\theta(t)] \subset L^{4/3}(\nu)$, for some odd function θ.

§1. Preliminaries

It is well known and easy to see that, if k is a non-zero integer, then a minimal projection from $L^p[-1, 1]$ onto $[\cos k\frac{\pi}{2}t, \sin k\frac{\pi}{2}t]$ is orthogonal. In the following, we establish a partial converse to that result. (This question is also relevant with respect to almost locally minimal projections (see [4,5] and the discussion in §3).)

Let ν be a symmetric measure on $[-1, 1]$. In this paper we show that any 2-dimensional unconditional subspace V of $L^{4/3}(\nu)$, admitting a minimal projection (from $L^{4/3}$) onto itself which is also orthogonal, must have the form $[\cos\theta(t), \sin\theta(t)] \subset L^{4/3}(\nu)$, for some odd function θ. (Recall that V is **unconditional** means that V has a basis (v_1, v_2) such that $\|a_1v_1 + a_2v_2\| = \||a_1|v_1 + |a_2|v_2\|$.)

In order to lay the foundation for our discussion, we recall first a theory providing a characterization theorem for such minimal operators.

Let X be a Banach space. Let $\mathcal{B} = \mathcal{B}(X, V)$ be the space of all bounded linear operators from a real or complex normed space X into a separable subspace V. Let $\mathcal{P}$ be the family of all operators in $\mathcal{B}$ with a given fixed action on V (e.g., the identity action corresponds to the family of bounded projections onto V), and assume that $\mathcal{P}$ has a minimal element.

Approximation Theory X: Abstract and Classical Analysis
Charles K. Chui, Larry L. Schumaker, and Joachim Stöckler (eds.), pp. 113–116.
Copyright © 2002 by Vanderbilt University Press, Nashville, TN.
ISBN 0-8265-1415-4.

Definition 1. $(x, y) \in S(X^{**}) \times S(X^*)$ *will be called an* extremal pair *for* $Q \in \mathcal{B}$ *if* $\langle Q^{**}x, y \rangle = \|Q\|$, *where* $Q^{**} : X^{**} \to V$ *is the second adjoint extension of* Q *TO* X^{**} (S *denotes the unit sphere*).

Definition 2. *Let* $\mathcal{E}(Q)$ *be the set of all extremal pairs for* Q. *To each* $(x, y) \in \mathcal{E}(Q)$ *associate the rank one operator* $y \otimes x$ *from* X *to* X^{**} *given by* $(y \otimes x)(z) = \langle z, y \rangle x$ *for* $z \in X$.

Theorem 1. (Characterization [2]) P *has minimal norm in* $\mathcal{P}$ *if and only if the closed convex hull of* $\{y \otimes x\}_{(x,y) \in \mathcal{E}(P)}$ *contains an operator for which* V *is an invariant subspace, i.e., if and only if there exists an operator* E_P *(from* X *into* X^{**}*) such that*

$$E_P = \int_{\mathcal{E}(P)} y \otimes x \, d\mu(x, y) : \quad V \to V \tag{1}$$

(μ *is a non-negative measure with total mass* 1).

In the discussion below it is helpful to introduce a fixed basis $\vec{v} = (v_1, v_2, \ldots)$ for V; we will write $V = [\vec{v}] = [v_1, v_2, \ldots]$. Then

$$\mathcal{P} = \left\{ \sum_i u_i \otimes v_i : \ \langle v_i, u_j \rangle = A_{ij} \quad \text{for } A \text{ a given fixed matrix} \right\}$$

and the necessary and sufficient condition (1) can be rewritten as equations

$$\int_{\mathcal{E}(P)} \langle \vec{v}, y \rangle x \, d\mu(x, y) = M\vec{v} \quad \text{for some matrix } M. \tag{2}$$

Notation. If $z \in Z$ and $z^* \in Z^*$ are such that $\langle z, z^* \rangle = \|z\| \|z^*\| \neq 0$, then z^* is an extremal for z and we write $z^* = \text{ext}z$. (Then also $z = \text{ext}z^*$.) Note that $\text{ext}z$ is determined only up to a non-zero scalar factor.

§2. Main Theorem

Theorem 2. *Let* ν *be a symmetric measure on* $[-1, 1]$. *Any 2-dimensional unconditional subspace* V *of* $L^{4/3}(\nu)$, *admitting a minimal projection (from* $L^{4/3}$*) onto itself which is also orthogonal, must have the form* $[\cos \theta(t), \sin \theta(t)] \subset L^{4/3}(\nu)$, *for some odd function* θ.

Proof: From Theorem 1, a projection P is minimal if and only if

$$\int_{\mathcal{E}(P)} \langle \vec{v}, y \rangle \, x \, d\mu(x, y) = M\vec{v}.$$

But now $\langle x, \vec{u} \rangle \cdot \langle \vec{v}, y \rangle = \|P\|$ implies that $x = \text{ext} \langle \vec{v}, y \rangle \cdot \vec{u}$ and $y = \text{ext} \langle x, \vec{u} \rangle \cdot \vec{v}$. Further, let $\vec{e} = \langle \vec{v}, y \rangle$ and let $f(\vec{e} \cdot \vec{u}) = \text{ext} \, \vec{e} \cdot \vec{u}$. Then (suppressing unnecessary notation, i.e., replacing $\langle \vec{v}, y \rangle$ by $\vec{e}$ and replacing $\mu(x, y)$ by $\mu(\vec{e})$) we have

$$\int \vec{e} f(\vec{e} \cdot \vec{u}(t)) \, d\mu(\vec{e}) = M\vec{v}(t). \tag{3}$$

Let $1 \leq p \leq \infty$. Then any 2-dimensional unconditional subspace V of $L^p(\nu)$ is given by some $[v_1, v_2]$, where v_1 is even and v_2 is odd. Furthermore, in this case it is easy to see that in (3) above the measure μ must be even in the second variable, M must be diagonal, and u_1 is even while u_2 is odd. Indeed $\mu(1, e) = \mu(1, -e)$ follows from the fact that u_1 is even while u_2 is odd in the case of a minimal projection. But if also the minimal P is orthogonal we have that $u_1 = v_1$ and $u_2 = v_2$.

Now note that, in the case $p = 4/3$ (whence $q = 4$), we have

$$f(u) = \text{ext}(u) = \text{sgn}(u)|u|^{q/p} = u^3.$$

Thus, in this case equation (3) above becomes

$$\int \vec{e}(\vec{e} \cdot \vec{v})^3 \, d\mu(\vec{e}) = M\vec{v},$$

i.e., assuming without loss that $\vec{e} = \kappa(1, e)$, where κ is some constant, and $M = \text{diag}(1, m)$, we have the following pair of equations (writing $d\mu(e)$ for $\kappa d\mu(\vec{e})$)

$$\int (v_1 + ev_2)^3 \, d\mu(e) = v_1, \qquad \int e(v_1 + ev_2)^3 \, d\mu(e) = mv_2. \tag{4}$$

Thus, using the notation $c_i = \int e^{2i} \, d\mu(e)$ and the fact that μ is symmetric (or even) (on $(-\infty, \infty)$), we have

$$c_0 v_1^3 + 3c_1 v_1 v_2^2 = v_1, \qquad 3c_1 v_1^2 v_2 + c_2 v_2^3 = mv_2. \tag{5}$$

Next (without loss, $v_1(t)v_2(t) \neq 0$) divide the first of the equations in (5) by v_1 and the second by v_2 to obtain

$$c_0 v_1^2 + 3c_1 v_2^2 = 1, \qquad 3c_1 v_1^2 + c_2 v_2^2 = m. \tag{6}$$

But now consider the matrix B with rows $(c_0, 3c_1)$ and $(3c_1, c_2)$. If $\det B$ is not zero, then it follows that $|v_1| = \alpha_1$ and $|v_1| = \alpha_2$, where α_1 and α_2 are constants. Thus by the orthogonality conditions (recall P is a projection) $\alpha_1 = \alpha_2 = 1/\sqrt{\nu[-1, 1]}$. Furthermore, if $\det B$ is zero then we have, again by use of the orthogonality conditions, that for each t in $[-1, 1]$, $v_1(t) = \kappa_1 \cos \theta(t)$ and $v_2(t) = \kappa_2 \sin \theta(t)$, for some $-\pi/2 \leq \theta(t) \leq \pi/2$ and some positive constants κ_i. Finally, note that θ must be odd on $[-1, 1]$. $\square$

§3. Remarks

Using the theory of [1], one can see that if $L^{4/3}$ is replaced by L^1 in the statement of Theorem 2, then the conclusion is the same with the extra necessary requirements that ν defined by $\nu(e^{i\frac{\pi}{2}t}) := \nu(t)$ is uniformly distributed on the (half) circle and $\theta(t) = k\frac{\pi}{2}t$, for some non-zero integer k. (These extra conditions follow from the requirement that the Lebesgue function be constant. There is no simple analogue of this requirement in the case $p \neq 1$.)

Conjecture. In Theorem 2 above the above requirements on ν and θ in the case of L^1 are also valid in the case of $L^{4/3}$. In addition Theorem 2 extends to all $1 \leq p \leq \infty$, $p \neq 2$.

Note, finally, that in [3] it was shown that, in the case of L^1 and the measure ν equally distributed (in the sense described above) on 3 points, the classes of almost locally minimal projections, as defined in [4], coincide with the orthogonal minimal projections.

References

1. Chalmers, B. L. and F. T. Metcalf, The determination of minimal projections and extensions in L^1, Trans. Am. Math. Soc. **329** (1992), 289–305.

2. Chalmers, B. L. and F. T. Metcalf, A characterization and equations for minimal projections and extensions, J. Oper. Theory. **32** (1994), 31–46.

3. Chalmers, B. L. and B. Shekhtman, On minimal, locally minimal and orthogonal projections, in *Trends in Approximation Theory*, K. Kopotun, T. Lyche, M. Neamtu (eds.), Vanderbilt University Press, Nashville (2000), 1–4.

4. Zippin, M., Almost locally minimal projections in finite dimensional Banach spaces, Israel J. Math. **110** (1999), 253–268.

5. Zippin, M., Orthogonal almost locally minimal projections on l_1^n, Israel J. Math. **115** (2000), 253–268.

Bruce L. Chalmers   Boris Shekhtman
Department of Mathematics   Department of Mathematics
University of California, Riverside   University of South Florida
Riverside, California 92521   Tampa, Florida 33620
blc@math.ucr.edu   boris@math.usf.edu

Hilbert Transforms and Orthonormal Expansions for Exponential Weights

S. B. Damelin

Abstract. We establish the uniform boundedness of the weighted Hilbert transform for a general class of symmetric and nonsymmetric weights on a finite or infinite interval $I := (c, d)$ with $c < 0 < d$. We then apply these results to study mean and uniform convergence of orthonormal expansions on the line.

§1. Introduction and Statement of Results

In this paper we shall study mean and uniform convergence of orthonormal expansions as well as uniform bounds for the weighted Hilbert transform for a large class of exponential decaying weights on an interval $I := (c, d)$ with $c < 0 < d$. For orthonormal expansions on $[-1, 1]$, there are many well known results for Chebyshev, Jacobi and generalized Jacobi weights starting from Riesz, and we do not review these here. Instead, we refer the reader to [19,20,24,28,32] and the many references cited therein for a detailed and comprehensive account of this vast and interesting subject. Our study involves exponentially decaying weights w on I and functions f which may grow at $\pm\infty$ or near the endpoints of I. More precisely, in Theorems 1.2 and 1.7 below, we study both mean and uniform convergence of orthonormal expansions for a class of symmetric exponential weights on the line of polynomial decay at infinity, the so called Freud class which will be defined more precisely in Definition 1.1 below. Our mean convergence results are motivated by a recent result of Jha and Lubinsky in [20, Theorem 1.2] and indeed we will show how to improve this latter result. Our uniform convergence result, see Theorem 1.7, relies on good uniform bounds for the weighted Hilbert transform, see Theorems 1.6(A-B). We establish these latter bounds for a general class of symmetric and nonsymmetric weights on I.

Approximation Theory X: Abstract and Classical Analysis
Charles K. Chui, Larry L. Schumaker, and Joachim Stöckler (eds.), pp. 117–135.
Copyright © 2002 by Vanderbilt University Press, Nashville, TN.
ISBN 0-8265-1415-4.

1.1 Statement of results

To set the scene for our investigations, let $I := (c, d)$ where $c < 0 < d$ with c and d finite or infinite. (Note that I need not be symmetric about 0 but should contain 0). Let w be a nonnegative weight on I with $x^n w(x) \in L^1$, $n = 0, 1, \dots$. The idea of this paper arose, partly, from an interesting paper of Geza Freud [18] who studied the dependence of the greatest zero of $p_n(w^2)$, the nth orthonormal polynomial for w^2, on the corresponding recurrence coefficients. More precisely, given w as above, we recall, see [19,30], that $p_n := p_n(w^2)$ admits the representation

$$p_n(x) = \gamma_n x^n + \cdots, \quad \gamma_n := \gamma_n(w^2) > 0$$

and satisfies

$$\int_I p_n(x) p_n(x) w^2(x) dx = \delta_{m,n}, \quad m, n \geq 0.$$

We denote by $x_{j,n} := x_{j,n}(w^2)$, $1 \leq j \leq n$, the jth simple zero of p_n in I, and we order these zeroes as

$$x_{n,n} < x_{n-1,n} \cdots < x_{2,n} < x_{1,n}.$$

If w is even, we write the three term recurrence of p_n in the form

$$x p_n(x) = A_n(w^2) p_{n+1}(x) + A_{n-1}(w^2) p_{n-1}(x), \quad n \geq 0.$$

Here, $p_{-1} = 0$, $p_0 = \left(\int w^2(x)\, dx \right)^{-1/2}$ and

$$A_n := A_n(w^2) = \gamma_{n-1}/\gamma_n > 0$$

are the corresponding recurrence coefficients. Given w as above, we may also form an orthonormal expansion

$$f(x) \to \sum_{j=0}^{\infty} b_j p_j(x), \, b_j := \int_I f p_j w, \, j \geq 0 \tag{1.1}$$

for any measurable function $f : I \to (-\infty, \infty)$ for which

$$\int_I |f(x) x^j| w(x) dx < \infty, \quad j = 1, 2, 3\dots \tag{1.2}$$

Let $S_n[.w] := S_n[.], n \geq 1$ denote the nth partial sums of the orthonormal expansions given by (1.2).

To state our main results, we require some additional notation. To this end, let us agree that henceforth C will denote a positive constant independent of x, y, n, f and j which may take on different values at different times. Moreover, for any two sequences b_n and c_n of nonzero real numbers, we shall write $b_n = O(c_n)$ if there exists a positive constant C, independent of n, such that

$$b_n \leq Cc_n, \qquad n \to \infty$$

and $b_n \sim c_n$ if

$$b_n = O(c_n) \qquad \text{and} \qquad c_n = O(b_n).$$

Henceforth, for functions and sequences of functions, O and $\sim$ will be uniform in x, y, n, f and j.

We begin with the definition of a large class of *admissible* weights, see Definitions 1.1 (A-C) below. For clarity of exposition, explicit and easily absorbed examples of admissible weights are presented immediately after Definitions 1.1(A-C).

Definition 1.1A. *A function* $g : (0, d) \to (0, \infty)$ *is said to be* quasi-increasing *if*

$$g(x) \leq Cg(y), \qquad 0 < x \leq y < d.$$

It is easy to see that any increasing function is quasi-increasing. Similarly, we may define the notion of quasi-decreasing.

Definition 1.1B. *A weight function* $w : I \to (0, \infty)$ *will be called* admissible *if each of the following conditions below is satisfied:*

a) $Q := \log(1/w)$ *is continuously differentiable and satisfies* $Q(0) = 0$;

b) Q' *is nondecreasing in* I *with*

$$\lim_{x \to c^+} Q(x) = \lim_{x \to d^-} Q(x) = \infty;$$

c) *The function*

$$T(x) := \frac{xQ'(x)}{Q(x)}, \, x \neq 0$$

is quasi-increasing in $(0, d)$ *and quasi decreasing in* $(c, 0)$ *with*

$$T(x) \geq \lambda > 1, \, x \in I \backslash 0;$$

d) *Given a sufficiently small* $B > 0$ *along with positive* ε *and* δ,

$$\frac{\log|Q'(x + \delta)|}{(|Q'(x - \varepsilon)|)^B} \leq C,$$

for all x *close enough to* c *and* d.

Definition 1.1B suffices for uniform bounds on the weighted Hilbert transform, see Theorems 1.6(A-B) below. For Theorems 1.2, 1.3, 1.5 and 1.7, we need some additional smoothness and regularity assumptions on Q'.

Definition 1.1C. *An admissible weight w will be called* strongly admissible *if the following additional assumptions on w hold:*

a) *There exists $\varepsilon_0 \in (0,1)$ such that for $y \in I\backslash\{0\}$*

$$T(y) \sim T\left(y\left[1 - \frac{\varepsilon_0}{T(y)}\right]\right);$$

b) *For every $\varepsilon > 0$, there exists $\delta > 0$ such that for every $x \in I\backslash\{0\}$,*

$$\int_{x - \frac{\delta x}{T(x)}}^{x + \frac{\delta x}{T(x)}} \frac{|Q'(s) - Q'(x)|}{|s - x|^{3/2}} ds \leq \varepsilon |Q'(x)| \sqrt{\frac{T(x)}{|x|}}.$$

1.2 Examples

The following are explicit examples of strongly admissible weights. Here and throughout, $\exp_k$ denotes the kth iterated exponential.

a) Symmetric exponential weights on the line of polynomial decay:

$$w_\alpha(x) := \exp\left(-|x|^\alpha\right), \alpha > 1, \qquad x \in (-\infty, \infty); \qquad (1.3)$$

b) Nonsymmetric exponential weights on the line with varying rates of polynomial and faster than polynomial decay:

$$w_{k,l,\alpha,\beta}(x) := \exp(-Q_{k,l,\alpha,\beta}(x))$$

with

$$Q_{k,l,\alpha,\beta}(x) := \begin{cases} \exp_l\left(x^\alpha\right) - \exp_l(0), & x \in [0, \infty), \\ \exp_k\left(|x|^\beta\right) - \exp_k(0), & x \in (-\infty, 0) \end{cases} \qquad (1.4)$$

where $l, k \geq 1$ and $\alpha, \beta > 1$;

c) Nonsymmetric exponential weights on $(-1, 1)$ with varying rates of decay near ± 1:

$$w^{k,l,\alpha,\beta}(x) := \exp(-Q^{k,l,\alpha,\beta}(x))$$

with

$$Q^{k,l,\alpha,\beta}(x) := \begin{cases} \exp_l(1 - x^2)^{-\alpha} - \exp_l(1), & x \in [0, 1), \\ \exp_k(1 - x^2)^{-\beta} - \exp_k(1), & x \in (-1, 0), \end{cases} \qquad (1.5)$$

where $l, k \geq 1$ and $\alpha, \beta > 1$.

1.3 Remarks

(a) Definitions 1.1(A-C) define a very general class of possibly nonsymmetric exponential weights of minimal smoothness and with varying rates of decrease on $(c, 0)$ and $[0, d)$. The weights given by (a) are widely called "Freud weights" and are characterized by one rate of polynomial decay at $\pm\infty$. In this case $T \sim 1$ in Definition 1.1B(c). Because of their current popularity, we shall henceforth adopt the name **Freud weight** in what follows. However, we also allow nonsymmetric weights of polynomial and faster than polynomial decay at c and/or d respectively. Notice that our definition allows w to decrease with one rate on $[0, d)$, and with another on $(c, 0)$. For a detailed perspective on this class of weights and its applications to orthogonal polynomials and various weighted approximation problems of current interest, we refer the reader to [5,6,7,8,9,12,13,15,20,22,23, 24,25,28,29,31] and the many references cited therein.

(b) Definitions 1.1B (a-d) involve smoothness and regularity conditions on w which suffice for uniform bounds on the weighted Hilbert transform, see Theorem 1.6(A-B) below. Note that when w decays as a polynomial, $T \sim 1$, whereas otherwise T increases without bound. Definition 1.1B(d) is needed to control the behavior of Q' close to c and d when Q' grows very quickly. It is trivially satisfied when Q' grows as a polynomial.

(c) The definition of a strongly admissible weight, see Definition 1.1C, is necessary in proving Theorems 1.2, 1.3, 1.5 and 1.7. Here, we need bounds on p_n and estimates for the spacing of its zeroes, which in turn require additional smoothness and regularity assumptions on Q'. Definition 1.1C(b) is a local lip $1/2$ condition on Q', and appeared first in [22]. Notice that we do not require Q'' to exist everywhere. Definition 1.1C(d) first appeared in [5,15], although it is motivated by a much older growth lemma of E. Borel. We apply it heavily in the proof of Theorem 1.3 below.

We need some additional notation: If w is strongly admissible and even, it is well known, see [22], that the asymptotic behavior of the recurrence coefficients A_n is expressed in terms of the scaled endpoints of the support of the equilibrium measure for w which is one interval. It is also well known, see [22,24,29,31] and the references cited therein, that firstly these scaled endpoints can be explicitly calculated and are given as the positive roots of the equation

$$u = \frac{2}{\pi} \int_0^1 \frac{a_u t Q'(a_u t)}{\sqrt{1 - t^2}} dt,$$

and secondly that

$$\lim_{n \to \infty} \frac{A_n}{a_n} = 1/2$$

and

$$\lim_{n\to\infty}\frac{A_{n+1}}{A_n}=1.$$

Here, the number a_u is, as a real valued function of u, uniquely defined and strictly increasing in $(-\infty,\infty)$ with

$$\lim_{u\to-\infty}a_u=c,\qquad\lim_{u\to\infty}a_u=d.$$

We refer the interested reader to [11,14,16,31] and the references cited therein for recent and related analogues of the above circle of ideas for weights whose equilibrium measure is supported on more than one interval.

1.4 Mean convergence of orthonormal expansions

We are ready to state our first result.

Theorem 1.2. *Let $I=(-\infty,\infty)$, w be a symmetric strongly admissible Freud weight, and let $1<p<\infty$. Suppose $b,B\in I$ satisfy*

$$b<1-1/p,\qquad B>-1/p,\qquad b\le B$$

and that

$$\frac{A_{n+1}}{A_n}=1+O\left(\frac{1}{n}\right),\qquad n\to\infty.\tag{1.6}$$

Then there exists an infinite subsequence $n=n_j$ of natural numbers such that for $j\ge C$ and for all f satisfying (1.2)

$$||S_n[f]wu_b||_{L_p(I)}\le C||fwu_B||_{L_p(I)}.\tag{1.7}$$

Moreover, if (1.7) holds for some infinite subsequence n_j and real b and B, then necessarily

$$b<1-1/p,\qquad B>-1/p,\qquad b\le B.$$

In particular, we have

$$\lim_{j\to\infty}||(S_n[f]-f)wu_b||_{L_p(I)}=0\tag{1.8}$$

for all f with

$$\lim_{|x|\to\infty}|fwu_B|(x)=0.$$

In [20, Theorem 1.2], Theorem 1.2 is established for every $n\ge 1$ assuming in addition to (1.6) the assumption that

$$\frac{A_n}{a_n}=\frac{1}{2}\left[1+O\left(\frac{1}{n^{2/3}}\right)\right],\qquad n\to\infty.\tag{1.9}$$

For the Hermite weight, [20, Theorem 1.2] essentially appears in earlier papers of Askey and Waigner, see [1] and Muckenhoupt, see [26] and [27]. Interesting generalizations of the work of Muckenhoupt in [26] and [27] have also been proved by Mhaskar and Xu in [25]. For a detailed discussion of the conditions (1.6) and (1.9), we refer the reader to Remark 1.4 below.

In order to establish Theorem 1.2, we use the following theorem of independent interest.

Theorem 1.3. *Let w be strongly admissible and symmetric. If*

$$\frac{A_{n+1}}{A_n} = 1 + O\left(\frac{1}{(nT(a_n))^{2/3}}\right), \qquad n \to \infty \qquad (1.10)$$

then there exists a subsequence $n = n_j$ of natural numbers such that

$$\frac{A_n}{a_n} = \frac{1}{2}\left[1 + O\left(\frac{1}{(nT(a_n))^{2/3}}\right)\right], \qquad j \to \infty. \qquad (1.11)$$

The following remark suffices:

Remark 1.4.

(a) Observe that (1.11) is a strong asymptotic whereas (1.10) is a ratio asymptotic. Thus applying the well-known identity

$$\left|\frac{a_{n+1}}{a_n} - 1\right| \sim \frac{1}{nT(a_n)},$$

see [22], it is clear that (1.11) implies (1.10) for every $n \geq 1$. It is the other direction which is new and nontrivial for strongly admissible weights w.

(b) Fortunately, the hypotheses on the recurrence coefficients given by (1.6) and (1.10) are not always vacuous ones. We list some related results on the line, and refer the interested reader to [7,16,21,22,23,28] and the references cited therein for further references and insights.

(A) Let $w = \exp(-Q)$, where Q is a polynomial of fixed degree. Then (see [16]),

$$\frac{A_n}{a_n} = \frac{1}{2}\left[1 + O\left(\frac{1}{n^2}\right)\right], \qquad n \to \infty.$$

(B) Let $w = W\exp(-Q)$, where Q is an even polynomial of fixed degree with nonnegative coefficients and $W(x) = |x|^\rho$ for some real ρ greater than -1. Then (see [7] and [23]),

$$\frac{A_n}{a_n} = \frac{1}{2}\left[1 + O\left(\frac{1}{n}\right)\right], \qquad n \to \infty.$$

Note that when $\rho \neq 0$, w is not always admissible.

(C) Let $w = w_\alpha$ be given by (1.3). Then (see [21]),

$$\frac{A_n}{a_n} = \frac{1}{2}\left[1 + O\left(\frac{1}{n}\right)\right], \qquad n \to \infty.$$

(D) Let $m \geq 1$ and $w = Ww_{1,1,2m,2m}$ given by (1.4). Then (see [7]),

$$\frac{A_n}{a_n} = \frac{1}{2}\left[1 + O\left(\frac{T(a_n)}{n}\right)\right], \qquad n \to \infty.$$

As a consequence of Theorem 1.3, we now state

Theorem 1.5. Let $I = (-\infty, \infty)$, w be a strongly admissible symmetric Freud weight and assume that the recurrence coefficients A_n satisfy (1.6). Then there exists N_0 and a infinite set of natural numbers Ω such that

$$\sup_{x \in I} |p_{n+1}(x) - p_n(x)|\, w(x) \times \left\{\left|1 - \frac{|x|}{a_n}\right| + n^{-2/3}\right\}^{-1/4} \sim a_n^{-1/2} \quad (1.12)$$

for all $1 \leq n \leq N_0$ and $n \geq N_0$, $n \in \Omega$.

The importance of (1.12) lies in the fact that for $|x|$ close to a_n, it improves the known bound (see [22])

$$\sup_{x \in I} |p_n(x)|\, w(x) \left\{\left|1 - \frac{|x|}{a_n}\right| + n^{-2/3}\right\}^{1/4} \sim a_n^{-1/2} \quad (1.13)$$

by a factor of $1/4$ as it should. Under the additional assumption (1.9), (1.12) is [20, Theorem 1.1] for $n \geq 1$.

We turn our attention to uniform convergence of orthonormal expansions for strongly admissible symmetric Freud weights on $(-\infty, \infty)$. To this end, we will need bounds on weighted Hilbert transforms. Define formally for continuous $f : I \to (-\infty, \infty)$

$$H[f](x) := \lim_{\varepsilon \to 0^+} \int_{|x-t| \geq \varepsilon} \frac{f(t)}{x - t}\, dt$$

where the integral above is understood as a Cauchy-Principal valued integral. It is known, see [26,27], that if $b < 1 - 1/p$, $B > -1/p$, $b \leq B$ and $1 < p < \infty$, we have

$$\|H[f]u_b\|_{L_p((-\infty,\infty))} \leq C\|fu_B\|_{L_p((-\infty,\infty))}, \quad (1.14)$$

provided the right hand side of (1.14) is finite. Indeed, relations such as (1.14) are essential in studying convergence of orthonormal expansions. This is mainly due to the following identity which follows from the Christoffel-Darboux formula for orthonormal polynomials:

$$S_n[f] = A_n \left\{ p_n H[f p_{n-1}] - p_{n-1} H[f p_n] \right\}. \tag{1.15}$$

For uniform convergence of orthonormal expansions, it thus seems natural to look for L_∞ analogues of (1.14), but until recently, even for finite intervals, such analogues have been scarce in the literature. We mention that uniform bounds for weighted Hilbert transforms are also important for the numerical solution and stability of integral equations on finite and infinite intervals, see [10] and the references cited therein. To this end, we now state two theorems which hold for any admissible weight on I. We refer the reader to [2,3,4,8,9] and the references cited therein for related results. We recall that for $h > 0$,

$$\Delta_h^1(f, I)(x) := f(x + h/2) - f(x - h/2), \ x \pm h/2 \in I \tag{1.16}$$

is the first symmetric difference operator of f. Then we have:

Theorem 1.6A. *Let $f : I \to (-\infty, \infty)$ be continuous, w be admissible, $B, \varepsilon > 0$ and suppose $\|fw(1 + |Q'|)^B\|_{L_\infty(I)} < \infty$ and $\frac{\|\Delta_u^1(f,I)\|_{L_\infty(I)}}{u} \in L_1[0, \varepsilon]$. Then*

$$\|H[fw]\|_{L_\infty(I)}$$
$$\leq C \left[\|fw(1 + |Q'|)^B\|_{L_\infty(I)} + \int_0^\varepsilon \frac{\|\Delta_u^1(f,I)\|_{L_\infty(I)}}{u} du \right]. \tag{1.17}$$

Theorem 1.6B. *Let $f : I \to (-\infty, \infty)$ be differentiable, w be admissible, $B > 0$ and suppose $\|fw(1 + |Q'|)^B\|_{L_\infty(I)} < \infty$ and $\|f'w\|_{L_\infty(I)} < \infty$. Then*

$$\|H[fw]\|_{L_\infty(I)} \leq C \left[\|fw(1 + |Q'|)^B\|_{L_\infty(I)} + \|f'w\|_{L_\infty(I)} \right]. \tag{1.18}$$

Note the factor $(1 + |Q'|)^B$ on the right-hand side of (1.17) and (1.18). For Q even and of polynomial growth, (the case $T \sim 1$), we may take it to be essentially u_{B_1} for some $B_1 > 0$. Theorems 1.6(A-B) improve [9, Theorem 1.1] and [4, Theorem 1] in several respects. Firstly they replace the weighting factor of w^2 in [9] by the correct factor $w(1 + |Q'|)^B$. This later observation is crucial in the formulation of Theorem 1.7 below. Secondly, Theorems 1.6(A-B) hold simultaneously for a much larger class of possibly nonsymmetric weights with varying rates of decay on $(c, 0)$ and $[0, d)$. Using Theorems 1.6(A-B), we are able to announce:

Theorem 1.7. *Let $I = (-\infty, \infty)$, w be a strongly admissible symmetric Freud weight, $B > 0$, and assume (1.6). Let $f : I \to (-\infty, \infty)$ be differentiable, and suppose $\|f w u_B\|_{L_\infty(I)} < \infty$ and $\|f'w\|_{L_\infty(I)} < \infty$. Then there exists an infinite subsequence $n = n_j$ of natural numbers such that for all $j \geq C$*

$$\|S_n[f]w\|_{L_\infty(I)} \leq C \left[\|f w u_B\|_{L_\infty(I)} + \|f'w\|_{L_\infty(I)} \right]. \tag{1.19}$$

Moreover, if (1.19) holds for some infinite subsequence n_j and real B, then necessarily $B > 0$. If f' is also continuous and

$$\lim_{|x|\to\infty} |f w u_B|(x) = 0$$

and

$$\lim_{|x|\to\infty} |f'w^\alpha|(x) = 0$$

for some $0 < \alpha < 1$, then

$$\lim_{j\to\infty} \|(S_{n_j}[f] - f)w\|_{L_\infty(I)} = 0. \tag{1.20}$$

If we assume (1.9), then (1.20) holds for every $n \geq 1$.

Theorem 1.7 provides necessary and sufficient conditions for uniform convergence of orthogonal expansions for strongly admissible Freud weights on the line. Its proof will appear in a future paper.

The remainder of this paper is devoted to the proofs of Theorem 1.2, Theorem 1.3, Theorem 1.5 and Theorems 1.6(A-B).

§2. Proofs of Theorems 1.2, 1.3, 1.5 and 1.6(A-B)

In this section, we prove Theorems 1.2, 1.3, 1.5 and 1.6(A-B).

2.1 The proof of Theorem 1.3.

Let $n \geq C$, and define

$$m = m(n) = [n^{1/3}(T(a_n))^{1/3}]$$

where $[x]$ denotes the greatest integer $\leq x$. It follows using Definition 1.1C (a) and Remark 1.4(a) that uniformly for $r = 1, ..., m$ and $n \geq 1$,

$$T(a_{n+r}) \sim T(a_n).$$

Armed with this identity, we apply (1.10) repeatedly and deduce that there exists $N = N(m)$ and $D > 0$ such that for $n \geq N$,

$$A_{n+r} \geq \left(1 - \frac{D}{(nT(a_n))^{2/3}}\right) A_n, \qquad r = 1, 2, ..., m. \qquad (2.1)$$

Here $D > 0$ does not depend on n or m so we fix it. Now set

$$\varepsilon = \varepsilon(n) = \frac{D}{(nT(a_n))^{2/3}}.$$

A careful adaption of the proof of [18, Theorem 6] then shows that

$$\begin{aligned}
x_{1,n}(w) &\geq 2(1-\varepsilon)A_n \cos\frac{\pi}{m+1} \\
&\geq 2(1-\varepsilon)A_n(1 - C/m^2) \\
&\geq 2\left(1 - \frac{D}{(nT(a_n))^{2/3}}\right) A_n \left(1 - \frac{C}{(nT(a_n))^{2/3}}\right).
\end{aligned} \qquad (2.2)$$

Now recall that

$$\left|\frac{x_{1,n}}{a_n} - 1\right| = O\left(\frac{1}{n^{2/3}T(a_n)^{2/3}}\right). \qquad (2.3)$$

Thus (2.2) and (2.3) give

$$\frac{A_n}{a_n} \leq \frac{1}{2}\left[1 + O\left(\frac{1}{(nT(a_n))^{2/3}}\right)\right], \qquad n \geq C. \qquad (2.4)$$

Another careful inspection of [18, Theorem 7], reveals that we have

$$\lim_{n\to\infty} \frac{x_{1,n}}{\max_{k\leq n}\alpha_k} \leq 2.$$

Thus, we may apply this with (2.3) to obtain

$$\frac{\max_{k\leq n} A_k}{a_n} \geq \frac{1}{2}\left(1 - \frac{1/C}{(nT(a_n))^{2/3}}\right). \qquad (2.5)$$

Choosing an increasing sequence n_r with

$$\max_{k\leq n_r} A_k = A_{n_r}, \qquad (2.6)$$

and applying (2.4)–(2.6) gives the theorem.

2.2 Proof of Theorems 1.2 and 1.5.

We sketch the important ideas of the proof. The remaining technical details are very similar to [20, Theorem 1.1], and so we refer the reader to that paper.

Step 1: We reduce the proof of Theorem 1.5 to one important case.

(a) By symmetry we may assume that $x \geq 0$.

(b) It suffices to prove (1.12) for n sufficiently large.

(c) Using infinite finite inequalities it suffices to prove (1.12) for

$$0 \leq x \leq a_n \left(1 - \frac{C}{n^{2/3}}\right).$$

(d) For $0 \leq x \leq a_{n/2}$,

$$\left|1 - \frac{|x|}{a_n}\right| + n^{-2/3} \sim 1$$

so (1.12) follows from (1.13).

Step 2: Without loss of generality we may thus assume that

$$a_{n/2} \leq x \leq a_n \left(1 - \frac{C_1}{n^{2/3}}\right)$$

for some $C_1 > 0$ which will be chosen later. Let us define for $y \geq 0$

$$\tau_n(y) := \frac{1}{A_n^2} \sum_{k=0}^{n-1} (\alpha_{k+1}^2 - \alpha_k^2) p_k^2(y).$$

Then the Dombrowski-Fricke identity, see [17], gives

$$|p_{n+1}(y) - p_{n-1}(y)|w(x) \leq \left\{2\tau_{n+1}(y)w^2(y)\right\}^{1/2} + \left\{2\tau_n(y)w^2(y)\right\}^{1/2},$$
$$(2.7)$$

for

$$0 \leq y \leq 2\min\{A_n, A_{n+1}\}.$$

Applying Theorem 1.3, we deduce that there exists $D > 0$ and an infinite set of natural numbers Ω such that (2.7) holds for all

$$0 \leq y \leq a_n \left(1 - Dn^{-2/3}\right), \qquad n \in \Omega.$$

Choose $C_1 = D$ in the above and assume, as we may, that (2.7) holds for x and $n \in \Omega$.

Step 3: Estimation of $\tau_n(x)$ for $n \in \Omega$: We write

$$\tau_n(x) := \frac{1}{A_n^2} \left\{ \sum_{k=0}^{[n/4]} + \sum_{k=[n/4]+1}^{n-1} \right\} (\alpha_{k+1}^2 - \alpha_k^2) p_k^2(x)$$

$$= \tau_{n,1}(x) + \tau_{n,2}(x).$$

Now applying (1.6), we have that

$$\tau_{n,2}(x) w^2(x) \le C \frac{1}{n} w^2(x) \lambda_n(x)^{-1},$$

where λ_n are the Cotes numbers for w^2. But for this range of x, it is well known that

$$\lambda_n(x)^{-1} w^2(x) \sim \frac{n}{a_n} \sqrt{\left| 1 - \frac{|x|}{a_n} \right|} + n^{-2/3}.$$

Thus,

$$\tau_{n,2}(x) w^2(x) \le C a_n^{-1} \sqrt{\left| 1 - \frac{|x|}{a_n} \right|} + n^{-2/3}.$$

Moreover, using infinite-finite range inequalities yields

$$\tau_{n,1}(x) w^2(x) = O\left(\exp(-Cn)\right).$$

Combining the above estimates for both $\tau_{n,1}$ and $\tau_{n,2}$ yields Theorem 1.5. Theorem 1.2 then follows using [20, Theorem 1.2], Theorem 1.5 and Theorem 1.3 setting $n = n_j$. The crucial point in the proofs of both Theorem 1.2 and Theorem 1.5 is the removal of the strong asymptotic (1.9), and this is achieved by means of Theorem 1.3.

2.3 The proof of Theorems 1.6(A-B).

We begin with Theorem 1.6(A). We may assume that $x \in [0, d)$. The other case is similar. We consider several subcases. We suppose first that $d = \infty$, $c = -\infty$, and that x is sufficiently close to d. More precisely, let us choose a constant $D > 0$ so close to d so that for $D < x < d$

$$\frac{1}{Q'(x+\varepsilon)} < \varepsilon.$$

(We will in practice always take ε small since clearly if Theorem 1.6(A-B) holds for such ε, then it holds for larger ε). Fix x and define

$$A(x) = A_\varepsilon(x) := \frac{1}{2Q'(x+\varepsilon)}.$$

Note $|Q'|(y) = Q'(y)$ for y close enough to d, and $|Q'|(y) = -Q'(y)$ for y close enough to c. This follows from the definition of T and Definitions 1.1(B-C). Let us write

$$H[fw](x) = \left(\int_{|t|>2x} + \int_{-2x}^0 + \int_0^{2x} \right) \frac{f(t)w(t)}{t-x} dt \qquad (2.8)$$

$$= I_1(x) + I_2(x) + I_3(x).$$

We first estimate $I_1(x)$. Here

$$|t - x| \ge |t| - |x| \ge |t| - |t|/2 = |t|/2$$

so

$$|I_1(x)| \le 2 \int_{|t|>2x} \frac{|f(t)|w(t)}{|t|} dt$$

$$\le C\|fw(1+|Q'|)^B\|_{L_\infty(I)} \int_{|t|\ge 2} \frac{1}{|t|(1+|Q'|(t))^B} dt$$

$$\le C\|fw(1+Q')^B\|_{L_\infty(I)}.$$

Similarly,

$$|I_2(x)| = \left| \int_{-2x}^0 \frac{f(t)w(t)}{t-x} dt \right| \le \|fw(1+|Q'|)^B\|_{L_\infty(I)} \int_{-2x}^0 \frac{1}{x-t} dt$$

$$\le C\|fw(1+|Q'|)^B\|_{L_\infty(I)} \int_x^{3x} \frac{du}{u}$$

$$\le C\|fw(1+|Q'|)^B\|_{L_\infty(I)}.$$

We proceed with $I_3(x)$. Note that by choice of $A(x)$ and using [22], Lemma 3.2a,

$$w(y) \sim w(x)$$

uniformly for every $y \in I$ with $|x-y| \le 2A(x)$. (The latter lemma implies that $Q(s) \ge Q(r)$, $s/r \ge 1$.) We then split $I_3(x)$ as follows: Write for $\beta := \beta(\varepsilon) > 0$,

$$I_3(x) = \int_0^{2x} \frac{f(t)w(t)}{t-x} dt$$

$$= \left(\int_0^{x/\beta} + \int_{x/\beta}^{x-\varepsilon} + \int_{x-\varepsilon}^{x-A(x)} + \int_{x-A(x)}^{x+A(x)} + \int_{x+A(x)}^{2x} \right) \frac{f(t)w(t)}{t-x} dt$$

$$= I_{31}(x) + I_{32}(x) + I_{33}(x) + I_{34}(x) + I_{35}(x).$$

Then,

$$|I_{31}(x)| \leq C\|fw(1+|Q'|)^B\|_{L_\infty(I)} \int_0^{x/\beta} \frac{(1+|Q'|(t))^{-B}}{x-t}dt$$
$$\leq C\|fw(1+|Q'|)^B\|_{L_\infty(I)} \int_0^{x/\beta} \frac{1}{x-t}dt$$
$$\leq C\|fw(1+|Q'|)^B\|_{L_\infty(I)}.$$

Similarly,

$$|I_{32}(x)| \leq C\|fw(1+|Q'|)^B\|_{L_\infty(I)} \frac{\log x}{(1+Q'(x/\beta))^B}$$
$$\leq C\|fw(1+|Q'|)^B\|_{L_\infty(I)}.$$

Next,

$$|I_{33}(x)| \leq C\|fw(1+|Q'|)^B\|_{L_\infty(I)} \int_{x-\varepsilon}^{x-A(x)} \frac{(1+|Q'|(t))^{-B}}{x-t}dt$$
$$\leq C\|fw(1+|Q'|)^B\|_{L_\infty(I)} \frac{\log Q'(x+\varepsilon)}{(1+Q'(x-\varepsilon))^B}$$
$$\leq C\|fw(1+|Q'|)^B\|_{L_\infty(I)}.$$

Similarly,

$$|I_{35}(x)| = \left|\int_{x+A(x)}^{2x} \frac{f(t)w(t)}{t-x}dt\right|$$
$$\leq C\|fw(1+|Q'|)^B\|_{L_\infty(I)}(1+Q'(x+A(x)))^{-B} \int_{A(x)}^{x} \frac{1}{u}du$$
$$\leq C\|fw(1+|Q'|)^B\|_{L_\infty(I)}(1+Q'(x+A(x)))^{-B}(\log Q'(x+\varepsilon)+\log x)$$
$$\leq C\|fw(1+|Q'|)^B\|_{L_\infty(I)}.$$

Finally we consider $I_{34}(x)$. We write

$$|I_{34}(x)| = \left|\int_{x-A(x)}^{x+A(x)} \frac{f(t)w(t)}{t-x}dt\right|$$
$$\leq \left|\int_{x-A(x)}^{x+A(x)} f(t)\frac{w(t)-w(x)}{t-x}dt\right| + w(x)\left|\int_{x-A(x)}^{x+A(x)} \frac{f(t)-f(x)}{t-x}dt\right|$$
$$= |I_{341}(x)| + |I_{342}(x)|.$$

We begin by making the substitution $t = u/2 + x$ into $I_{342}(x)$. Then we have

$$|I_{342}(x)| \leq Cw(x) \int_0^{2A(x)} \left| \frac{f(x + u/2) - f(x - u/2)}{u} \right| du$$

$$\leq C \int_0^{2A(x)} \left\| \Delta_u^1(f, I)w \right\|_{L_\infty(I)} \frac{1}{u} du$$

$$\leq C \int_0^{\varepsilon} \left\| \Delta_u^1(f, I)w \right\|_{L_\infty(I)} \frac{1}{u} du.$$

Also

$$|I_{341}(x)| = \left| \int_{x-A(x)}^{x+A(x)} f(t) \frac{w(t) - w(x)}{t - x} dt \right|$$

$$\leq C \| fw(1 + |Q'|)^B \|_{L_\infty(I)} \int_{x-A(x)}^{x+A(x)} w^{-1}(t) |w'(\eta)| dt$$

$$\leq C A(x)w(x - A(x))w^{-1}(x + A(x))Q'(x + A(x)) \| fw(1 + |Q'|)^B \|_{L_\infty(I)}$$

$$\leq C \| fw(1 + |Q'|)^B \|_{L_\infty(I)}.$$

We observe that if $1 \leq x < D$, then the estimates for $I_1(x)$ go through as before as we only needed the fact that $D > 1$ to ensure that the integral converged for t much larger than x. $I_2(x)$ follows without change except that we use the boundedness of $|Q'|$ rather than its sign. For $I_3(x)$, x is bounded and $w \sim 1$, so the proof is easier than before. If $0 < x < 1$, then we write

$$H[f; w](x) = \int_{-\infty}^{\infty} \frac{f(t)w(t)}{t - x} dt = \left(\int_{-\infty}^{x-1} + \int_{x-1}^{x+1} + \int_{x+1}^{\infty} \right) \frac{w(t)f(t)}{t - x} dt.$$

For the first two integrals, t is bounded away from x, and for the third we proceed as above, but the proof is easier since x is bounded and $w \sim 1$. Suppose now that d and c are finite. Let us first suppose that x is close to d. Then choose $D > 0$ such that $D \leq x < d$, and write

$$H[f; w](x) = \left(\int_c^0 + \int_0^d \right) \frac{w(t)f(t)}{t - x} dt.$$

For the first integral, $t - x$ is bounded away from 0, so bounding this term gives us the required estimate. For the second integral, split as in $I_3(x)$. (Note that here we choose ε at the start small enough so that $x \pm \varepsilon \in I$). A very similar and easier argument works if also $\gamma \leq x \leq D$ for some fixed and positive γ, for here $w \sim 1$. If for this γ, $0 \leq x < \gamma$, then split

$$H[f; w](x) = \int_c^d \frac{f(t)w(t)}{t - x} dt = \left(\int_c^{x-\gamma} + \int_{x-\gamma}^{x+\gamma} + \int_{x+\gamma}^d \right) \frac{w(t)f(t)}{t - x} dt.$$

Then the estimate goes through very much as above. Finally, suppose that d is finite and $c = -\infty$. (The other case is similar). Then choose $A > 1$, and write

$$H[f;w](x) = \left(\int_{-\infty}^{-Ax} + \int_{-Ax}^{0} + \int_{0}^{d} \right) \frac{w(t)f(t)}{t-x} dt.$$

This proves Theorem 1.6(A). Theorem 1.6(B) follows by noting that in the proof of Theorem 1.6(A), for a given x, we may take ε small enough so that $w(x \pm u/2) \sim w(x)$ for all $u \leq \varepsilon$. This easily yields the result.

Acknowledgments. . The author is indebted to Larry Schumaker and an anonymous referee who helped to improve this paper in many ways.

References

1. Askey, R., and S. Wainger, Mean convergence of expansions in Laguerre and Hermite series, J. Math **87**, 695–708.

2. De Bonis, M. C, B. Della Vecchia, and G. Mastroianni, Approximation of the Hilbert transform on the real semiaxis using Laguerre zeros, preprint.

3. De Bonis, M. C, and G. Mastroianni, Some simple quadrature rules to evaluate the Hilbert transform on the real line, preprint.

4. De Bonis, M. C, B. Della Vecchia, and G. Mastroianni, Approximation of the Hilbert transform on the real line using Hermite zeroes, Mathematics of Computation, to appear.

5. Damelin, S. B., Converse and smoothness theorems for Erdös weights in L_p, $(0 < p \leq \infty)$, J. Approx. Theory **93** (1998), 349–398.

6. Damelin, S. B., On the distribution of interpolation arrays for exponential weights, Electronic Transactions of Numerical Analysis, to appear.

7. Damelin, S. B., Asymptotics of recurrence coefficients for exponential weights, Magnus's method revisited, manuscript.

8. Damelin, S. B., and K. Diethelm, Interpolatory product quadratures for Cauchy principal value integrals with Freud weights, Numer. Math **83** (1999), 87–105.

9. Damelin, S. B., and K. Diethelm, Boundness and numerical approximation of the weighted Hilbert transform on the real line, Numerical Functional Analysis and Optimization **22(1,2)**, (2000), 13–54.

10. Damelin, S.B., and K. Diethelm, The numerical approximation and stability of certain integral equations on the line, manuscript.

11. Damelin, S.B., P. Dragnev, and A. B. J. Kuijlaars, The support of the equilibrium measure for a class of external fields on a finite interval, Pacific J. Math **199(2)** (2001), 303–321.

12. Damelin, S. B., H. S. Jung, and K. H. Kwon, Mean convergence of Hermite Fejèr and Hermite interpolation of higher order for Freud type weights, J. Approx. Theory **113** (2001), 21–58.

13. Damelin, S. B., H. S. Jung, and K. H. Kwon, On mean convergence of Lagrange and extended Lagrange interpolation for exponential weights: Methods, recent and new results, manuscript.

14. Damelin, S. B., and A. B. J. Kuijlaars, The support of the extremal measure for monomial external fields on $[-1, 1]$, Trans. Amer. Math. Soc. **351** (1999), 4561–4584.

15. Damelin, S. B., and D. S. Lubinsky, Jackson theorems for Erdös weights in L_p $(0 < p \leq \infty)$, J. Approx. Theory **94** (1998), 333–382.

16. Deift, P., T. Kriecherbauer, K. McLaughlin, S. Venakides, and X. Zhou, Strong asymptotics of orthonormal polynomials with respect to exponential weights, Commun. Pure Appl. Math. **52** (1999), 1491–1552.

17. Dombrowski, J. M., and G. H. Fricke, The absolute continuity of phase operators, Trans. Amer. Math. Soc, **213**, 363–372.

18. Freud, G., On the greatest zero of orthonormal polynomials, J. Approx. Theory **46** (1986), 15–23.

19. Freud, G., *Orthonormal Polynomials*, Akadémiai Kiadó, Budapest, 1971.

20. Jha, S. W., and D. S. Lubinsky, Necessary and sufficient conditions for mean convergence of orthonormal expansions for Freud weights, Constr. Approx. **11** (1995), 331–363.

21. Kriecherbauer, T., and K. McLaughlin, Strong asymptotics of polynomials orthonormal with respect to Freud weights, International Mathematics Research Notices **6** (1999), 299–333.

22. Levin, A. L., and D. S. Lubinsky, *Orthonormal Polynomials for Exponential Weights*, Springer Verlag 2001.

23. Magnus, A. P., On Freud's equations for exponential weights, J. Approx. Theory **46** (1986), 65–99.

24. Mhaskar, H. N., *Introduction to the Theory of Weighted Polynomial Approximation*, World Scientific, Singapore, 1996.

25. Mhaskar, H. N., and Y. Xu, Mean convergence of expansions in Freud type orthogonal polynomials, SIAM J. Math. Anal **22**, 847–855.

26. Muckenhoupt, B., Mean convergence of Hermite and Laguerre series I, Trans. Amer. Math. Soc **147**, 419–431.

27. Muckenhoupt, B., Mean convergence of Hermite and Laguerre series II, Trans. Amer. Math. Soc **147**, 433–460.

28. Nevai, P., Geza Freud, orthogonal polynomials and Christoffel functions (a case study), J. Approx. Theory **48** (1986), 3–167.

29. Rakhmanov, E. A., On asymptotic properties of polynomials orthogonal on the real axis, Mat. Sb. **119** (1982), 163–203. English transl: Math. USSR-Sb. **47** (1984), 155–193.

30. Stahl, H., and V. Totik, *General Orthogonal Polynomials*, Cambridge University Press, Cambridge, 1992.

31. Totik, V., *Weighted Approximation with Varying Weight*, Lect. Notes in Math. Vol. 1569, Springer-Verlag, Berlin, 1994.

32. Xu, Y., Mean convergence of generalised Jacobi series and interpolating polynomials, J. Approx Theory **72**, 237–251.

Steven B. Damelin
Department of Mathematics and Computer Science
Georgia Southern University
Post Box 8093
Statesboro, GA 30460-8093
USA
damelin@gasou.edu

On the Separation of Logarithmic Points on the Sphere

P. D. Dragnev

Abstract. In this article we consider the distribution of N points on the unit sphere, whose mutual distances have maximal geometric mean. The problem is reduced to a weighted energy problem for circular symmetric weights. As a result, a new improved separation condition on the optimal configuration is derived.

§1. Introduction

The distribution of points on the sphere which satisfy some optimal conditions is a fascinating subject that has interested researchers for a long time. Ever since the ancient Greek and the Platonic solids, passing through Isaak Newton and the Thirteen Sphere Problem, and in recent times the discovery of fullerenes, patterns on the sphere have been a source of many comprehensive investigations throughout the centuries.

In this article we shall focus our attention on a particular problem concerning the so called logarithmic points on the sphere. Sometimes referred to as elliptic Fekete points [20], these points maximize the geometric mean of their mutual distances. Indeed, if we define the logarithmic energy of a collection of N points $\omega_N := \{\mathbf{x}_1, \mathbf{x}_2, \dots, \mathbf{x}_N\}$ on the unit sphere $S^2 := \{\mathbf{x} \in \mathbb{R}^3 \ : \ |\mathbf{x}| = 1\}$ to be

$$E_0(\omega_N) := \sum_{i<j} \log \frac{1}{|\mathbf{x}_i - \mathbf{x}_j|}, \tag{1}$$

then the logarithmic points are the ones that minimize this energy (here $|\cdot|$ denotes the Euclidean norm). More generally, when $\alpha > 0$ $(\alpha < 0)$ the α-extremal points minimize (maximize) the α-energy

$$E_\alpha(\omega_N) := \sum_{i<j} \frac{1}{|\mathbf{x}_i - \mathbf{x}_j|^\alpha}. \tag{2}$$

Approximation Theory X: Abstract and Classical Analysis
Charles K. Chui, Larry L. Schumaker, and Joachim Stöckler (eds.), pp. 137–144.

The logarithmic points are then the limiting case of the α-extremal points as $\alpha \to 0$. When $\alpha \to \infty$ we obtain the best packing problem, where the smallest distance is maximized. We shall refer to these points as best-packing points on the sphere.

The distribution of logarithmic points was posed as a problem by L. L. Whyte [25] in 1952 and has been a subject of extensive study, see [1,2,8,9,10,11,12,13,14,16,17,18,23,24]. In this paper we shall restrict ourselves to the separation of the logarithmic points.

Definition 1. *For fixed $N \in \mathbb{N}$ and $\alpha \in \mathbb{R}$, the minimal distance*

$$d_{N,\alpha} := \min_{i \neq j} |\mathbf{x}_i - \mathbf{x}_j| \tag{3}$$

among any N-configuration of α-extremal points is called a separation distance *for N and α.*

It is intuitively clear that $d_{N,\alpha}$ cannot become too small. For the best-packing points, it is known that $\lim_{N \to \infty} \sqrt{N} d_{N,\infty} = \sqrt{8\pi/\sqrt{3}}$ (see [4,11,13]). For the so-called Fekete points ($\alpha = 1$), Dahlberg [5] showed that the extremal points on a smooth surface in $\mathbb{R}^3$, and in particular on S^2, have a separation distance of order $1/\sqrt{N}$, and therefore there is a constant C such that $d_{N,1} \geq C/\sqrt{N}$. On the other hand for $\alpha < 0$, Stolarsky [21] derived that

$$d_{N,\alpha} \geq \left(\frac{4|\alpha|2^\alpha}{(2-\alpha)N} \right)^{1/(2+\alpha)}, \quad -2 < \alpha < 0.$$

We recall here that when $\alpha \leq -2$ some of the points may coincide (see [3]).

Regarding the logarithmic points, it was Rakhmanov, Saff, and Zhou [18], who first showed that they are well separated, namely that

$$d_{N,0} \geq \frac{3/5}{\sqrt{N}}.$$

Dubickas, using a refinement of their method, improved upon this estimate and showed that

$$d_{N,0} \geq \frac{7/4}{\sqrt{N}}.$$

It is not known, whether $\lim_{N \to \infty} \sqrt{N} d_{N,0}$ exists.

In this paper we reduce Whyte's problem to a minimal energy problem for circular symmetric weights. A general theory on the position of the weighted Fekete points and the support of the equilibrium measure associated with circular symmetric weights is then applied to derive a new bound on the separation distance $d_{N,0}$.

Theorem 2. *Suppose that $N \geq 2$. Then the separation distance of the logarithmic points satisfies*

$$d_{N,0} \geq \frac{2}{\sqrt{N-1}}. \tag{4}$$

Remark 3. In both [18] and [9] the approach is to fix one of the logarithmic points in an arbitrary optimal configuration to be at the south pole, and then apply a stereographical projection of the sphere with a pole at the north pole onto the equatorial plane. Then the image points are shown to be bounded away from the origin, which is the image of the south pole, hence leading to the desired inequalities. In this article we also apply a stereographical projection, but the difference is that the north pole is chosen to be one of the logarithmic points. This (small at first glance) modification dramatically changes the picture, allowing us to reduce the problem to a weighted Fekete points problem with *admissible* weight, which eventually implies that the stereographical images of the other $N-1$ logarithmic points stay away from ∞, which is exactly what we want.

Observe that equality holds in (4) for $N = 2$, therefore it is not possible to improve on the constant in (4). However, it is not clear from the present proof whether in the limit one can obtain a better constant than 2. The difficulty is that the weights, which we obtain after the stereographical projection depend on N, and a very sensitive analysis of the weighted Fekete points will be needed to further advance in this direction. Let us also note that the approach might be useful for numerical investigations of this important problem.

In Section 2 we briefly review the necessary potential-theoretical tools and the properties of circular symmetric weights. The proof of Theorem 2 is given in Section 3.

§2. Potential-Theoretical Preliminaries

Here we briefly introduce some notions from the theory of logarithmic potentials (for details see [6,19]). A positive continuous function $w : \mathbf{C} \to \mathbb{R}^+$ is called an admissible weight if $|z|w(z) \to 0$ as $|z| \to \infty$. For a measure μ in the class $\mathcal{M}$ of positive unit Borel measures, the weighted logarithmic energy is defined as

$$I_w(\mu) := \int \int \log \left[\frac{1}{|y - z| w(y) w(z)} \right] d\mu(y) d\mu(z)$$
$$= I(\mu) + 2 \int Q(z) \, d\mu(z),$$

where $I(\mu)$ is the logarithmic energy, and $Q(z) := \log(1/w(z))$ is called an external field. The logarithmic potential of μ is denoted by

$$U^\mu(z) := \int \log \frac{1}{|z-y|}\, d\mu(y).$$

There exists a unique measure μ_w, such that $I_w(\mu_w) = \inf_{\mathcal{M}} I_w(\mu)$. This measure, called the **weighted equilibrium measure** associated with the weight w, has a compact support S_w, and is determined completely by its variational inequalities

$$
\begin{aligned}
U^{\mu_w}(z) + Q(z) &\geq F_w \quad \text{quasi} - \text{everywhere on} \ \ \mathbb{C} \\
U^{\mu_w}(z) + Q(z) &\leq F_w \quad \text{on} \ \ S_w,
\end{aligned}
\tag{5}
$$

where F_w is called the **equilibrium constant**, and quasi-everywhere means everywhere except on a set of logarithmic capacity zero. To find μ_w, or to solve the **weighted energy problem**, is usually a very difficult task, often associated with a comprehensive study of the equilibrium support S_w. The weighted energy problem has wide applications in various areas of analysis, like orthogonal polynomials, random matrices, integrable systems, etc. (see [6,7,15,19,22]). Here we establish a nice connection with logarithmic points on the unit sphere.

We are interested in the specific case when $w(z)$ is a circular symmetric weight, that is $Q(z)$ is a radially symmetric function, i.e. it has the property that $Q(z) = Q(|z|)$ for all $z \in \mathbb{C}$. Then the problem is easily solved and the equilibrium measure and support can be explicitly found. We summarize the results that we need and refer the reader to [19, Section IV.6] for details. Let Q, when restricted to $\mathbb{R}^+$, be a differentiable function with absolutely continuous derivative, bounded below, and such that

$$\lim_{r \to \infty} (Q(r) - \log r) = \infty. \tag{6}$$

If (i) $rQ'(r)$ is increasing or (ii) Q is convex, then the equilibrium support is the annulus

$$S_w = \{z \mid r_0 \leq |z| \leq R_0\}, \tag{7}$$

where $r_0 \geq 0$ is the smallest number for which $Q'(r) > 0$ for all $r > r_0$, and R_0 is the smallest solution of $R_0 Q'(R_0) = 1$. The equilibrium measure μ_w is given by

$$d\mu_w(z) = \frac{1}{2\pi}(rQ'(r))'dr d\phi, \quad z = re^{i\phi}, \ \ 0 \leq \phi \leq 2\pi, \ r_0 \leq r \leq R_0, \tag{8}$$

where $F_w = Q(R_0) - \log(R_0)$, and the equilibrium potential is

$$U^{\mu_w}(z) = \begin{cases} F_w - Q(r_0), & \text{if } |z| < r_0 \\ F_w - Q(z), & \text{if } r_0 \leq |z| \leq R_0 \\ \log 1/|z|, & \text{if } |z| > R_0. \end{cases} \tag{9}$$

For more details on the theory and the applications of the weighted energy problem, we refer to the recent seminal monograph on logarithmic potentials by Saff and Totik [19] and the references therein.

§2. Proof of Theorem 2

This section is devoted to the proof of the separation condition in Theorem 2. Let $\omega_N^* := \{\mathbf{x}_1^*, \mathbf{x}_2^*, \dots, \mathbf{x}_N^*\}$ be a global minimizer of the problem

$$E_0(\omega_N^*) = \min_{\omega_N \subset S^2} E_0(\omega_N). \tag{10}$$

Choose a coordinate system such that $\mathbf{x}_N^*$ is the north Pole, and let $z_i^* \in \mathbb{C}$, $i = 1, \dots, N-1$ be the stereographical projections of $\mathbf{x}_i^*$ on the equatorial plane (observe that the image of $\mathbf{x}_N^*$ is ∞). The conversion of the mutual distances for any points $\mathbf{x}_i \in S^2, i = 1, \dots, N-1$ is the following

$$|\mathbf{x}_i - \mathbf{x}_j| = \frac{2|z_i - z_j|}{\sqrt{1+|z_i|^2}\sqrt{1+|z_j|^2}}, \ j \neq N, \ \text{and} \ |\mathbf{x}_N - \mathbf{x}_i| = \frac{2}{\sqrt{1+|z_i|^2}}. \tag{11}$$

For the product of these distances, we evaluate

$$\prod_{i<j} |\mathbf{x}_i - \mathbf{x}_j| = \prod_{1 \leq i < j \leq N-1} |\mathbf{x}_i - \mathbf{x}_j| \prod_{i=1}^{N-1} |\mathbf{x}_N - \mathbf{x}_i|$$

$$= \prod_{1 \leq i < j \leq N-1} \frac{2|z_i - z_j|}{\sqrt{1+|z_i|^2}\sqrt{1+|z_j|^2}} \prod_{i=1}^{N-1} \frac{2}{\sqrt{1+|z_i|^2}}$$

$$= 2^{N(N-1)/2} \prod_{1 \leq i < j \leq N-1} \frac{|z_i - z_j|}{(1+|z_i|^2)^{\frac{N-1}{2(N-2)}}(1+|z_j|^2)^{\frac{N-1}{2(N-2)}}}. \tag{12}$$

Then $\{z_i^*\}_{i=1}^{N-1}$ are solutions to the following minimization problem (compare (10) and (12)):

$$\min_{\{z_i\} \subset \mathbb{C}} \sum_{1 \leq i < j \leq N-1} \log \frac{1}{|z_i - z_j| W_N(z_i) W_N(z_j)}, \tag{13}$$

where

$$W_N(z) := (1+|z|^2)^{-(N-1)/(2(N-2))}.$$

We remind the reader that for any k the weighted Fekete points associated with a positive weight w are defined to be any k points that maximize the product

$$\prod_{1 \leq i < j \leq k} |z_i - z_j| w(z_i) w(z_j).$$

In particular, if we consider $w = W_N$ and $k = N - 1$, the points $\{z_i^*\}_{i=1}^{N-1}$ turn out to be weighted Fekete points. This implies some very nice properties about where z_i^* could possibly lie, which we now examine. Since $|z|W_N(z) \to 0$ as $z \to \infty$, the weight W_N is admissible. Then we can apply [19, Theorem III.1.2], which states that for any k the weighted Fekete points lie in the set

$$S_w^* := \{z \in \mathbf{C} \mid U^{\mu_w}(z) + Q(z) \leq F_w\}. \tag{14}$$

It is clear from (5) that $S_w \subset S_w^*$. Since in this case the weight w is also circular symmetric, from (8) we are able to determine μ_w explicitly, find its support, and show that $S_w = S_w^*$. We find that $R_0 = \sqrt{N - 2}$, and using (7) we determine an upper bound on all weighted Fekete points, and in particular on $\{z_i^*\}$, from which we conclude that

$$|z_i^*| \leq \sqrt{N - 2}, \ i = 1, \ldots, N - 1. \tag{15}$$

Then from (11) we see that $\min_i |\mathbf{x}_N^* - \mathbf{x}_i^*| \geq 2/\sqrt{N - 1}$, and since the choice of $\mathbf{x}_N^*$ was arbitrary (4) follows.

It remains to prove (15). For the weight $w(z) := W_N(z)$ the external field

$$Q(r) = \frac{N - 1}{2(N - 2)} \log(1 + r^2) \tag{16}$$

is a radially symmetric function. Property $\lim_{r \to \infty}(Q(r) - \log r) = \infty$ is clear. That $rQ'(r)$ is increasing on $(0, \infty)$ can be easily seen from

$$rQ'(r) = \left(\frac{N - 1}{N - 2}\right) \frac{r^2}{1 + r^2}.$$

In our case we easily derive that $r_0 = 0$ and $R_0 = \sqrt{N - 2}$. The equilibrium measure is

$$d\mu_w(z) = \frac{1}{\pi} \left(\frac{N - 1}{N - 2}\right) \frac{r}{(1 + r^2)^2} dr d\phi, \qquad z = re^{i\phi}, \ 0 \leq \phi \leq 2\pi,$$

for $r \leq \sqrt{N - 2}$, and the equilibrium constant is

$$F_w = Q(R_0) - \log(R_0) = \frac{(N - 1)\log(N - 1) - (N - 2)\log(N - 2)}{2(N - 2)}.$$

Recall from (9) that for $|z| > R_0$

$$U^{\mu_w}(z) + Q(z) = Q(z) - \log|z| = Q(r) - \log r \quad \text{where} \quad z = re^{i\theta}.$$

Since $rQ'(r)$ is increasing and R_0 is the only positive solution of $rQ'(r) = 1$, we have that $rQ'(r) > 1$ for all $r > R_0$. Therefore, $Q'(r) > 1/r$, or $Q'(r) - 1/r > 0$ for all $r > R_0$. This implies that $U^{\mu_w}(z) + Q(z) > F_w$ for $|z| > R_0$. Thus, $S_w^* = S_w$, which combined with (7) and (14) implies (15). This completes the proof of the theorem. $\square$

Acknowledgments. This work was supported in part by PFW and PRF Summer Research Grants.

References

1. Andreev, N. N., An extremal property of the icosahedron, East J. Approx. **2**, no. 4 (1996), 459–462.

2. Bergersen, B., D. Boal, and P. Palffy-Muhoray, Equilibrium configurations of particles on the sphere: the case of logarithmic interactions, J. Phys. A: Math. Gen. **27** (1994), 2579–2586.

3. Björk, G., Distribution of positive mass which maximize a certain generalized energy integral, Ark. Mat. **3** (1956), 255–269.

4. Conway, J. H., and N. J. A. Sloane, *Sphere Packings, Lattices and Groups*, 2nd ed., New York: Springer-Verlag (1993).

5. Dahlberg, B. E. J., On the distribution of Fekete points, Duke Math. J. **45** (1978), 537–542.

6. Deift P., *Orthogonal Polynomials and Random Matrices: a Riemann-Hilbert Approach*, Courant Lecture Notes in Mathematics, Courant Institute, New York, 1999.

7. Deift, P., and K. T-R McLaughlin, A continuum limit of the Toda lattice, Mem. Amer. Math. Soc. **624** (1998).

8. Dragnev, P. D., D. A. Legg, and D. W. Townsend, Discrete logarithmic energy on the sphere, Pacific J. Math., to appear.

9. Dubickas, A., On the maximal product of distances between points on the sphere, Lithuanian Math. J. **36**, no. 3 (1996), 241–248.

10. Fejes Tóth, L., *Regular Figures*, Pergamon-Macmillan, New York, 1964.

11. Habicht, W., and B. L. van der Waerden, Lagerung von Punkten auf der Kugel, Math. Ann. **123** (1951), 223–234.

12. Kolushov, A. V., and V. A. Yudin, Extremal dispositions of points on the sphere, Anal. Math. **23**, no. 1 (1997), 25–34.

13. Kuijlaars, A. B. J., and E. B. Saff, Distributing many points on a sphere, Math. Intelligencer **19**, no. 1 (1997), 5–11.

14. Kuijlaars, A. B. J., and E. B. Saff, Asymptotics for minimal discrete energy on the sphere, Trans. Amer. Math. Soc. **350**, no. 2 (1998), 523–538.

15. Kuijlaars, A. B. J., and K. T-R McLaughlin, Generic behavior of the density of states in random matrix theory and equilibrium problems in the presence of real analytic external fields, Comm. Pure Appl. Math. **53** (2000), 736–785.

16. Melnik, T. W., O. Knop, and W. R. Smith, Extremal arrangements of points and unit charges on a sphere: equilibrium configurations revised, Can. J. Chem. **55** (1977), 1745–1761.

17. Rakhmanov, E. A., E. B. Saff, and Y. M. Zhou, Minimal discrete energy on the sphere, Math. Res. Lett. **1** (1994), 647–662.

18. Rakhmanov, E. A., E. B. Saff, and Y. M. Zhou, Electrons on the sphere, *Computational Methods and Functional Theory*, R. M. Ali, St. Ruscheweyh, and E. B. Saff, (eds.), Singapore: World Scientific (1995), 111–127.

19. Saff, E. B., and V. Totik, *Logarithmic Potentials with External Fields*, Springer-Verlag, New York (1997).

20. Shub, M., and S. Smale, Complexity of Bezout's theorem III, J. of Complexity **9** (1993), 4–14.

21. Stolarsky, K. B., Spherical distributions of N points with maximal distance sums are well spaced, Proc. Amer. Math. Soc. **48** (1975), 203–206.

22. Totik, V., *Weighted Approximation with Varying Weight*, Lect. Notes in Math. vol. 1569, Springer-Verlag, Berlin, 1994.

23. Wagner, G., On the product of distances to a point set on a sphere, J. Austral. Math. Soc. (Series A), **47** (1989), 466–489.

24. Wagner, G., On the mean of distances on the surface of a sphere, Pacific J. Math. **144** (1990), 389–398.

25. Whyte, L. L., Unique arrangements of points on a sphere, Amer. Math. Monthly **59** (1952), 606–611.

P. D. Dragnev
Department of Mathematics
Indiana-Purdue University
Fort Wayne, IN 46825
dragnevp@ipfw.edu

Best One–Sided L^1–Approximation by Blending Functions and Approximate Cubature

Dimiter Dryanov, Werner Haußmann
and Petar Petrov

Abstract. In the case of numerical quadrature an integral is approximately computed by point evaluation at certain points, i.e. by lower dimensional data, see e.g. Davis–Rabinowitz [1]. Similarly, by a cubature formula, multiple integrals are evaluated by lower dimensional data. In this paper we consider a compound cubature procedure on $I^2 := [-1,1]^2$ which allows to calculate double integrals using line integrals, i.e. bivariate integrals are approximated by univariate ones. Our cubature formula is based on one–sided L^1–approximation to the integrand from above and from below by blending functions. We will get an inclusion of the value of the double integral. The line integrals involved are taken over the diagonals of I^2 as well as over lines parallel to the diagonals. These diagonals are canonical point sets for the best one–sided L^1–approximation by blending functions.

§1. One–Sided L^1–Approximation by Blending Functions

We consider bivariate one–sided L^1–approximation by blending functions on I^2, where $I := [-1,1]$, and corresponding approximate cubatures. The vector space of blending functions is defined by

$$B^{m,n}(I^2) := \{\, g \in C^{m,n}(I^2) \ : \ D^{m,n}g := \frac{\partial^{m+n}}{\partial x^m \partial y^n}\, g = 0 \ \text{ on } \ I^2\},$$

where

$$C^{m,n}(I^2) := \{f : I^2 \to \mathbb{R} \ : \ D^{k,\ell}f \text{ continuous}, \ 0 \le k \le m, \ 0 \le \ell \le n \,\}.$$

Approximation Theory X: Abstract and Classical Analysis
Charles K. Chui, Larry L. Schumaker, and Joachim Stöckler (eds.), pp. 145–152.
Copyright © 2002 by Vanderbilt University Press, Nashville, TN.
ISBN 0-8265-1415-4.

A function $h^* \in B^{m,n}$ is defined to be a best one–sided L^1–approximant from above to f with repect to $B^{m,n}$ if $h^* - f \geq 0$ and

$$\int_{I^2} (h^* - f) \leq \int_{I^2} (h - f), \quad \text{for all } h \in B^{m,n} \text{ with } h - f \geq 0 \text{ on } I^2.$$

Analogously, we define a best one–sided L^1–approximant h_* from below.

First, let $m = n = 1$. For a given function $f \in C^{1,1}(I^2)$ with $D^{1,1}f \geq 0$, we shall consider existence and uniqueness of the best one–sided L^1–approximants h^* (from above) and h_* (from below) with respect to $B^{1,1}$. Note that $B^{1,1}$ consists of all bivariate functions which are sums of univariate continuously differentiable functions. We shall give explicit formulas for h^* resp. h_* based on simultaneous interpolation of the function f and its gradient grad f on a canonical point set. In the case of one–sided L^1–approximation *from above* the canonical point set is the diagonal

$$\Delta^* := \{(t,t) \ : \ t \in I\}$$

of I^2, in the case of approximation *from below* the canonical point set is the anti–diagonal

$$\Delta_* := \{(t,-t) \ : \ t \in I\}.$$

Indeed, we have

Theorem 1. (see [2]). *Let $f \in C^{1,1}(I^2)$ satisfy $D^{1,1}f \geq 0$ on I^2. Then*

(a) *f possesses a unique best one–sided L^1–approximant h^* from above with respect to $B^{1,1}$. h^* is characterized by the following simultaneous interpolation conditions:*

    (i) $h^*_{|\Delta^*} = f_{|\Delta^*}$,

    (ii) grad $h^*_{|\Delta^*} =$ grad $f_{|\Delta^*}$.

(b) *The best one–sided L^1–approximant h_* from below also exists and is unique, and is characterized by*

    (i) $h_{*|\Delta_*} = f_{|\Delta_*}$,

    (ii) grad $h_{*|\Delta_*} -$ grad $f_{|\Delta_*}$.

§2. Interpolation by $B^{1,1}$–Blending Functions on Diagonals

Theorem 1 is based on the possibility of interpolation of the function f and its gradient grad f on Δ^* and Δ_*. In addition, we consider error formulas for the interpolation remainder which control the sign of $f - h^*$ and $f - h_*$, respectively.

Theorem 2. *Let $f \in C^{1,1}(I^2)$ be given. Then*

(a) *The function*

$$h^*(x,y) := f(-1,-1) + \int\limits_{-1}^{x} D^{1,0}f(t,t)\, dt + \int\limits_{-1}^{y} D^{0,1}f(t,t)\, dt$$

is the unique $B^{1,1}$–function which satisfies the interpolation conditions $h^*_{|\Delta^*} = f_{|\Delta^*}$ and grad $h^*_{|\Delta^*} = $ grad $f_{|\Delta^*}$ on the diagonal Δ^*. Moreover, the following error representation holds:

$$f(x,y) - h^*(x,y) = -\frac{(x-y)^2}{2} D^{1,1}f(\xi,\eta), \tag{1}$$

where $(\xi,\eta) = (\xi(x,y),\eta(x,y)) \in I^2$.

(b) *Similarly, the function*

$$h_*(x,y) := f(-1,1) + \int\limits_{-1}^{x} D^{1,0}f(t,-t)\, dt - \int\limits_{-1}^{-y} D^{0,1}f(t,-t)\, dt$$

is the unique $B^{1,1}$–function which satisfies the interpolation conditions $h_{*|\Delta_*} = f_{|\Delta_*}$ and grad $h_{*|\Delta_*} = $ grad $f_{|\Delta_*}$ on the anti–diagonal Δ_*. The error expression in this case is given by

$$f(x,y) - h_*(x,y) = \frac{(x+y)^2}{2} D^{1,1}f(\rho,\sigma) \tag{2}$$

with $(\rho,\sigma) = (\rho(x,y),\sigma(x,y)) \in I^2$.

For the proof of Theorem 2 we refer to [2].

§3. One–Line Segment Cubature Formulas

By the interpolation formulas (1) and (2) we obtain the following cubature formulas which approximate integrals over I^2 by lower dimensional integrals, namely

$$\int\limits_{I^2} f \approx 2\int\limits_{I} f(x,x)dx, \tag{3}$$

and

$$\int\limits_{I^2} f \approx 2\int\limits_{I} f(x,-x)dx, \tag{4}$$

which are exact for functions from $B^{1,1}$.

Theorem 3. *The only one–line segment cubatures of the form*

$$\int_{I^2} f \approx a \cdot \int_J f, \tag{5}$$

where the right hand side is integrated over a line segment of the form

$$J := \{\, (x,y) \in I^2 \;:\; Ax + By + C = 0, \; A^2 + B^2 \neq 0 \,\},$$

which are exact for all functions in $B^{1,1}$, are the cubatures (3) and (4).

Proof: If $A = 0$, then $B \neq 0$, and $J = \{(x,y) \;:\; y = y_0\}$. Then for $h \in B^{1,1}$ the cubature (5) becomes

$$\int_{I^2} h = a \cdot \int_{-1}^{1} h(x, y_0)\, dx, \quad a \neq 0. \tag{6}$$

Using the test functions $h_1(x,y) = x + y$ and $h_2(x,y) = 1$, we get that the cubature formula must have the form

$$\int_{I^2} h = 2 \cdot \int_{-1}^{1} h(x,0)\, dx, \quad \text{for } h \in B^{1,1}.$$

Now it is obvious that for $h_3(x,y) = y^2$ the cubature is not exact, but $h_3 \in B^{1,1}$. The case $B = 0$ and $A \neq 0$ is similar.

Thus, let $A \cdot B \neq 0$. Without loss of generality we can assume that $B = -1$, hence $J = \{(x,y) \in I^2 \;:\; y = Ax + C\}$, where $A > 0$ or $A < 0$. Consider the case when $A > 0$. Then the cubature has the form

$$\int_{I^2} h = a \cdot \int_{\max\left(-1, \frac{-1-C}{A}\right)}^{\min\left(1, \frac{1-C}{A}\right)} h(x, Ax + C)\, dx, \quad a \neq 0. \tag{7}$$

The test function $h_4(x,y) = y - Ax - C$ shows that $C = 0$.

If $A \in (0,1]$, then $\min\left(1, \frac{1}{A}\right) = 1$ and $\max\left(-1, -\frac{1}{A}\right) = -1$. For $h_5(x,y) = y^2$ and $h_6(x,y) = y^4$ we get $aA^2 \cdot 2/3 = 4/3$, and $aA^4 \cdot 2/5 = 4/5$. Hence $A^2 = 1$, i.e. $A = 1$. The exactness for h_2 gives $a = 2$.

For $A \in [1, \infty)$ we have $\max\left(-1, -\frac{1}{A}\right) = -\frac{1}{A}$ and $\min\left(1, \frac{1}{A}\right) = \frac{1}{A}$. Testing with $h_7(x,y) = x^2$ and $h_8(x,y) = x^4$, we get $a/A^3 = 2$, and $a/A^5 = 2$. Thus $A^2 = 1$, i.e. again $A = 1$. Hence the cubature formula is of the desired form. It is immediate that (3) is exact for functions $f \in B^{1,1}$. In a similar way we obtain $C = 0$, $a = 2$ and $A = -1$ in the case $A < 0$. $\square$

§4. Compound Cubatures

In this section, we shall present compound cubature formulas based on the best one–sided L^1–approximants h^* resp. h_* as given in Theorem 1. First we give a Peano kernel representation of the error expression for the interpolants considered in Theorem 2.

$$f(x,y) - h^*(x,y) = f(x,y) - \int_{-1}^{x} \frac{d}{dt} f(t,t)\, dt - \int_{x}^{y} D^{0,1} f(t,t)\, dt - f(-1,-1)$$

$$= f(x,y) \; - \; f(x,x) \; - \; \int_{x}^{y} D^{0,1} f(t,t)\, dt$$

$$= \int_{x}^{y} \left[\, D^{0,1} f(x,t) \; - \; D^{0,1} f(t,t) \,\right] dt = \int_{x}^{y} \int_{t}^{x} D^{1,1} f(v,t)\, dv dt$$

$$= - \int_{-1}^{1} \int_{-1}^{1} \left[\, (y-t)_+^0 (t-x)_+^0 (t-v)_+^0 (v-x)_+^0 \right.$$
$$\left. + (x-t)_+^0 (t-y)_+^0 (v-t)_+^0 (x-v)_+^0 \,\right] \cdot D^{1,1} f(v,t)\, dv dt.$$

The expression in square brackets is the corresponding Peano kernel. The subscript $+$ designates the truncated power function. Note that the Peano kernel is positive. Integrating the above equation over I^2, we get the error expression for the cubature formula (3) as follows:

$$\int_{I^2} f \; - \; 2 \int_{I} f(x,x)\, dx = - \int_{-1}^{1} \int_{-1}^{1} \left[\, (1-t)(1+v)(t-v)_+^0 + \right. \tag{8}$$
$$\left. + (1+t)(1-v)(v-t)_+^0 \,\right] \cdot D^{1,1} f(v,t)\, dv dt.$$

Now the mean value theorem yields

$$\int_{I^2} f - 2 \int_{I} f(x,x)\, dx \; = \; -\frac{4}{3} D^{1,1} f(\xi, \eta). \tag{9}$$

Based on the best one–sided L^1–approximant h_* from below, we get an error expression for the cubature formula (4):

$$\int_{I^2} f - 2 \int_{I} f(x,-x)\, dx = \frac{4}{3} D^{1,1} f(\rho, \sigma). \tag{10}$$

Remark. Note that the distance $\inf\{\int_{I^2}(h-f) \ : \ h \in B^{1,1}, h \geq f\}$ is given by the absolute value of the integral (8).

We now turn to compound cubature formulas based on the cubatures (9) and (10). For $n \in \mathbb{N}$ we denote

$$x_i = -1 + \frac{2i}{n}, \quad i = 0, 1, ..., n; \quad y_j = -1 + \frac{2j}{n}, \quad j = 0, 1, ..., n.$$

Further, for $0 \leq i, j \leq n-1$, define

$$\square_{i,j} := \{(x,y) \in I^2 : x \in [x_i, x_{i+1}], \ y \in [y_j, y_{j+1}]\}.$$

Applying the cubature formula (9) with respect to $\square_{i,j}$, we obtain

$$\int\limits_{\square_{i,j}} f = \frac{2}{n} \int\limits_{x_i}^{x_{i+1}} f\left(x, x+(y_j - x_i)\right) dx - \frac{4}{3n^4} D^{1,1} f(\xi_{i,j}, \eta_{i,j}),$$

where $(\xi_{i,j}, \eta_{i,j}) \in \square_{i,j}$. After summation over all $\square_{i,j}$, we obtain the compound cubature formula

$$\int\limits_{I^2} f = \sum\limits_{i,j=0}^{n} \int\limits_{\square_{i,j}} f$$

$$= \frac{2}{n} \sum\limits_{k=-n+1}^{n-1} \int\limits_{\max\left(-1,-1+\frac{2k}{n}\right)}^{\min\left(1,1+\frac{2k}{n}\right)} f\left(x, x-\frac{2k}{n}\right) dx - \frac{4}{3n^2} \cdot D^{1,1} f(\xi, \eta)$$

$$=: CF_n^*(f) - \frac{4}{3n^2} \cdot D^{1,1} f(\xi, \eta), \qquad (\xi, \eta) \in I^2.$$

$$(11)$$

In a similar way one can obtain a compound cubature formula based on the cubature (10):

$$\int\limits_{I^2} f = \frac{2}{n} \sum\limits_{k=-n+1}^{n-1} \int\limits_{\max\left(-1,-1+\frac{2k}{n}\right)}^{\min\left(1,1+\frac{2k}{n}\right)} f\left(x, -x+\frac{2k}{n}\right) dx + \frac{4D^{1,1} f(\rho, \sigma)}{3n^2}$$

$$=: CF_*^n(f) + \frac{4}{3n^2} \cdot D^{1,1} f(\rho, \sigma), \qquad (\rho, \sigma) \in I^2.$$

$$(12)$$

With the aid of (11) and (12), we get a two–sided cubature for approximate integration of double integrals using univariate integrals as follows:

Theorem 4. *Given* $g \in C^{1,1}(I^2)$, *let* $D^{1,1}g \geq m$ *on* I^2. *Then for*
$g_m(x,y) := g(x,y) - mxy$,

$$CF^n_*(g_m) \leq \int_{I^2} g \leq CF^*_n(g_m).$$

This is the two–sided compound cubature formula *for approximate integration with order of convergence* $O(n^{-2})$ *as* $n \to \infty$.

Proof: Applying (11) and (12) to the perturbed function

$$g_m(x,y) = g(x,y) - mxy \qquad \text{satisfying } D^{1,1}g_m = D^{1,1}g - m \geq 0 \text{ on } I^2,$$

we obtain

$$CF^n_*(g_m) + \frac{4[D^{1,1}g(\rho,\sigma) - m]}{3n^2} = \int_{I^2} g_m$$

$$= CF^*_n(g_m) - \frac{4[D^{1,1}g(\xi,\eta) - m]}{3n^2}.$$

Hence, since $\int_{I^2} g_m = \int_{I^2} g$, we get

$$CF^n_*(g_m) \leq \int_{I^2} g \leq CF^*_n(g_m). \quad \square$$

§5. One–Sided L^1–Approximation by $B^{2,1}$–Blending Functions

We conclude the paper with a result concerning one–sided L^1–approximation by $B^{2,1}$–blending functions. Here we have the following V–shaped resp. Λ–shaped canonical point sets

$$V^* := \{(x,y) \in I^2 \ : \ y = 2|x| - 1\}$$

when we approximate from above resp.

$$\Lambda_* := \{(x,y) \in I^2 \ : \ y = -2|x| + 1\}$$

in the case of approximation from below.

The characterization theorem for best one–sided L^1–approximation by $B^{2,1}$–blending functions is given in the next theorem.

Theorem 5. (see [3]). *Let* $f \in C^{2,1}(I^2)$ *satisfy* $D^{2,1}f \geq 0$ *on* I^2. *Then* f *possesses a unique best one–sided* L^1–*approximant* g^* *from above with respect to* $B^{2,1}$. *The unique best one–sided* L^1–*approximant* $g^* \in B^{2,1}$ *is characterized by the simultaneous Hermite interpolation* $h^*_{|V^*} = f_{|V^*}$ *and* grad $h^*_{|V^*} = $ grad $f_{|V^*}$.

A corresponding result holds for approximation from below. In this case, the canonical point set is Λ_*. We remark that based on the corresponding one–sided L^1–approximation theorems, a compound cubature formula for double integrals can be derived as well.

Acknowledgments. This work was supported by Volkswagen Foundation under grant I/70523.

References

1. Davis, P. J., and P. Rabinowitz, *Methods of Numerical Integration*, Academic Press, New York–San Francisco–London, 1975.

2. Dryanov, D., W. Haußmann, and P. Petrov, Best one–sided L^1 approximation of bivariate functions by sums of univariate ones, Arch. Math. (Basel) **75** (2000), 125–131.

3. Dryanov, D., W. Haußmann, and P. Petrov, Best one–sided L^1 approximation by $B^{2,1}$–blending functions, in *Recent Progress in Multivariate Approximation*, W. Haußmann, K. Jetter and M. Reimer (eds.), Internat. Ser. Numer. Math. **137**, Birkhäuser, Basel, 2001, 115–134.

Dimiter Dryanov
Département des Mathématiques et de Statistique
Université de Montréal, C. P. 6128
Montréal, Québec H3C 3J7
Canada
dryanovd@dms.umontreal.ca

Werner Haußmann
Department of Mathematics
Gerhard–Mercator–University
47048 Duisburg
Germany
haussmann@math.uni-duisburg.de

Petar Petrov
Department of Mathematics
University of Sofia
1164 Sofia
Bulgaria
peynov@fmi.uni-sofia.bg

Polynomials with Littlewood-Type Coefficient Constraints

Tamás Erdélyi

Abstract. This survey paper focuses on my contributions to the area of polynomials with Littlewood-type coefficient constraints. It summarizes the main results from many of my recent papers some of which are joint with Peter Borwein.

§1. Introduction

Let D be the open unit disk of the complex plane. Its boundary, the unit circle of the complex plane, is denoted by ∂D. Let

$$\mathcal{K}_n := \left\{ p_n : p_n(z) = \sum_{k=0}^{n} a_k z^k, \ a_k \in \mathbb{C}, \ |a_k| = 1 \right\}.$$

The class $\mathcal{K}_n$ is often called the collection of all (complex) unimodular polynomials of degree n. Let

$$\mathcal{L}_n := \left\{ p_n : p_n(z) = \sum_{k=0}^{n} a_k z^k, \ a_k \in \{-1, 1\} \right\}.$$

The class $\mathcal{L}_n$ is often called the collection of all (real) unimodular polynomials of degree n. By Parseval's formula,

$$\int_0^{2\pi} |P_n(e^{it})|^2 \, dt = 2\pi(n+1)$$

for all $P_n \in \mathcal{K}_n$. Therefore

$$\min_{z \in \partial D} |P_n(z)| \le \sqrt{n+1} \le \max_{z \in \partial D} |P_n(z)| \qquad (1.1)$$

for all $P_n \in \mathcal{K}_n$. An old problem (or rather an old theme) is the following.

Approximation Theory X: Abstract and Classical Analysis   153
Charles K. Chui, Larry L. Schumaker, and Joachim Stöckler (eds.), pp. 153–196.
Copyright © 2002 by Vanderbilt University Press, Nashville, TN.
ISBN 0-8265-1415-4.

Problem 1.1. (Littlewood's Flatness Problem). *How close can a unimodular polynomial $P_n \in \mathcal{K}_n$ or $P_n \in \mathcal{L}_n$ come to satisfying*

$$|P_n(z)| = \sqrt{n+1}, \qquad z \in \partial D ? \tag{1.2}$$

Obviously (1.2) is impossible if $n \geq 1$, so one must look for less than (1.2). But then there are various ways of seeking such an "approximate situation". One way is the following. In his paper [51], Littlewood had suggested that, conceivably, there might exist a sequence (P_n) of polynomials $P_n \in \mathcal{K}_n$ (possibly even $P_n \in \mathcal{L}_n$) such that $(n+1)^{-1/2}|P_n(e^{it})|$ converge to 1 uniformly in $t \in \mathbb{R}$. We shall call such sequences of unimodular polynomials "ultraflat". More precisely, we give the following definitions. In the rest of the paper, we assume that (n_k) is a strictly increasing sequence of positive integers.

Definition 1.2. *Given a positive number ε, we say that a polynomial $P_n \in \mathcal{K}_n$ is ε-flat if*

$$(1 - \varepsilon)\sqrt{n+1} \leq |P_n(z)| \leq (1 + \varepsilon)\sqrt{n+1}, \qquad z \in \partial D, \tag{1.3}$$

or equivalently

$$\max_{z \in \partial D} \left||P_n(z)| - \sqrt{n+1}\right| \leq \varepsilon\sqrt{n+1}.$$

Definition 1.3. *Given a sequence (ε_{n_k}) of positive numbers tending to 0, we say that a sequence (P_{n_k}) of unimodular polynomials $P_{n_k} \in \mathcal{K}_{n_k}$ is (ε_{n_k})-ultraflat if each P_{n_k} is ε_{n_k}-flat, that is*

$$(1 - \varepsilon_{n_k})\sqrt{n_k+1} \leq |P_{n_k}(z)| \leq (1 + \varepsilon_{n_k})\sqrt{n_k+1}, \qquad z \in \partial D. \tag{1.4}$$

We say that a sequence (P_{n_k}) of unimodular polynomials $P_{n_k} \in \mathcal{K}_{n_k}$ is ultraflat *if there is a sequence (ε_{n_k}) of positive numbers tending to 0 for which (P_{n_k}) is (ε_{n_k})-ultraflat.*

The existence of an ultraflat sequence of unimodular polynomials seemed very unlikely, in view of a 1957 conjecture of P. Erdős (Problem 22 in [28]) asserting that, for all $P_n \in \mathcal{K}_n$ with $n \geq 1$,

$$\max_{z \in \partial D} |P_n(z)| \geq (1 + \varepsilon)\sqrt{n+1}, \tag{1.5}$$

where $\varepsilon > 0$ is an absolute constant (independent of n). Yet, combining some probabilistic lemmas from Körner's paper [47] with some constructive methods (Gauss polynomials, etc., which are completely unrelated to the deterministic part of Körner's paper), Kahane [43] proved that there exists a sequence (P_n) of unimodular polynomials $P_n \in \mathcal{K}_n$ which is (ε_n)-ultraflat, where

$$\varepsilon_n = O\left(n^{-1/17}\sqrt{\log n}\right).\qquad(1.6)$$

Thus the Erdős conjecture (1.5) was disproved for the classes $\mathcal{K}_n$. For the more restricted class $\mathcal{L}_n$, the analogous Erdős conjecture is unsettled to this date. It is a common belief that the analogous Erdős conjecture for $\mathcal{L}_n$ is true, and consequently there is no ultraflat sequence of unimodular polynomials $P_n \in \mathcal{L}_n$. I thank H. Queffelec for providing more details about the existence of ultraflat sequences (P_n) of unimodular polynomials $P_n \in \mathcal{K}_n$. The story is roughly the following.

Littlewood [51] had constructed polynomials $P_n \in \mathcal{K}_n$ so that on one hand $|P_n(z)| \le B\sqrt{n+1}$ for every $z \in \partial D$, and on the other hand $|P_n(z)| \ge A\sqrt{n+1}$ with an absolute constant $A > 0$ for every $z \in \partial D$ except for a small arc. In the light of this result he asked how close we can get to satisfying $|P_n(z)| = \sqrt{n+1}$ for every $z \in \partial D$ if $P_n \in \mathcal{K}_n$. The first result in this direction is due to Körner [47]. By using a result of Byrnes, he showed that there are absolute constants $0 < A < B$ such that $A\sqrt{n+1} \le |P_n(z)| \le B\sqrt{n+1}$ for every $z \in \partial D$. Then Kahane [43] constructed a sequence (P_n) of polynomials $P_n \in \mathcal{K}_n$ for which

$$(1-\varepsilon_n)\sqrt{n+1} \le |P_n(z)| \le (1+\varepsilon_n)\sqrt{n+1}, \qquad z \in \partial D,$$

with a sequence (ε_n) of positive real numbers converging to 0. Such a sequence is called (ε_n)-ultraflat.

Kahane's construction seemed to indicate a very rigid behavior for the phase function α_n, where

$$P_n(e^{it}) = R_n(t)e^{i\alpha_n(t)}, \qquad R_n(t) = |P_n(e^{it})|.$$

Saffari [66] had conjectured in 1991 that for every ultraflat sequence (P_n), $\alpha_n'(t)/n$ converges in measure to the uniform distribution on $[0,1]$, that is,

$$m\{t \in [0,2\pi] : 0 \le \alpha_n'(t) \le nx\} \to 2\pi x, \qquad 0 \le x \le 1,\qquad(1.7)$$

where m is the Lebesgue measure on the Borel subsets of $[0,2\pi]$. Since it can be seen easily that $X_n := \alpha_n'(t)/n$ is uniformly bounded, the method of moments applies and everything could be obtained from

$$\int_0^1 X_n^q(t)\,dt = \frac{1}{q+1} + o_{n,q}, \qquad q = 0,1,\ldots,\qquad(1.8)$$

where the numbers $o_{n,q}$ converge to 0 for every fixed q as $n \to \infty$. This was proved by Saffari [66] for $q = 0,1,2$. Then in 1996 Queffelec and Saffari [65] used Kahane's method with a slight modification to show the existence of an ultraflat sequence (P_n) which satisfies (1.7). They also showed that (1.8) is true for $q = 3$ (and almost for $q = 4$) for any ultraflat sequence (P_n) of polynomials $P_n \in \mathcal{K}_n$. When their work was submitted to Journal of Fourier Analysis and Applications, the editor in chief, J. Benedetto, and one of his students discovered an error in Byrnes work which, as a result, invalidated Körner's work. It was discovered that the deterministic part of Körner's [47] work was incorrect, and it was based on the incorrect "Theorem 2" of Byrnes' paper [21]. For details of the story see the forthcoming paper by J.S. Byrnes and Saffari [22].

Fortunately Kahane's work was independent of Byrnes'. It contained though another slight error which was corrected in [65]. Ultraflat sequences (P_n) of polynomials $P_n \in \mathcal{K}_n$ do exist! It is important to note this, otherwise the work of a number of papers would be without object. In [32] we answer Saffari's Problem affirmatively, namely we show that (1.7) (or equivalently (1.8)) is true for every ultraflat sequence (P_n) of unimodular polynomials $P_n \in \mathcal{K}_n$.

An interesting related result to Kahane's breakthrough is given by Beck [4]. He proved that for every sufficiently large integer k (he states the result for $k = 400$) there are polynomials P_n of degree n $(n = 1, 2, \ldots)$ so that each coefficient of each P_n is a k-th root of unity, and with some absolute constants $c_1, c_2 > 0$ we have

$$c_1 \sqrt{n} \le |P_n(e^{it})| \le c_2 \sqrt{n}, \qquad t \in \mathbb{R}, \quad n = 1, 2, \ldots.$$

For an account of some of the work done till the mid 1960's, see Littlewood's book [52] and [65].

§2. On the Phase Problem of Saffari

Let (P_n) be an ultraflat sequence of unimodular polynomials $P_n \in \mathcal{K}_n$. We write

$$P_n(e^{it}) = R_n(t)e^{i\alpha_n(t)}, \qquad R_n(t) = |P_n(e^{it})|. \qquad (2.1)$$

It is a simple exercise to show that α_n can be chosen to be an element of $C^\infty(\mathbb{R})$. This is going to be our understanding throughout this section. The following result was conjectured by Saffari [66] and proved in [32]:

Theorem 2.1. (Uniform Distribution Theorem for the Angular Speed.) Let (P_n) be an ultraflat sequence of unimodular polynomials $P_n \in \mathcal{K}_n$. Then, with the notation (2.1), in the interval $[0, 2\pi]$, the distribution of the normalized angular speed $\alpha'_n(t)/n$ converges to the uniform distribution as $n \to \infty$. More precisely, we have

$$\mathrm{m}\{t \in [0, 2\pi] : 0 \le \alpha'_n(t) \le nx\} = 2\pi x + o_n(x) \qquad (2.2)$$

for every $x \in [0, 1]$, where $\lim_{n \to \infty} o_n(x) = 0$ for every $x \in [0, 1]$, As a consequence, $|P'_n(e^{it})|/n^{3/2}$ also converges to the uniform distribution as $n \to \infty$. More precisely, we have

$$\mathrm{m}\{t \in [0, 2\pi] : 0 \leq |P'_n(e^{it})| \leq n^{3/2}x\} = 2\pi x + o_n(x) \qquad (2.3)$$

for every $x \in [0, 1]$, where $\lim_{n \to \infty} o_n(x) = 0$ for every $x \in [0, 1]$. In both statements the convergence of $o_n(x)$ is uniform on $[0, 1]$.

The basis of Saffari's conjecture was that for the special ultraflat sequences of unimodular polynomials produced by Kahane [43], (2.2) is indeed true. In Section 4 of [32] we prove this conjecture in general.

In the general case, (2.2) can, by integration, be reformulated (equivalently) in terms of the moments of the angular speed $\alpha'_n(t)$. This was observed and recorded by Saffari [66]. We present the proof of this equivalence in Section 4 of [32] and we settle Theorem 2.1 by proving the following result.

Theorem 2.2. (Reformulation of the Uniform Distribution Conjecture.) *Let (P_n) be an ultraflat sequence of unimodular polynomials $P_n \in \mathcal{K}_n$. Then, for any $q > 0$ we have*

$$\frac{1}{2\pi} \int_0^{2\pi} |\alpha'_n(t)|^q \, dt = \frac{n^q}{q+1} + o_{n,q} n^q, \qquad (2.4)$$

with suitable constants $o_{n,q}$ converging to 0 as $n \to \infty$ for every fixed $q > 0$.

An immediate consequence of (2.4) is the remarkable fact that for large values of $n \in \mathbb{N}$, the $L_q(\partial D)$ Bernstein factors

$$\frac{\int_0^{2\pi} |P'_n(e^{it})|^q \, dt}{\int_0^{2\pi} |P_n(e^{it})|^q \, dt}$$

of the elements of ultraflat sequences (P_n) of unimodular polynomials are essentially independent of the polynomials. More precisely (2.4) implies the following result.

Theorem 2.3. (The Bernstein Factors.) *Let q be an arbitrary positive real number. Let (P_n) be an ultraflat sequence of unimodular polynomials $P_n \in \mathcal{K}_n$. We have*

$$\frac{\int_0^{2\pi} |P'_n(e^{it})|^q \, dt}{\int_0^{2\pi} |P_n(e^{it})|^q \, dt} = \frac{n^q}{q+1} + o_{n,q} n^q \, ,$$

and as a limit case,

$$\frac{\max_{0 \le t \le 2\pi} |P_n'(e^{it})|}{\max_{0 \le t \le 2\pi} |P_n(e^{it})|} = n + o_n n,$$

with suitable constants $o_{n,q}$ and o_n converging to 0 as $n \to \infty$ for every fixed q.

In Section 3 of [32] we show the following result which turns out to be stronger than Theorem 2.2.

Theorem 2.4. (Negligibility Theorem for Higher Derivatives.) *Let* (P_n) *be an ultraflat sequence of unimodular polynomials* $P_n \in \mathcal{K}_n$. *For every integer* $r \ge 2$, *we have*

$$\max_{0 \le t \le 2\pi} |\alpha_n^{(r)}(t)| \le o_{n,r} n^r$$

with suitable constants $o_{n,r} > 0$ *converging to 0 for every fixed* $r = 2, 3, \ldots.$.

We show in Section 4 of [32] how Theorem 2.1 follows from Theorem 2.4. Finally in Section 4 of [32] we give the following extension of Theorem 2.1 (Uniform Distribution Conjecture) to higher derivatives.

Theorem 2.5. (Distribution of the Modulus of Higher Derivatives of Ultraflat Sequences of Unimodular Polynomials.) *Let* (P_n) *be an ultraflat sequence of unimodular polynomials* $P_n \in \mathcal{K}_n$. *Then*

$$\left(\frac{|P_n^{(r)}(e^{it})|}{n^{r+1/2}} \right)^{1/r}$$

converges to the uniform distribution as $n \to \infty$. *More precisely, we have*

$$\mathrm{m}\left\{ t \in [0, 2\pi] : 0 \le |P_n^{(r)}(e^{it})| \le n^{r+1/2} x^r \right\} = 2\pi x + o_{r,n}(x)$$

for every $x \in [0, 1]$, *where* $\lim_{n \to \infty} o_{r,n}(x) = 0$ *for every fixed* $r = 1, 2, \ldots$ *and* $x \in [0, 1]$. *The convergence of* $o_n(x)$ *is uniform on* $[0, 1]$.

Remark 2.6. Assume that (P_n) is an ultraflat sequence of unimodular polynomials $P_n \in \mathcal{K}_n$. As before, we use notation (2.1). We denote the number of zeros of P_n inside the open unit disk D by $Z(P_n)$. We claim that

$$Z(P_n) = \frac{n}{2}(1 + o_n),$$

where o_n is a sequence converging to 0 as $n \to \infty$. To see this we argue as follows. By Theorem 2.1 (proved in [32]) we have

$$\alpha_n(2\pi) - \alpha_n(0) = \frac{1}{2}(1 + o_n)(2\pi) = (1 + o_n)n\pi$$

with constants o_n converging to 0 as $n \to \infty$. So the "Argument Principle" yields the result we stated.

For continuous functions f defined on $[0, 2\pi]$, and for $q \in (0, \infty)$, we define

$$\|f\|_q := \left(\int_0^{2\pi} |f(t)|^q \, dt \right)^{1/q}.$$

We also define

$$\|f\|_\infty := \lim_{q \to \infty} \|f\|_q = \max_{t \in [0, 2\pi]} |f(t)|.$$

In [65] the following conjecture is made.

Conjecture 2.7. *Assume that* (P_n) *is an ultraflat sequence of unimodular polynomials* $P_n \in \mathcal{K}_n$ *and* $f_n(t) = \mathrm{Re}(P_n(e^{it}))$. *Let* $q \in (0, \infty)$. *Then*

$$\|f_n\|_q \sim \left(\frac{\Gamma\left(\frac{q+1}{2}\right)}{\Gamma\left(\frac{q}{2}+1\right)\sqrt{\pi}} \right)^{1/q} n^{1/2},$$

and

$$\|f_n'\|_q \sim \left(\frac{\Gamma\left(\frac{q+1}{2}\right)}{(q+1)\Gamma\left(\frac{q}{2}+1\right)\sqrt{\pi}} \right)^{1/q} n^{3/2},$$

where Γ *denotes the usual gamma function and the* $\sim$ *symbol means that the ratio of the left and right hand sides converges to 1 as* $n \to \infty$.

The above conjecture follows from Theorems 2.1 and 2.4. The arguments will be presented in my forthcoming paper [36].

§3. On Saffari's Near Orthogonality Conjectures

The structure of ultraflat sequences of unimodular polynomials is studied in [32] and [34] where several conjectures of Saffari are proved. In [35], based on the results in [32], we proved yet another Saffari conjecture formulated in [66].

Theorem 3.1. (Saffari's Near-Orthogonality Conjecture.) *Assume that* (P_n) *is an ultraflat sequence of unimodular polynomials* $P_n \in \mathcal{K}_n$. *Let*

$$P_n(z) := \sum_{k=0}^{n} a_{k,n} z^k.$$

Then

$$\sum_{k=0}^{n} a_{k,n} a_{n-k,n} = o(n).$$

Here, as usual, $o(n)$ denotes a quantity for which $\lim_{n\to\infty} o(n)/n = 0$. The statement remains true if the ultraflat sequence (P_n) of unimodular polynomials $P_n \in \mathcal{K}_n$ is replaced by an ultraflat sequence (P_{n_k}) of unimodular polynomials $P_{n_k} \in \mathcal{K}_{n_k}$, $0 < n_1 < n_2 < \cdots$.

If Q_n is a polynomial of degree n of the form $Q_n(z) = \sum_{k=0}^{n} a_k z^k$ with $a_k \in \mathbb{C}$, then its conjugate reciprocal polynomial is defined by $Q_n^*(z) := z^n \overline{Q}_n(1/z) := \sum_{k=0}^{n} \overline{a}_{n-k} z^k$. In terms of the above definition, Theorem 3.1 may be rewritten as

Corollary 3.2. Assume that (P_n) is an ultraflat sequence of unimodular polynomials $P_n \in \mathcal{K}_n$. Then

$$\int_{\partial D} |P_n(z) - P_n^*(z)|^2 \, |dz| = 2n + o(n) \, .$$

Remark 3.3. Theorem 3.1 clearly shows that there is no ultraflat sequence (P_n) of unimodular polynomials $P_n \in \mathcal{K}_n$ that are conjugate reciprocal. Otherwise, using the fact that $a_{k,n} = \overline{a}_{n-k,n}$, we would have

$$\sum_{k=0}^{n} a_{k,n} a_{n-k,n} = \sum_{k=0}^{n} |a_{k,n}|^2 = n + 1,$$

which contradicts Theorem 3.1. In fact, Theorem 3.1 tells us much more. It measures how far an ultraflat sequence of unimodular polynomials is from being conjugate reciprocal.

Remark 3.4. In [66] another "near orthogonality" relation has been conjectured. Namely it was suspected that if (P_{n_m}) is an ultraflat sequence of unimodular polynomials $P_{n_m} \in \mathcal{K}_{n_m}$ and

$$P_n(z) := \sum_{k=0}^{n} a_{k,n} z^k, \qquad n = n_m, \qquad m = 1, 2, \dots,$$

then

$$\sum_{k=0}^{n} a_{k,n} \overline{a}_{n-k,n} = o(n), \qquad n = n_m, \qquad m = 1, 2, \dots,$$

where, as usual, $o(n_m)$ denotes a quantity for which $\lim_{n_m \to \infty} o(n_m)/n_m$ equals 0. However, it was Saffari himself, together with Queffelec [65], who showed that this could not be any farther away from being true. Namely they constructed an ultraflat sequence (P_{n_m}) of plain-reciprocal unimodular polynomials $P_{n_m} \in \mathcal{K}_{n_m}$ such that

$$P_n(z) := \sum_{k=0}^{n} a_{k,n} z^k, \qquad a_{k,n} = a_{n-k,n}, \qquad k = 0, 1, 2, \dots n,$$

and hence

$$\sum_{k=0}^{n} a_{k,n}\overline{a}_{n-k,n} = n+1$$

for the values $n = n_m$, $m = 1, 2, \ldots$.

Remark 3.5. One can ask how flat a conjugate reciprocal unimodular polynomial can be. We present a simple result here. Let $P_n \in \mathcal{K}_n$ be a conjugate reciprocal polynomial of degree n. Then

$$\max_{z \in \partial D} |P_n(z)| \geq (1 + \varepsilon)\sqrt{n}$$

with $\varepsilon := \sqrt{\frac{4}{3}} - 1$. This is an observation made by Erdős [29] but his constant $\varepsilon > 0$ is unspecified.

To prove the statement, observe that Malik's inequality [57], p. 676 gives

$$\max_{z \in \partial D} |P_n'(z)| \leq \frac{n}{2} \max_{z \in \partial D} |P_n(z)|.$$

(Note that the fact that P_n is conjugate reciprocal improves the Bernstein factor on ∂D from n to $n/2$.) Using $P_n \in \mathcal{K}_n$, Parseval's formula, and Malik's inequality, we obtain

$$2\pi \frac{n^3}{3} \leq 2\pi \frac{n(n+1)(2n+1)}{6} = \int_{\partial D} |P_n'(z)|^2 \, |dz| \leq 2\pi \left(\frac{n}{2}\right)^2 \max_{z \in \partial D} |P_n(z)|^2,$$

and

$$\max_{z \in \partial D} |P_n(z)| \geq \sqrt{4/3}\sqrt{n}$$

follows.

§4. Some Littlewood-type Results

This section is essentially copied from [17]. We examine a number of problems concerning polynomials with coefficients restricted in various ways. We are particularly interested in how small such polynomials can be on the interval $[0, 1]$. For example, in [17] we prove that there are absolute constants $c_1 > 0$ and $c_2 > 0$ such that

$$\exp\left(-c_1\sqrt{n}\right) \leq \inf_{0 \neq p \in \mathcal{F}_n} \|p\|_{[0,1]} \leq \exp\left(-c_2\sqrt{n}\right)$$

for every $n \geq 2$, where $\mathcal{F}_n$ denotes the set of polynomials of degree at most n with coefficients from $\{-1, 0, 1\}$.

Littlewood considered minimization problems of this variety on the unit disk, hence, the title of the section. His most famous, now solved, conjecture (see [27] on pages 285 – 288) was that the L_1 norm of an element $f \in \mathcal{F}_n$ on the unit circle grows at least as fast as $c \log N$, where N is the number of non-zero coefficients in f and $c > 0$ is an absolute constant. This was proved by Konjagin [45] and independently by McGehee, Pigno, and Smith [56].

When the coefficients are required to be integers, the questions have a Diophantine nature and have been studied from a variety of points of view. See [2,3,8,18,39,61].

One key to the analysis is a study of the related problem of how large an order zero these restricted polynomials can have at 1. In [17] we answer this latter question precisely for the class of polynomials of the form

$$p(x) = \sum_{j=0}^{n} a_j x^j\,, \qquad |a_j| \le 1\,, \quad a_j \in \mathbb{C}$$

with fixed $|a_0| \neq 0$.

Variants of these questions have attracted considerable study, though rarely have precise answers been possible to give. See in particular [1,7,38, 68,69,71,41,6]. Indeed the classical, much studied, and presumably very difficult problem of Prouhet, Tarry, and Escott rephrases as a question of this variety. (Precisely: what is the maximal vanishing at 1 of a polynomial with integer coefficients with l_1 norm $2n$? It is conjectured to be n. See [18] and [6].)

We introduce the following classes of polynomials. Let

$$\mathcal{P}_n^c := \left\{ \sum_{i=0}^{n} a_i x^i : a_i \in \mathbb{C} \right\}$$

denote the set of algebraic polynomials of degree at most n with complex coefficients. Let

$$\mathcal{P}_n := \left\{ \sum_{i=0}^{n} a_i x^i : a_i \in \mathbb{R} \right\}$$

denote the set of algebraic polynomials of degree at most n with real coefficients. Let

$$\mathcal{Z}_n := \left\{ \sum_{i=0}^{n} a_i x^i : a_i \in \mathbb{Z} \right\}$$

denote the set of algebraic polynomials of degree at most n with integer coefficients. Let

$$\mathcal{F}_n := \left\{ \sum_{i=0}^{n} a_i x^i : a_i \in \{-1, 0, 1\} \right\}$$

denote the set of polynomials of degree at most n with coefficients from $\{-1, 0, 1\}$. Let

$$\mathcal{A}_n := \left\{ \sum_{i=0}^{n} a_i x^i : a_i \in \{0, 1\} \right\}$$

denote the set of polynomials of degree at most n with coefficients from $\{0, 1\}$. Finally, let

$$\mathcal{L}_n := \left\{ \sum_{i=0}^{n} a_i x^i : a_i \in \{-1, 1\} \right\}$$

denote the set of polynomials of degree at most n with coefficients from $\{-1, 1\}$.

So obviously

$$\mathcal{L}_n, \ \mathcal{A}_n \subset \mathcal{F}_n \subset \mathcal{Z}_n \subset \mathcal{P}_n \subset \mathcal{P}_n^c.$$

Throughout this section the uniform norm on a set $A \subset \mathbb{R}$ is denoted by $\|.\|_A$.

In his monograph [52], Littlewood discusses the class $\mathcal{L}_n$ and its complex analogue when the coefficients are complex numbers of modulus 1. On page 25 he writes "These raise fascinating questions." It is easy to see that the L_2 norm of any polynomial of degree n with complex coefficients of modulus one on the unit circle is $\sqrt{n+1}$. (Here we have normalized so that the unit circle has length 1.) Hence the minimum supremum norm of any such polynomial on the unit circle is at least $\sqrt{n+1}$.

The Rudin-Shapiro polynomials (see [51], for example) show that there are polynomials from $\mathcal{L}_n$ with maximum modulus less than $c\sqrt{n+1}$ on the unit circle. Littlewood remarks in [52] that although it has been known for more than 50 years that $g_n(\theta) := \sum_{m=0}^{n} e^{im \log m} e^{im\theta}$ satisfies $|g_n(\theta)| < c\sqrt{n+1}$ on the real line, the existence of polynomials $p_n \in \mathcal{L}_n$ with $|p_n(z)| < c\sqrt{n+1}$ on the unit circle has only fairly recently been shown. He adds "As a matter of cold fact, many people had doubted its truth." Rudin and Shapiro had the following simple idea:

$$P_0(z) = Q_0(z) = 1,$$
$$P_{n+1}(z) = P_n(z) + z^{2^n} Q_n(z),$$
$$Q_{n+1}(z) = P_n(z) - z^{2^n} Q_n(z).$$

We have at once

$$\begin{aligned}
|P_n|^2 + |Q_n|^2 &= 2(|P_{n-1}|^2 + |Q_{n-1}|^2) \\
&= 2^2(|P_{n-2}|^2 + |Q_{n-2}|^2) \\
&= \cdots \\
&= 2^n(|P_0|^2 + |Q_0|^2) = 2(\mu_n + 1)
\end{aligned}$$

on the unit circle, where

$$\mu_n := \deg(P_n) = \deg(Q_n) = 2^n - 1.$$

So $P_n, Q_n \in \mathcal{L}_{\mu_n}$ and $|P_n(z)| \le \sqrt{2}\sqrt{\deg(P_n) + 1}$ on the unit circle. From this it is a routine work to construct $P_n \in \mathcal{L}_n$ such that $|P_n(z)| \le c\sqrt{n}$ for every $n = 1, 2, \ldots$ with an absolute constant $c > 0$.

However, it is not known whether or not there are such polynomials from $p_n \in \mathcal{L}_n$ with minimal modulus also at least $c\sqrt{n}$ on the unit circle, where $c > 0$ is an absolute constant. Littlewood conjectures that there are such polynomials.

Littlewood also makes the above conjecture in [51] as well as several others. In [48] he writes that the problem of finding polynomials of degree n with coefficients of modulus 1 and with modulus on the unit disk bounded below by $c\sqrt{n}$ "seems singularly elusive and intriguing."

Erdős conjectured that the maximum modulus of a polynomial from $\mathcal{L}_n$ is always at least $c\sqrt{n+1}$ with an absolute constant $c > 1$. Erdős offers \$100 for a solution to this problem in [30]. Both Littlewood's and Erdős' conjectures are still open.

In the paper [48] Littlewood also considers $\sum_{m=0}^{n-1} \omega^{m(m+1)/2} z^m$ and shows that this polynomial has almost constant modulus (in an asymptotic sense) except on a set of measure $cn^{-1/2+\delta}$. Here ω is a primitive nth root of unity. Further related results are to be found in [4, 5, 9, 19, 25, 26, 40, 43, 45, 47, 55, 60].

Carrol, Eustice, and T. Figiel [40] show that

$$\liminf \frac{\log(m(n))}{\log(n+1)} > .431,$$

where $m(n)$ denotes the largest value that the minimum modulus of a polynomial from $\mathcal{L}_n$ can be on the unit circle. They also prove that

$$\sup \frac{\log(m(n))}{\log(n+1)} = \lim \frac{\log(m(n))}{\log(n+1)}.$$

They further conjecture that $m(n)n^{-1/2}$ tends to zero (contrary to Littlewood).

The average maximum modulus is computed by Salem and Zygmund [67] who show that for all but $o(2^n)$ polynomials from $\mathcal{L}_n$ the maximum modulus on the unit disk lies between $c_1\sqrt{n \log n}$ and $c_2\sqrt{n \log n}$.

The expected L^4 norm of a polynomial $p \in \mathcal{L}_n$ is $(2n^2 - n)^{1/4}$. This is due to Newman and Byrnes [59]. They also compute the L^4 norm of the Rudin-Shapiro polynomials.

In the case of complex coefficients these problems are mostly solved. A very interesting result of Kahane [43] proves the existence of polynomials of degree n with *complex* coefficients of modulus 1 and with minimal and maximal modulus both asymptotically $\sqrt{n+1}$ on the unit circle. See also Section 2 and [32].

The study of the location of zeros of the classes $\mathcal{F}_n$, $\mathcal{L}_n$, and $\mathcal{A}_n$ begins with Bloch and Pólya [6]. They prove that the average number of real zeros of a polynomial from $\mathcal{F}_n$ is at most $c\sqrt{n}$. They also prove that a polynomial from $\mathcal{F}_n$ cannot have more than

$$\frac{cn \log \log n}{\log n}$$

real zeros. This quite weak result appears to be the first on this subject. Schur [69] and by different methods Szegő [71] and Erdős and Turán [38] improve this to $c\sqrt{n \log n}$ (see also [10]). (Their results are more general, but in this specialization not sharp.)

Our Theorem 7.2 gives the right upper bound of $c\sqrt{n}$ for the number of real zeros of polynomials from a much larger class, namely for all polynomials of the form

$$p(x) = \sum_{j=0}^{n} a_j x^j, \qquad |a_j| \le 1, \quad |a_0| = |a_n| = 1, \quad a_j \in \mathbb{C}.$$

Schur [69] claims that Schmidt gives a version of part of this theorem. However, it does not appear in the reference he gives, namely [68], and we have not been able to trace it to any other source. Also, our method is able to give $c\sqrt{n}$ as an upper bound for the number of zeros of a polynomial $p \in \mathcal{P}_n^c$ with $|a_0| = 1, |a_i| \le 1$, inside any polygon with vertices in the unit circle (of course, c depends on the polygon). This is discussed in Section 8.

Bloch and Pólya [6] also prove that there are polynomials $p \in \mathcal{F}_n$ with

$$\frac{cn^{1/4}}{\sqrt{\log n}}$$

distinct real zeros of odd multiplicity. (Schur [69] claims they do it for polynomials with coefficients only from $\{-1, 1\}$, but this appears to be incorrect.)

In a seminal paper Littlewood and Offord [53] prove that the number of real roots of a $p \in \mathcal{L}_n$, on average, lies between

$$\frac{c_1 \log n}{\log \log \log n} \qquad \text{and} \qquad c_2 \log^2 n$$

and it is proved by Boyd [20] that every $p \in \mathcal{L}_n$ has at most $c \log^2 n / \log \log n$ zeros at 1 (counting multiplicities).

Kac [42] shows that the expected number of real roots of a polynomial of degree n with random uniformly distributed coefficients is asymptotically $(2/\pi) \log n$. He writes "I have also stated that the same conclusion holds if the coefficients assume only the values 1 and -1 with equal probabilities. Upon closer examination it turns out that the proof I had in mind is inapplicable... . This situation tends to emphasize the particular interest of the discrete case, which surprisingly enough turns out to be the most difficult." In a recent related paper Solomyak [70] studies the random series $\sum_{n=0}^{\infty} \pm \lambda^n$.

§5. Number of Zeros at 1

Theorems 5.1 and 5.2 below (see [17] for the proofs) offer upper bounds for the number of zeros at 1 of certain classes of polynomials with restricted coefficients. The first result sharpens and generalizes results of Amoroso [1], Bombieri and Vaaler [7], and Hua [41], who give versions of this result for polynomials with integer coefficients.

Theorem 5.1. *There is an absolute constant $c > 0$ such that every polynomial p of the form*

$$p(x) = \sum_{j=0}^{n} a_j x^j, \qquad |a_j| \leq 1, \quad a_j \in \mathbb{C},$$

has at most

$$c \left(n(1 - \log |a_0|) \right)^{1/2}$$

zeros at 1.

Applying Theorem 5.1 with $q(x) := x^n p(x^{-1})$ immediately gives the following.

Theorem 5.2. *There is an absolute constant $c > 0$ such that every polynomial p of the form*

$$p(x) = \sum_{j=0}^{n} a_j x^j, \qquad |a_j| \leq 1, \quad a_j \in \mathbb{C},$$

has at most

$$c \left(n(1 - \log |a_n|) \right)^{1/2}$$

zeros at 1.

The sharpness of the above theorems is shown by

Theorem 5.3. *Suppose $n \in \mathbb{N}$. Then there exists a polynomial p of the form*

$$p(x) = \sum_{j=0}^{n} a_j x^j, \qquad |a_j| \le 1, \quad a_j \in \mathbb{R},$$

such that p has a zero at 1 with multiplicity at least

$$\min \left\{ \frac{1}{6} \left(\left(n \left(1 - \log |a_0| \right) \right)^{1/2} - 1, n \right\} .$$

The following two theorems can be obtained from the results above with slightly worse constants. However, we have distinct attractive proofs of Theorems 5.4 and 5.5 below and in [17] we give them also.

Theorem 5.4. *Every polynomial p of the form*

$$p(x) = \sum_{j=0}^{n} a_j x^j, \qquad |a_0| = 1, \quad |a_j| \le 1, \quad a_j \in \mathbb{C},$$

has at most $\lfloor \frac{16}{7} \sqrt{n} \rfloor + 4$ zeros at 1.

Theorem 5.5. *For every $n \in \mathbb{N}$, there exists a polynomial*

$$p_n(x) = \sum_{j=0}^{n^2-1} a_j x^j$$

such that $a_{n^2-1} = 1$; $a_0, a_1, \ldots, a_{n^2-2}$ are real numbers of modulus less than 1; and p_n has a zero at 1 with multiplicity at least $n - 1$.

Theorem 5.5 immediately implies

Corollary 5.6. *For every $n \in \mathbb{N}$, there exists a polynomial*

$$p_n(x) = \sum_{j=0}^{n} a_j x^j, \qquad a_n = 1, \quad |a_j| \le 1, \quad a_j \in \mathbb{R},$$

and p_n has a zero at 1 with multiplicity at least $\lfloor \sqrt{n} - 1 \rfloor$.

The next related result (see [18]) is well known (in a variety of forms) but its proof is simple and we include it in [17].

Theorem 5.7. *There is an absolute constant $c > 0$ such that for every $n \in \mathbb{N}$ there is a $p \in \mathcal{F}_n$ having at least $c\sqrt{n/\log(n+1)}$ zeros at 1.*

Theorems 5.4 and 5.7 show that the right upper bound for the number of zeros a polynomial $p \in \mathcal{F}_n$ can have at 1 is somewhere between $c_1 \sqrt{n/\log(n+1)}$ and $c_2 \sqrt{n}$ with absolute constants $c_1 > 0$ and $c_2 > 0$. Completely closing the gap in this problem looks quite difficult.

Our next theorem from [17] slightly generalizes Theorem 5.1 and offers an *explicit* constant.

Theorem 5.8. *If $|a_0| \geq \exp(-L^2)$ and $|a_j| \leq 1$ for each $j = L^2 + 1, L^2 + 2, \ldots, n$, then the polynomial*

$$p(x) = \sum_{j=0}^{n} a_j x^j, \qquad a_j \in \mathbf{C}$$

has at most $\frac{44}{7}(L+1)\sqrt{n} + 5$ zeros at 1.

The next result from [17] is a simple observation about the maximal number of zeros a polynomial $p \in \mathcal{A}_n$ can have.

Theorem 5.9. *There is an absolute constant $c > 0$ such that every $p \in \mathcal{A}_n$ has at most $c \log n$ zeros at -1.*

Remark to Theorem 5.9. Let R_n be defined by

$$R_n(x) := \prod_{i=1}^{n} \left(1 + x^{a_i}\right),$$

where $a_1 := 1$ and a_{i+1} is the smallest odd integer that is greater than $\sum_{k=1}^{i} a_k$. It is tempting to speculate that R_n is the lowest degree polynomial with coefficients from $\{0, 1\}$ and a zero of order n at -1. This is true for $n = 1, 2, 3, 4, 5$ but fails for $n = 6$ and hence for all larger n.

Our final result in this section shows that a polynomial $Q \in \mathcal{F}_n$ with k zeros at 1 has many other zeros on the unit circle (at certain roots of unity). The proof can be found in [17], but a version of it may also be deduced from results in [7].

Theorem 5.10. *Let $p \leq n$ be a prime. Suppose $Q \in \mathcal{F}_n$ and Q has exactly k zeros at 1 and exactly m zeros at a primitive pth root of unity. Then*

$$p(m+1) \geq k \frac{\log p}{\log(n+1)}.$$

§6. The Chebyshev Problem on $[0,1]$

If p is a polynomial of the form

$$p(x) = \sum_{j=0}^{n} a_j x^j$$

with $a_1 = a_2 = \cdots = a_{m-1} = 0$ and $a_m \neq 0$, then we call $I(p) := a_m$ the first non-zero coefficient of p.

Our first theorem in this section (see [17] for a proof) shows how small the uniform norm of a polynomial $0 \neq p$ on $[0, 1]$ can be under some restriction on its coefficients.

Theorem 6.1. *Let* $\delta \in (0, 1]$. *There are absolute constants* $c_1 > 0$ *and* $c_2 > 0$ *such that*

$$\exp\left(-c_1(n(1 - \log \delta))^{1/2}\right) \leq \inf_p \|p\|_{[0,1]} \leq \exp\left(-c_2(n(1 - \log \delta))^{1/2}\right),$$

where the infimum is taken over all polynomials p *of the form*

$$p(x) = \sum_{j=0}^n a_j x^j, \qquad |a_j| \leq 1, \quad a_j \in \mathbb{C},$$

with $|I(p)| \geq \delta \geq \exp\left(\frac{1}{2}(6 - n)\right)$.

The following result is a special case of Theorem 6.1.

Theorem 6.2. *There are absolute constants* $c_1 > 0$ *and* $c_2 > 0$ *such that*

$$\exp\left(-c_1\sqrt{n}\right) \leq \inf_p \|p\|_{[0,1]} \leq \exp\left(-c_2\sqrt{n}\right)$$

for every $n \geq 2$, *where the infimum is taken over all polynomials* p *of the form*

$$p(x) = \sum_{j=0}^n a_j x^j, \qquad |a_j| \leq 1, \quad a_j \in \mathbb{C},$$

with $|I(p)| = 1$.

For the class $\mathcal{F}_n$ we have

Theorem 6.3. *There are absolute constants* $c_1 > 0$ *and* $c_2 > 0$ *such that*

$$\exp\left(-c_1\sqrt{n}\right) \leq \inf_{0 \neq p \in \mathcal{F}_n} \|p\|_{[0,1]} \leq \exp\left(-c_2\sqrt{n}\right)$$

for every $n \geq 2$.

See [17] for a proof. Note that the lower bound in the above theorem is a special case of Theorem 6.2. The proof of the upper bound, however, requires new ideas.

The approximation rate in Theorems 6.2 and 6.3 should be compared with

$$\inf_p \|p\|_{[0,1]}^{1/n} = \frac{2^{1/n}}{4},$$

where the infimum is taken for all monic $p \in \mathcal{P}_n$, and also with

$$\frac{1}{2.376\ldots} < \inf_{0 \neq p \in \mathcal{Z}_n} \|p\|_{[0,1]}^{1/n} < \frac{1 + \varepsilon_n}{2.3605}, \qquad \varepsilon_n \to 0.$$

The first equality above is attained by the normalized Chebyshev polynomial shifted linearly to $[0,1]$ and is proved by a simple perturbation argument. The second inequality is much harder (the exact result is open) and is discussed in [9]. It is an interesting fact that the polynomials $0 \neq p \in \mathcal{Z}_n$ with the smallest uniform norm on $[0,1]$ are very different from the usual Chebyshev polynomial of degree n. For example, they have at least 52% of their zeros at either 0 or 1. Relaxation techniques do not allow for their approximate computation.

Likewise, polynomials $0 \neq p \in \mathcal{F}_n$ with small uniform norm on $[0,1]$ are again quite different from polynomials $0 \neq p \in \mathcal{Z}_n$ with small uniform norm on $[0,1]$.

The story is roughly as follows. Polynomials $0 \neq p \in \mathcal{P}_n$ with leading coefficient 1 and with smallest possible uniform norm on $[0,1]$ are characterized by equioscillation and are given explicitly by the Chebyshev polynomials. In contrast, finding polynomials from $\mathcal{Z}_n$ with small uniform norm on $[0,1]$ is closely related to finding irreducible polynomials with all their roots in $[0,1]$.

The construction of non-zero polynomials from $\mathcal{F}_n$ with small uniform norm on $[0,1]$ is more or less governed by how many zeros such a polynomial can have at 1. Indeed, non-zero polynomials from $\mathcal{F}_n$ with minimal uniform norm on $[0,1]$ are forced to have close to the maximal possible number of zeros at 1.

This problem of the maximum order of a zero at 1 for a polynomial in $\mathcal{F}_n$, and closely related problems for polynomials of small height have attracted considerable attention but there is still a gap in what is known (see Theorems 5.4 and 5.7).

For the class $\mathcal{A}_n$ we have the following Chebyshev-type theorem. This result should be compared with Theorem 6.3. See [17] for a proof.

Theorem 6.4. *There are absolute constants $c_1 > 0$ and $c_2 > 0$ such that*

$$\exp\left(-c_1 \log^2(n+1)\right) \leq \inf_{0 \neq p \in \mathcal{A}_n} \|p(-x)\|_{[0,1]} \leq \exp\left(-c_2 \log^2(n+1)\right)$$

for every $n \geq 2$.

Our last theorem in this section is a sharp Chebyshev-type inequality for $\mathcal{F} := \cup_{n=1}^{\infty} \mathcal{F}_n$ and $\mathcal{S}$, where $\mathcal{S}$ denotes the collection of all analytic functions f on the open unit disk $D := \{z \in \mathbf{C} : |z| < 1\}$ that satisfy

$$|f(z)| \leq \frac{1}{1 - |z|}, \qquad z \in D.$$

See [17] for a proof.

Theorem 6.5. *There are absolute constants $c_1 > 0$ and $c_2 > 0$ such that*

$$\exp(-c_1/a) \leq \inf_{p \in \mathcal{S}, \, |p(0)|=1} \|p\|_{[1-a,1]} \leq \inf_{p \in \mathcal{F}, \, |p(0)|=1} \|p\|_{[1-a,1]} \leq \exp(-c_2/a)$$

for every $a \in (0,1)$.

§7. More on the Number of Real Zeros

Theorems 7.2 and 7.3 below give upper bounds for the number of *real* zeros of polynomials p when their coefficients are restricted in various ways.

The prototype for these theorems is given below. It was apparently first proved, at least up to the correct constant, by E. Schmidt in the early thirties. His complicated proof was not published – the first published proof is due to Schur [69]. Later new and simpler proofs and generalizations were published by Szegő [71] and Erdős and Turán [38] and others. A version of the approach of Erdős and Turán is presented in [8].

Theorem 7.1. *Suppose*

$$p(z) := \sum_{j=0}^{n} a_j z^j, \qquad a_j \in \mathbb{C},$$

has m positive real roots. Then

$$m^2 \leq 2n \log \left(\frac{|a_0| + |a_1| + \cdots + |a_n|}{\sqrt{|a_0 a_n|}} \right).$$

Our Theorem 7.2 below (see [17] for a proof) improves the above bound of $c\sqrt{n \log n}$ in the cases we are interested in where the coefficients are of similar size. Up to the constant c it is the best possible result.

Theorem 7.2. *There is an absolute constant $c > 0$ such that every polynomial p of the form*

$$p(x) = \sum_{j=0}^{n} a_j x^j, \qquad |a_j| \leq 1, \quad |a_0| = 1, \quad a_j \in \mathbb{C},$$

has at most $c\sqrt{n}$ zeros in $[-1, 1]$.

There is an absolute constant $c > 0$ such that every polynomial p of the form

$$p(x) = \sum_{j=0}^{n} a_j x^j, \qquad |a_j| \leq 1, \quad |a_n| = 1, \quad a_j \in \mathbb{C},$$

has at most $c\sqrt{n}$ zeros in $\mathbb{R} \setminus (-1, 1)$.

There is an absolute constant $c > 0$ such that every polynomial p of the form

$$p(x) = \sum_{j=0}^{n} a_j x^j, \qquad |a_j| \leq 1, \quad |a_0| = |a_n| = 1, \quad a_j \in \mathbb{C},$$

has at most $c\sqrt{n}$ real zeros.

In [17] we also prove

Theorem 7.3. *There is an absolute constant $c > 0$ such that every polynomial p of the form*

$$p(x) = \sum_{j=0}^{n} a_j x^j, \qquad |a_j| \leq 1, \quad |a_0| = 1, \quad a_j \in \mathbf{C}, \qquad (7.1)$$

has at most c/a zeros in $[-1 + a, 1 - a]$ whenever $a \in (0, 1)$.

Theorem 7.3 is sharp up to the constant. It is possible to construct a polynomial (of degree $n \leq ck^2$) of the form (4.1) with a zero of order k in the interval $(0, 1 - 1/k]$. This is discussed in [3].

The next theorem from [13] gives an upper bound for the number of zeros of a polynomial p lying on a subarc of the unit circle when the coefficients of p are restricted as in the first statement of Theorem 7.2.

Theorem 7.4. *There is an absolute constant $c > 0$ such that every polynomial p of the form*

$$p(x) = \sum_{j=0}^{n} a_j x^j, \qquad |a_j| \leq 1, \quad |a_0| = 1, \quad a_j \in \mathbf{C},$$

has at most $cn\alpha$ zeros on a subarc I_α of length α of the unit circle if $\alpha \geq n^{-1/2}$, while it has at most $c\sqrt{n}$ zeros on I_α if $\alpha \leq n^{-1/2}$. The polynomial $p(z) := z^n - 1$ ($\alpha \geq n^{-1/2}$) and Theorem 5.4 ($\alpha \leq n^{-1/2}$) show that these bounds are essentially sharp.

One can observe that Jensen's inequality implies that every function f analytic in the open unit disk $D := \{z \in \mathbf{C} : |z| < 1\}$ and satisfying the growth condition

$$|f(0)| = 1, \qquad |f(z)| \leq \frac{1}{1 - |z|}, \qquad z \in D$$

has at most $(c/a) \log(1/a)$ zeros in the disk $D_a := \{z \in \mathbf{C} : |z| < 1 - a\}$, where $0 < a < 1$ and $c > 0$ is an absolute constant. This observation plays a crucial role in the next section.

§8. Further Results on the Zeros

There is a huge literature on the zeros of polynomials with restricted coefficients. See, for example, [1,6,3,7,41,38,8,13,17,52,57,61,68,69,71]. In [13] we prove the three essentially sharp theorems below.

Theorem 8.1. *Every polynomial p of the form*

$$p(x) = \sum_{j=0}^{n} a_j x^j, \qquad |a_0| = 1, \quad |a_j| \leq 1, \quad a_j \in \mathbb{C}, \qquad (8.1)$$

has at most $c\sqrt{n}$ zeros inside any polygon with vertices on the unit circle, where the constant $c > 0$ depends only on the polygon.

Theorem 8.2. *There is an absolute constant $c > 0$ such that every polynomial p of the form*

$$p(x) = \sum_{j=0}^{n} a_j x^j, \qquad |a_0| = |a_n| = 1, \quad |a_j| \leq 1, \quad a_j \in \mathbb{C},$$

has at most $c(n\alpha + \sqrt{n})$ zeros in the strip

$$\{z \in \mathbb{C} : |\operatorname{Im}(z)| \leq \alpha\},$$

and in the sector

$$\{z \in \mathbb{C} : |\arg(z)| \leq \alpha\}.$$

Theorem 8.3. *Let $\alpha \in (0,1)$. Every polynomial p of the form*

$$p(x) = \sum_{j=0}^{n} a_j x^j, \qquad |a_0| = 1, \quad |a_j| \leq 1, \quad a_j \in \mathbb{C},$$

has at most c/α zeros inside any polygon with vertices on the circle

$$\{z \in \mathbb{C} : |z| = 1 - \alpha\},$$

where the constant $c > 0$ depends only on the number of the vertices of the polygon.

For $z_0 \in \mathbb{C}$ and $r > 0$, let

$$D(z_0, r) := \{z \in \mathbb{C} : |z - z_0| < r\}.$$

In [33] we show that a polynomial p of the form

$$p(x) = \sum_{j=0}^{n} a_j x^j, \qquad |a_0| = 1, \quad |a_j| \leq 1, \quad a_j \in \mathbb{C},$$

has at most $(c_1/\alpha) \log(1/\alpha)$ zeros in the disk $D(0, 1 - \alpha)$ for every $\alpha \in (0, 1)$, where $c_1 > 0$ is an absolute constant. This is a simple consequence of Jensen's formula. However it is not so simple to show that this estimate for the number of zeros in $D(0, 1-\alpha)$ is sharp. In [33] we present two examples to show the existence of polynomials p_α ($\alpha \in (0,1)$) of the form (8.1) (with a suitable $n \in \mathbb{N}$ depending on α) with at least $\lfloor (c_2/\alpha) \log(1/\alpha) \rfloor$ zeros in $D(0, 1-\alpha)$ ($c_2 > 0$ is an absolute constant). In fact, we show the existence of such polynomials from much smaller classes with more restrictions on the coefficients. Our first example has probabilistic background and shows the existence of polynomials p_α ($\alpha \in (0,1)$) with *complex* coefficients of modulus *exactly* 1 and with at least $\lfloor (c_2/\alpha) \log(1/\alpha) \rfloor$ zeros in $D(0, 1-\alpha)$ ($c_2 > 0$ is an absolute constant). Our second example is constructive and defines polynomials p_α ($\alpha \in (0,1)$) with *real* coefficients of modulus *at most* 1, with constant term 1, and with at least $\lfloor (c_2/\alpha) \log(1/\alpha) \rfloor$ zeros in $D(0, 1 - \alpha)$ ($c_2 > 0$ is an absolute constant). So, in particular, the constant in Theorem 8.3 cannot be made independent of the number of vertices of the polygon.

Some other observations on polynomials with restricted coefficients are also formulated in [33]. More precisely in [33] we prove Theorems 8.4 - 8.9, 8.11 and 8.13 below.

Theorem 8.4. *Let $\alpha \in (0,1)$. Every polynomial of the form*

$$p(x) = \sum_{j=0}^{n} a_j x^j, \qquad |a_0| = 1, \quad |a_j| \leq 1, \quad a_j \in \mathbb{C},$$

has at most $(2/\alpha) \log(1/\alpha)$ zeros in the disk $D(0, 1 - \alpha)$.

Theorem 8.5. *For every $\alpha \in (0,1)$ there is a polynomial $Q := Q_\alpha$ of the form*

$$Q_\alpha(x) = \sum_{j=0}^{n} a_{j,\alpha} x^j, \qquad |a_{j,\alpha}| = 1, \quad a_{j,\alpha} \in \mathbb{C},$$

such that Q_α has at least $\lfloor (c_2/\alpha) \log(1/\alpha) \rfloor$ zeros in the disk $D(0, 1 - \alpha)$, where $c_2 > 0$ is an absolute constant.

Theorem 8.5 follows from

Theorem 8.6. *For every $n \in \mathbb{N}$ there is a polynomial p_n of the form*

$$p_n(x) = \sum_{j=0}^{n} a_{j,n} x^j, \qquad |a_{j,n}| = 1, \quad a_{j,n} \in \mathbb{C},$$

such that p_n has no zeros in the annulus

$$\left\{ z \in \mathbb{C} : 1 - \frac{c_3 \log n}{n} < |z| < 1 + \frac{c_3 \log n}{n} \right\},$$

where $c_3 > 0$ is an absolute constant.

To formulate some interesting corollaries of Theorems 8.4 and 8.6 we introduce some notation. Let $\mathcal{E}_n$ be the collection of polynomials of the form

$$p(x) = \sum_{j=0}^{n} a_j x^j, \qquad |a_0| = |a_n| = 1, \quad a_j \in [-1, 1].$$

Let $\mathcal{E}_n^c$ be the collection of polynomials of the form

$$p(x) = \sum_{j=0}^{n} a_j x^j, \qquad |a_0| = |a_n| = 1, \quad a_j \in \mathbf{C}, \quad |a_j| \le 1.$$

As before, let $\mathcal{L}_n$ be the collection of polynomials of the form

$$p(x) = \sum_{j=0}^{n} a_j x^j, \qquad a_j \in \{-1, 1\}.$$

Finally let $\mathcal{K}_n$ be the collection of polynomials of the form

$$p(x) = \sum_{j=0}^{n} a_j x^j, \qquad a_j \in \mathbf{C}, \quad |a_j| = 1.$$

For a polynomial p, let

$$d(p) := \min\{|1 - |z|| : \ z \in \mathbf{C}, \ p(z) = 0\}.$$

For a class of polynomials $\mathcal{A}$ we define

$$\gamma(\mathcal{A}) := \sup\{d(p) : \ p \in \mathcal{A}\}.$$

Theorem 8.7. *There are absolute constants $c_4 > 0$ and $c_5 > 0$ such that*

$$\frac{c_4 \log n}{n} \le \gamma(\mathcal{K}_n^c) \le \gamma(\mathcal{E}_n^c) \le \frac{c_5 \log n}{n}.$$

Theorem 8.8. *There is an absolute constant $c_6 > 0$ such that*

$$\gamma(\mathcal{L}_n) \le \gamma(\mathcal{E}_n) \le \frac{c_6 \log n}{n}.$$

There is an absolute constant $c_7 > 0$ such that for infinitely many positive integer values of n we have

$$\frac{c_7}{n} \le \gamma(\mathcal{L}_n) \le \gamma(\mathcal{K}_n).$$

Theorem 8.9. *For every* $\alpha \in (0,1)$ *there is a polynomial* $P := P_\alpha$ *of the form*

$$P(x) = \sum_{j=0}^{n} a_{j,\alpha} x^j \,, \qquad a_{0,\alpha} = 1 \,, \quad a_{j,\alpha} \in [-1,1] \,,$$

that has at least $\lfloor (c_8/\alpha) \log(1/\alpha) \rfloor$ *zeros in the disk* $D(0, 1 - \alpha)$*, where* $c_8 > 0$ *is an absolute constant.*

Conjecture 8.10. *Every polynomial* $p \in \mathcal{L}_n$ *has at least one zero in the annulus*

$$\left\{ z \in \mathbb{C} : 1 - \frac{c_9}{n} < |z| < 1 + \frac{c_9}{n} \right\} \,,$$

where $c_9 > 0$ *is an absolute constant.*

In the case when a polynomial $p \in \mathcal{L}_n$ is self-reciprocal, we can prove more than the conclusion of Conjecture 8.10. Namely

Theorem 8.11. *Every self-reciprocal polynomial* $p \in \mathcal{L}_n$ *has at least one zero on the unit circle* $\{z \in \mathbb{C} : |z| = 1\}$*.*

In [33] we also show that Conjecture 8.10 implies the conjecture below.

Conjecture 8.12. *There is no ultraflat sequence* $(p_{n_m})_{m=1}^{\infty}$ *of polynomials* $p_{n_m} \in \mathcal{L}_{n_m}$ *satisfying*

$$(1 - \varepsilon_{n_m})(n_m + 1)^{1/2} \leq |p_{n_m}(z)| \leq (1 + \varepsilon_{n_m})(n_m + 1)^{1/2}$$

for all $z \in \mathbb{C}$ *with* $|z| = 1$ *and for all* $m \in \mathbb{N}$*, where* $(\varepsilon_{n_m})_{m=1}^{\infty}$ *is a sequence of positive numbers converging to* 0*.*

Theorem 8.13. *Conjecture 8.10 implies Conjecture 8.12.*

§9. Littlewood-Type Problems on Subarcs of the Unit Circle

Littlewood's well-known and now resolved conjecture of around 1948 concerns polynomials of the form

$$p(z) := \sum_{j=1}^{n} a_j z^{k_j} \,,$$

where the coefficients a_j are complex numbers of modulus at least 1 and the exponents k_j are distinct non-negative integers. It states that such polynomials have L_1 norms on the unit circle

$$\partial D := \{z \in \mathbb{C} : |z| = 1\}$$

that grow at least like $c \log n$ with an absolute constant $c > 0$. This was proved by Konjagin [45] and independently by McGehee, Pigno, and Smith [56].

Pichorides, who contributed essentially to the proof of the Littlewood conjecture, observed in [62] that the original Littlewood conjecture (when all the coefficients are from $\{0, 1\}$ would follow from a result on the L_1 norm of such polynomials on sets $E \subset \partial D$ of measure π. Namely if

$$\int_E \left| \sum_{j=0}^n z^{k_j} \right| |dz| \geq c$$

for any subset $E \subset \partial D$ of measure π with an absolute constant $c > 0$, then the original Littlewood conjecture holds. Throughout this section the measure of a set $E \subset \partial D$ is the linear Lebesgue measure of the set

$$\{ t \in [-\pi, \pi) : e^{it} \in E \} .$$

Konjagin [46] gives a lovely probabilistic proof that this hypothesis fails. He does however conjecture the following: for any *fixed* set $E \subset \partial D$ of positive measure there exists a constant $c = c(E) > 0$ depending only on E such that

$$\int_E \left| \sum_{j=0}^n z^{k_j} \right| |dz| \geq c(E) .$$

In other words the sets $E_\varepsilon \subset \partial D$ of measure π in his example where

$$\int_{E_\varepsilon} \left| \sum_{j=0}^n z^{k_j} \right| |dz| < \varepsilon$$

must vary with $\varepsilon > 0$.

In [11] we show, among other things, that Konjagin's conjecture holds on subarcs of the unit circle ∂D.

Additional material on Littlewood's conjecture and related problems concerning the growth of polynomials with unimodular coefficients in various norms on the unit disk is to be found, for example, in [19, 4, 44, 52, 55, 59, 61, 70].

All the results of [11] concern how small polynomials of the above and related forms can be in the L_p norms on subarcs of the unit disk. For $1 \leq p \leq \infty$ the results are sharp, at least up to a constant in the exponent.

An interesting related result is due to Nazarov [58]. One of its simpler versions states that there is an absolute constant $c > 0$ such that

$$\max_{z \in I} |p(z)| \leq \left(\frac{c \, m(I)}{m(A)} \right)^n \max_{z \in A} |p(z)|$$

for every polynomial p of the form $p(z) = \sum_{j=0}^{n} a_j z^{k_j}$ with $k_j \in \mathbb{N}$ and $a_j \in \mathbb{C}$ and for every $A \subset I$, where I is a subarc of ∂D with length $m(I)$, and A is measurable with Lebesgue measure $m(A)$. This extends a result of Turán [72] called *Turán's Lemma*, where $I = \partial D$ and A is a subarc.

We introduce some notation. For $M > 0$ and $\mu \geq 0$, let $\mathcal{S}_M^\mu$ denote the collection of all analytic functions f on the open unit disk $D := \{z \in \mathbb{C} : |z| < 1\}$ that satisfy

$$|f(z)| \leq \frac{M}{(1 - |z|)^\mu}, \qquad z \in D.$$

We define the following subsets of $\mathcal{S}_1^1$. Let

$$\mathcal{F}_n := \left\{ f : f(z) = \sum_{j=0}^{n} a_j z^j, \ a_j \in \{-1, 0, 1\} \right\},$$

and denote the set of all polynomials with coefficients from the set $\{-1, 0, 1\}$ by

$$\mathcal{F} := \bigcup_{n=0}^{\infty} \mathcal{F}_n.$$

More generally we define the following classes of polynomials. For $M > 0$ and $\mu \geq 0$ let

$$\mathcal{K}_M^\mu := \left\{ f : f(z) = \sum_{j=0}^{n} a_j z^j, \ a_j \in \mathbb{C}, \ |a_j| \leq M j^\mu, \ |a_0| = 1, \ n \in \mathbb{N} \right\}.$$

On occasion we let $\mathcal{S} := \mathcal{S}_1^1$, $\mathcal{S}_M := \mathcal{S}_M^1$, and $\mathcal{K}_M := \mathcal{K}_M^0$.

We also employ the following standard notations. We denote by $\mathcal{P}_n$ the set of all polynomials of degree at most n with real coefficients. We denote by $\mathcal{P}_n^c$ the set of all polynomials of degree at most n with complex coefficients. The height of a polynomial

$$p_n(z) := \sum_{j=0}^{n} a_j z^j, \qquad a_j \in \mathbb{C}, \quad a_n \neq 0,$$

is defined by

$$H(p_n) := \max \left\{ \frac{|a_j|}{|a_n|} : \ j = 0, 1, \ldots, n \right\}.$$

Also,

$$\|p\|_A := \sup_{z \in A} |p(z)|$$

and

$$\|p\|_{L_q(A)} := \left(\int_A |p(z)|^q \, |dz| \right)^{1/q}$$

are used throughout this section for measurable functions (in this section usually polynomials) p defined on a measurable subset of the unit circle or the real line, and for $q \in (0, \infty)$.

Theorems 9.1 - 9.5, Corollaries 9.6 and 9.7, and Theorem 9.8 below are proved in [11].

Theorem 9.1. *Let* $0 < a < 2\pi$ *and* $M \geq 1$. *Let* A *be a subarc of the unit circle with length* $m(A) = a$. *Then there is an absolute constant* $c_1 > 0$ *such that*

$$\|f\|_A \geq \exp \left(\frac{-c_1(1 + \log M)}{a} \right)$$

for every $f \in \mathcal{S}_M(:= \mathcal{S}_M^1)$ *that is continuous on the closed unit disk and satisfies* $|f(z_0)| \geq \frac{1}{2}$ *for every* $z_0 \in \mathbb{C}$ *with* $|z_0| = \frac{1}{4M}$.

Corollary 9.2. *Let* $0 < a < 2\pi$ *and* $M \geq 1$. *Let* A *be a subarc of the unit circle with length* $m(A) = a$. *Then there is an absolute constant* $c_1 > 0$ *such that*

$$\|f\|_A \geq \exp \left(\frac{-c_1(1 + \log M)}{a} \right)$$

for every $f \in \mathcal{K}_M(:= \mathcal{K}_M^1)$.

The next two results from [11] show that the previous results are, up to constants, sharp.

Theorem 9.3. *Let* $0 < a < 2\pi$. *Let* A *be the subarc of the unit circle with length* $m(A) = a$. *Then there are absolute constants* $c_1 > 0$ *and* $c_2 > 0$ *such that*

$$\inf_{0 \neq f \in \mathcal{F}} \|f\|_A \leq \exp \left(\frac{-c_1}{a} \right)$$

whenever $m(A) = a \leq c_2$.

Theorem 9.4. *Let* $0 < a < 2\pi$ *and* $M \geq 1$. *Let* A *be the subarc of the unit circle with length* $m(A) = a$. *Then there are absolute constants* $c_1 > 0$ *and* $c_2 > 0$ *such that*

$$\inf_{0 \neq f \in \mathcal{K}_M} \|f\|_A \leq \exp \left(\frac{-c_1(1 + \log M)}{a} \right)$$

whenever $m(A) = a \leq c_2$.

The next two results from [11] extend the first two results to the L_1 norm (and hence to all L_p norms with $p \geq 1$).

Theorem 9.5. Let $0 < a < 2\pi$, $M \geq 1$, and $\mu = 1, 2, \ldots$. Let A be a subarc of the unit circle with length $m(A) = a$. Then there is an absolute constant $c_1 > 0$ such that

$$\|f\|_{L_1(A)} \geq \exp\left(\frac{-c_1(\mu + \log M)}{a}\right)$$

for every $f \in S_M^\mu$ that is continuous on the closed unit disk and satisfies $|f(z_0)| \geq \frac{1}{2}$ for every $z_0 \in \mathbb{C}$ with $|z_0| \leq \frac{1}{4M2^\mu}$.

Corollary 9.6. Let $0 < a < 2\pi$, $M \geq 1$, and $\mu = 1, 2, \ldots$. Let A be a subarc of the unit circle with length $m(A) = a$. Then there is an absolute constant $c_1 > 0$ such that

$$\|f\|_{L_1(A)} \geq \exp\left(\frac{-c_1(1 + \mu \log \mu + \log M)}{a}\right)$$

for every $f \in \mathcal{K}_M^\mu$.

The following is an interesting consequence of the preceding results.

Corollary 9.7. Let A be a subarc of the unit circle with length $m(A) = a$. If (p_k) is a sequence of monic polynomials that tends to 0 in $L_1(A)$, then the sequence $H(p_k)$ of heights tends to ∞.

The final result of this section shows that the theory does not extend to arbitrary sets of positive measure. This is shown in [11] as well.

Theorem 9.8. For every $\varepsilon > 0$ there is a polynomial $p \in \mathcal{K}_1$ such that $|p(z)| < \varepsilon$ everywhere on the unit circle except possibly in a set of linear measure at most ε.

§10. Markov- and Bernstein-type Inequalities

Erdős studied and raised many questions about polynomials with restricted coefficients. Both Erdős and Littlewood showed particular fascination about the class $\mathcal{L}_n$, where, as before, $\mathcal{L}_n$ denotes the set of all polynomials of degree n with each of their coefficients in $\{-1, 1\}$. A related class of polynomials is $\mathcal{F}_n$ that, as before, denotes the set of all polynomials of degree at most n with each of their coefficients in $\{-1, 0, 1\}$. Another related class is $\mathcal{G}_n$, that is the collection of all polynomials p of the form

$$p(x) = \sum_{j=m}^{n} a_j x^j, \qquad |a_m| = 1, \quad |a_j| \leq 1,$$

where m is an unspecified nonnegative integer not greater than n. For the sake of brevity, let

$$\|p\|_A := \sup_{z \in A} |p(z)|$$

for a complex-valued function p defined on A.

In [12] and [14] we establish the right Markov-type inequalities for the classes $\mathcal{F}_n$ and $\mathcal{G}_n$ on $[0,1]$. Namely there are absolute constants $c_1 > 0$ and $c_2 > 0$ such that

$$c_1 n \log(n+1) \leq \max_{0 \neq p \in \mathcal{F}_n} \frac{\|p'\|_{[0,1]}}{\|p\|_{[0,1]}} \leq c_2 n \log(n+1)$$

and

$$c_1 n^{3/2} \leq \max_{0 \neq p \in \mathcal{G}_n} \frac{|p'(1)|}{\|p\|_{[0,1]}} \leq \max_{0 \neq p \in \mathcal{G}_n} \frac{\|p'\|_{[0,1]}}{\|p\|_{[0,1]}} \leq c_2 n^{3/2} .$$

It is quite remarkable that the right Markov factor for $\mathcal{G}_n$ is much larger than the right Markov factor for $\mathcal{F}_n$. In [12] and [14] we also show that there are absolute constants $c_1 > 0$ and $c_2 > 0$ such that

$$c_1 n \log(n+1) \leq \max_{0 \neq p \in \mathcal{L}_n} \frac{|p'(1)|}{\|p\|_{[0,1]}} \leq \max_{0 \neq p \in \mathcal{L}_n} \frac{\|p'\|_{[0,1]}}{\|p\|_{[0,1]}} \leq c_2 n \log(n+1)$$

for every $p \in \mathcal{L}_n$. For polynomials $p \in \mathcal{F} := \bigcup_{n=0}^{\infty} \mathcal{F}_n$ with $|p(0)| = 1$ and for $y \in [0,1)$ the Bernstein-type inequality

$$\frac{c_1 \log\left(\frac{2}{1-y}\right)}{1-y} \leq \max_{\substack{p \in \mathcal{F} \\ |p(0)|=1}} \frac{\|p'\|_{[0,y]}}{\|p\|_{[0,1]}} \leq \frac{c_2 \log\left(\frac{2}{1-y}\right)}{1-y}$$

is also proved with absolute constants $c_1 > 0$ and $c_2 > 0$.

For continuous functions p defined on the complex unit circle, and for $q \in (0, \infty)$, we define

$$\|p\|_q := \left(\int_0^{2\pi} |p(e^{it})|^q \, dt \right)^{1/q} .$$

We also define

$$\|p\|_\infty := \lim_{q \to \infty} \|p\|_q = \max_{t \in [0, 2\pi]} |p(e^{it})| .$$

Based on the ideas of F. Nazarov, Qeffelec and Saffari [65] showed that

$$\sup_{p \in \mathcal{L}_n} \frac{\|p'\|_q}{\|p\|_q} = \gamma_{n,q} n , \qquad \lim_{n \to \infty} \gamma_{n,q} = 1 ,$$

for every $q \in (0, \infty]$, $q \neq 2$ (when $q = 2$, $\lim_{n \to \infty} \gamma_{n,q} = 3^{-1/2}$ by the Parseval Formula). It is interesting to compare this result with Theorem 2.3. It shows that Bernstein's classical inequality (extended by Arestov for all $q \in (0, \infty]$) stating that

$$\|p'\|_q \leq n\|p'\|_q$$

for all polynomials of degree at most n with complex coefficients, cannot be essentially improved for the class $\mathcal{L}_n$, except the trivial $q = 2$ case.

§11. Trigonometric Polynomials with Many Real Zeros

As before, let

$$\mathcal{L}_n := \left\{ p : p(z) = \sum_{j=0}^{n} a_j z^j, \ a_j \in \{-1, 1\} \right\}.$$

Let D denote the closed unit disk of the complex plane. Let ∂D denote the unit circle of the complex plane. Littlewood made the following conjecture about $\mathcal{L}_n$ in the fifties.

Conjecture 11.1. *(Littlewood). There are at least infinitely many values of $n \in \mathbb{N}$ for which there are polynomials $p_n \in \mathcal{L}_n$ so that*

$$c_1(n+1)^{1/2} \leq |p_n(z)| \leq c_2(n+1)^{1/2}$$

for all $z \in \partial D$. Here the constants c_1 and c_2 are independent of n.

There is a related conjecture of Erdős [29].

Conjecture 11.2. *(Erdős). There is a constant $\varepsilon > 0$ (independent of n) so that*

$$\max_{z \in \partial D} |p_n(z)| \geq (1 + \varepsilon)(n+1)^{1/2}$$

for every $p_n \in \mathcal{L}_n$ and $n \in \mathbb{N}$. That is, the constant C_2 in Conjecture 1.1 must be bounded away from 1 (independently of n).

This conjecture is also open. One of our results in [15] is formulated by Corollary 11.6. Littlewood gives a proof of this in [48] and explores related issues in [49, 50, 51]. The approach is via Theorem 11.3 which estimates the measure of the set where a real trigonometric polynomial of degree at most n with at least k zeros in $K := \mathbb{R} \pmod{2\pi}$ is small. There are two reasons for doing this. First the approach is, we believe, easier and secondly it leads to explicit constants.

Let $K := \mathbb{R} \pmod{2\pi}$. For the sake of brevity the uniform norm of a continuous function p on K will be denoted by $\|p\|_K := \|p\|_{L_\infty(K)}$. Let $\mathcal{T}_n$ denote the set of all real trigonometric polynomials of degree at most n, and let $\mathcal{T}_{n,k}$ denote the subset of those elements of $\mathcal{T}_n$ that have at least k zeros in K (counting multiplicities). In [15] we prove Theorems 11.3 – 11.5 and Corollary 11.6 below.

Theorem 11.3. *Suppose* $p \in \mathcal{T}_n$ *has at least* k *zeros in* K *(counting multiplicities). Let* $\alpha \in (0,1)$. *Then*

$$m\{t \in K : |p(t)| \leq \alpha \|p\|_K\} \geq \frac{\alpha}{e} \frac{k}{n},$$

where $m(A)$ *denotes the one-dimensional Lebesgue measure of* $A \subset K$.

Theorem 11.4. *We have*

$$2\pi \left(1 - \frac{c_2 k}{n}\right) \leq \sup_{p \in \mathcal{T}_{n,k}} \frac{\|p\|_{L_1(K)}}{\|p\|_{L_\infty(K)}} \leq 2\pi \left(1 - \frac{c_1 k}{n}\right)$$

for some absolute constants $0 < c_1 < c_2$.

Theorem 11.5. *Assume that* $p \in \mathcal{T}_n$ *satisfies*

$$\|p\|_{L_2(K)} \leq An^{1/2} \tag{11.1}$$

and

$$\|p'\|_{L_2(K)} \geq Bn^{3/2}. \tag{11.2}$$

Then there is a constant $\varepsilon > 0$ *depending only on* A *and* B *such that*

$$\|p\|_K^2 \geq (2\pi - \varepsilon)^{-1}\|p\|_{L_2(K)}^2. \tag{11.3}$$

Here

$$\varepsilon = \frac{\pi^3}{1024e}\frac{B^6}{A^6}$$

works.

Corollary 11.6. *Let* $p \in \mathcal{T}_n$ *be of the form*

$$p(t) = \sum_{k=1}^{n} a_k \cos(kt - \gamma_k), \qquad a_k = \pm 1, \quad \gamma_k \in \mathbb{R}, \quad k = 1, 2, \ldots, n.$$

Then there is a constant $\varepsilon > 0$ *such that*

$$\|p\|_K^2 \geq (2\pi - \varepsilon)^{-1}\|p\|_{L_2(K)}^2.$$

Here

$$\varepsilon := \frac{\pi^3}{1024e}\frac{1}{27}$$

works.

§12. On the Flatness of Trigonometric Polynomials

In [15] we give short and elegant proofs of some of the main results from Littlewood's papers [48, 49, 50, 51, 52]. There are two reasons for doing this. First our approaches are, we believe, much easier, and secondly they lead to explicit constants. Littlewood himself remarks that his methods were "extremely indirect".

We use the notation $K := \mathbb{R} \pmod{2\pi}$. Let

$$\|p\|_{L_\lambda(K)} := \left(\int_K |p(t)|^\lambda \, dt \right)^{1/\lambda}$$

and

$$M_\lambda(p) := \left(\frac{1}{2\pi} \int_K |p(t)|^\lambda \, dt \right)^{1/\lambda}.$$

In [16] we prove

Theorem 12.1. *Assume that p is a trigonometric polynomial of degree at most n with real coefficients that satisfies*

$$\|p\|_{L_2(K)} \le A n^{1/2} \tag{12.1}$$

and

$$\|p'\|_{L_2(K)} \ge B n^{3/2}. \tag{12.2}$$

Then there exists a constant $\varepsilon > 0$ so that

$$M_4(p) - M_2(p) \ge \varepsilon M_2(p)$$

where

$$\varepsilon := \left(\frac{1}{221} \right) \left(\frac{B}{A} \right)^{12}.$$

Let the Littlewood class $\mathcal{H}_n$ be the collection of all trigonometric polynomials of the form

$$p(t) := p_n(t) := \sum_{j=1}^{n} a_j \cos(jt + \alpha_j), \qquad a_j = \pm 1, \quad \alpha_j \in \mathbb{R}.$$

Note that for the Littlewood class $\mathcal{H}_n$ we have

$$\left(\frac{B}{A} \right)^{12} = 3^{-6}.$$

Corollary 12.2. *We have*

$$M_4(p) - M_2(p) \ge \frac{M_2(p)}{160874}$$

for every $p \in \mathcal{A}_n$. The merit factor

$$\left(\frac{M_4^4(p)}{M_2^4(p)} - 1 \right)^{-1}$$

is bounded above by 20110 for every $p \in \mathcal{H}_n$.

If Q_n is a polynomial of degree n of the form

$$Q_n(z) = \sum_{k=0}^{n} a_k z^k, \qquad a_k \in \mathbb{C},$$

and the coefficients a_k of Q_n satisfy

$$a_k = \bar{a}_{n-k}, \qquad k = 0, 1, \ldots n,$$

then we call Q_n a **conjugate-reciprocal polynomial of degree** n. We say that the polynomial Q_n is **unimodular** if $|a_k| = 1$ for each $k = 0, 1, 2, \ldots, n$. Note that if $p \in \mathcal{A}_n$, then

$$1 + p(t) = e^{int} Q_{2n}(e^{it})$$

with a conjugate-reciprocal unimodular polynomial Q_{2n} of degree at most $2n$. One can ask how flat a conjugate reciprocal unimodular polynomial can be. The inequality of Remark 3.5 implies the result below.

Theorem 12.3. *For every* $p \in \mathcal{H}_n$,

$$M_\infty(1+p) - M_2(1+p) \geq (\sqrt{4/3} - 1) M_2(1+p).$$

This improves the unspecified constant in a result of Erdős [29].

In [16] we give a numerical value of an unspecified constant in one of the main results of [50].

Theorem 12.4. *For every* $p \in \mathcal{A}_n$,

$$M_2(p) - M_1(p) \geq 10^{-31} M_2(p).$$

Based on the fact that for a fixed trigonometric polynomial p the function

$$\lambda \to \lambda \log(M_\lambda(p))$$

is an increasing convex function on $[0, \infty)$, we can state explicit numerical values of certain unspecified constants in some other related Littlewood results. For example, as a consequence of Theorem 12.4, we have

Theorem 12.5. *For every* $p \in \mathcal{H}_n$ *and* $\lambda > 2$,

$$\log(M_\lambda(p)) - \log(M_2(p)) \geq \frac{\lambda - 2}{\lambda} \log\left(\frac{1}{1 - 10^{-31}} \right), \qquad \lambda > 2,$$

and

$$\log(M_2(p)) - \log(M_\lambda(p)) \geq \frac{2 - \lambda}{\lambda} \log\left(\frac{1}{1 - 10^{-31}} \right), \qquad 1 \leq \lambda < 2.$$

§13. On the Norm of the Polynomial Truncation Operator

Let D and ∂D denote the open unit disk and the unit circle of the complex plane, respectively. We denote the set of all polynomials of degree at most n with real coefficients by $\mathcal{P}_n$. We denote the set of all polynomials of degree at most n with complex coefficients by $\mathcal{P}_n^c$. We define the truncation operator S_n for polynomials $P_n \in \mathcal{P}_n^c$ of the form

$$P_n(z) := \sum_{j=0}^{n} a_j z^j \,, \qquad a_j \in \mathbf{C} \,,$$

by

$$S_n(P_n)(z) := \sum_{j=0}^{n} \widetilde{a}_j z^j \,, \qquad \widetilde{a}_j := (a_j/|a_j|)\min\{|a_j|, 1\} \qquad (13.1)$$

(here $0/0$ is interpreted as 1). In other words, we take the coefficients $a_j \in \mathbf{C}$ of a polynomial P_n of degree at most n, and we truncate them. That is, we leave a coefficient a_j unchanged if $|a_j| < 1$, while we replace it by $a_j/|a_j|$ if $|a_j| \geq 1$. We form the new polynomial with the new coefficients $\widetilde{a}_j$ defined by (13.1), and we denote this new polynomial by $S_n(P_n)$. We define the norms of the truncation operators by

$$\|S_n\|_{\infty,\partial D}^{real} := \sup_{P_n \in \mathcal{P}_n} \frac{\max_{z \in \partial D} |S_n(P_n)(z)|}{\max_{z \in \partial D} |P_n(z)|}$$

and

$$\|S_n\|_{\infty,\partial D}^{comp} := \sup_{P_n \in \mathcal{P}_n^c} \frac{\max_{z \in \partial D} |S_n(P_n)(z)|}{\max_{z \in \partial D} |P_n(z)|} \,.$$

Our main theorem in [37] establishes the right order of magnitude of the norms of the operators S_n. This settles a question asked by S. Kwapien.

Theorem 13.1. *With the notation introduced above there is an absolute constant $c_1 > 0$ such that*

$$c_1 \sqrt{2n+1} \leq \|S_n\|_{\infty,\partial D}^{real} \leq \|S_n\|_{\infty,\partial D}^{comp} \leq \sqrt{2n+1} \,.$$

In fact we are able to establish an $L_p(\partial D)$ analogue of this as follows. For $p \in (0, \infty)$, let

$$\|S_n\|_{p,\partial D}^{real} := \sup_{P_n \in \mathcal{P}_n} \frac{\|S_n(P_n)\|_{L_p(\partial D)}}{\|P_n\|_{L_p(\partial D)}}$$

and

$$\|S_n\|_{p,\partial D}^{comp} := \sup_{S_n \in \mathcal{P}_n^c} \frac{\|S_n(P_n)\|_{L_p(\partial D)}}{\|P_n\|_{L_p(\partial D)}} \,.$$

Theorem 13.2. *With the notation introduced above there is an absolute constant $c_1 > 0$ such that*

$$c_1(2n+1)^{1/2-1/p} \leq \|S_n\|_{p,\partial D}^{real} \leq \|S_n\|_{p,\partial D}^{comp} \leq (2n+1)^{1/2-1/p}$$

for every $p \in [2, \infty)$.

Note that it remains open what is the right order of magnitude of $\|S_n\|_{p,\partial D}^{real}$ and $\|S_n\|_{p,\partial D}^{comp}$, respectively when $0 < p < 2$. In particular, it would be interesting to see if $\|S_n\|_{p,\partial D}^{comp} \leq c$ is possible for any $0 < p < 2$ with an absolute constant c. We record the following observation, due to S. Kwapien (see also [36]), in this direction.

Theorem 13.3. *There is an absolute constant $c > 0$ such that*

$$\|S_n\|_{1,\partial D}^{real} \geq c\sqrt{\log n}\,.$$

If the unit circle ∂D is replaced by the interval $[-1,1]$, we get a completely different order of magnitude of the polynomial truncation projector. In this case the norms of the truncation operators S_n are defined in the usual way. That is, let

$$\|S_n\|_{\infty,[-1,1]}^{real} := \sup_{P_n \in \mathcal{P}_n} \frac{\max_{x \in [-1,1]} |S_n(P_n)(x)|}{\max_{x \in [-1,1]} |P_n(x)|}$$

and

$$\|S_n\|_{\infty,[-1,1]}^{comp} := \sup_{P_n \in \mathcal{P}_n^c} \frac{\max_{x \in [-1,1]} |S_n(P_n)(x)|}{\max_{x \in [-1,1]} |P_n(x)|}\,.$$

In [36] we prove the following result.

Theorem 13.4. *With the notation introduced above we have*

$$2^{n/2-1} \leq \|S_n\|_{\infty,[-1,1]}^{real} \leq \|S_n\|_{\infty,[-1,1]}^{comp} \leq \sqrt{2n+1} \cdot 8^{n/2}\,.$$

In [36] we base the proof of the lower bound of Theorems 13.1 and 13.2 on the following lemma from [54].

Lemma 13.5. (Lovász, Spencer, Vesztergombi). *Let $a_{j,k}, j = 1, 2, \ldots, n_1,$ $k = 1, 2, \ldots, n_2$ be such that $|a_{j,k}| \leq 1$. Let also $p_1, p_2, \ldots, p_{n_2} \in [0,1]$. Then there are choices*

$$\varepsilon_k \in \{-p_k, 1 - p_k\}, \qquad k = 1, 2, \ldots, n_2\,,$$

such that for all j,

$$\left| \sum_{k=1}^{n_2} \varepsilon_k a_{j,k} \right| \leq c\sqrt{n_1}$$

with an absolute constant c.

§14. Problems

We use the notation introduced in Section 4. Some of the problems below are closely related to each other. Some of them have been mentioned before. Some of them have already been formulated in [10]. Here we summarize the simplest looking but most challenging ones.

Problem 14.1. (Erdős). Is it true that

$$\max_{z \in \partial D} |P_n(z)| \geq (1 + \varepsilon)\sqrt{n+1}$$

for every $P_n \in \mathcal{L}_n$ with $n \geq 1$, where $\varepsilon > 0$ is an absolute constant (independent of n)?

As a matter of fact in Problem 14.1

$$\max_{z \in \partial D} |P_n(z)| \geq \sqrt{n+1} + \varepsilon$$

with an absolute constant $\varepsilon > 0$ would already be remarkable to prove. A stronger version of Problem 14.1 is the following.

Problem 14.2. Is it true that

$$\|P_n\|_{L_4(\partial D)} \geq (1 + \varepsilon)\sqrt{n+1}$$

for every $P_n \in \mathcal{L}_n$ with $n \geq 1$, where $\varepsilon > 0$ is an absolute constant (independent of n)?

In Problem 14.2 even

$$\|P_n\|_{L_4(\partial D)} \geq \sqrt{n+1} + \varepsilon$$

with an absolute constant $\varepsilon > 0$ would already be remarkable to prove. Problem 14.2 can be reformulated as follows.

Problem 14.3. Suppose $n \geq 1$ and $a_0 = \pm 1, a_1 = \pm 1, \ldots, a_n = \pm 1$. Let

$$b_k := \sum_{j=0}^{n-k} a_j a_{j+k}, \qquad k = 1, 2, \ldots, n,$$

$$b_{-k} := \sum_{j=k}^{n} a_j a_{j-k}, \qquad k = 1, 2, \ldots, n.$$

Is it true that

$$\sum_{k=1}^{n} \left(b_k^2 + b_{-k}^2 \right) > \varepsilon(n+1)^2$$

with an absolute constant $\varepsilon > 0$ (independent of n)?

In Problem 14.3 even

$$\sum_{k=1}^{n} \left(b_k^2 + b_{-k}^2 \right) > \varepsilon (n+1)^{3/2}$$

with an absolute constant $\varepsilon > 0$ would already be remarkable to prove.

Problem 14.4. Is there an ultraflat sequence of unimodular polynomials $P_n \in \mathcal{L}_n$ (or at least $P_{n_k} \in \mathcal{L}_{n_k}$)?

A polynomial $P_n \in \mathcal{P}_n$ is called skew-reciprocal if

$$P_n(z) = z^n P_n(-1/z), \qquad z \in \mathbb{C}, \ z \neq 0.$$

Problem 14.5. Is there an ultraflat sequence of skew-reciprocal unimodular polynomials $P_{n_k} \in \mathcal{L}_{n_k}$, where each n_k is a multiple of 4?

Problem 14.6. Is there a sequence of unimodular polynomials $P_n \in \mathcal{L}_n$ (or at least $P_{n_k} \in \mathcal{L}_{n_k}$) for which

$$|P_n(z)| > c\sqrt{n+1}, \qquad z \in \partial D?$$

Problem 14.7. Is there a sequence of unimodular polynomials $P_n \in \mathcal{K}_n$ (or at least $P_{n_k} \in \mathcal{K}_{n_k}$) for which the derivative sequence (P_n') is ultraflat, that is

$$\lim_{n \to \infty} \frac{\max_{z \in \partial D} |P_n'(z)|}{\min_{z \in \partial D} |P_n'(z)|} = 1 ?$$

If the answer is yes, what can one say about the case when $P_n \in \mathcal{L}_n$?

Theorems 5.4 and 5.7 show that the right upper bound for the number of zeros a polynomial $p \in \mathcal{F}_n$ can have at 1 is somewhere between $c_1 \sqrt{n/\log(n+1)}$ and $c_2 \sqrt{n}$ with absolute constants $c_1 > 0$ and $c_2 > 0$.

Problem 14.8. How many zeros can a polynomial $p \in \mathcal{F}_n$ have at 1? Close the gap between Theorems 5.4 and 5.7. Any improvements would be interesting.

Problem 14.9. How many *distinct* zeros can a polynomial p_n of the form

$$p_n(x) = \sum_{j=0}^{n} a_j x^j, \qquad |a_0| = 1, \quad |a_j| \leq 1, \quad a_j \in \mathbb{C} \tag{14.1}$$

(or $p_n \in \mathcal{F}_n$) have in $[-1, 1]$? In particular, is it possible to give a sequence (p_n) of polynomials of the form (14.1) (or maybe $(p_n) \subset \mathcal{F}_n$) so that p_n has $c\sqrt{n}$ *distinct* zeros in $[-1, 1]$, where $c > 0$ is an absolute constant?

If not, what is the sharp analogue of Theorem 7.2 for *distinct* zeros in $[-1, 1]$?

Problem 14.10. How many *distinct* zeros can a polynomial $p_n \in \mathcal{F}_n$ have in the interval $[-1+a, 1-a]$, $a \in (0, 1)$? In particular, is it possible to give a sequence (p_n) of polynomials of the form (14.1) (or maybe $(p_n) \subset \mathcal{F}_n$) so that p_n has c/a *distinct* zeros in $[-1 + a, 1 - a]$, where $c > 0$ is an absolute constant and $a \in [n^{-1/2}, 1)$? If not, what is the sharp analogue of Theorem 7.3 for *distinct real* zeros?

It is easy to prove (see [17]) that a polynomial $p \in \mathcal{A}_n$ can have at most $\log_2 n$ zeros at -1.

Problem 14.11. Is it true that there is an absolute constant $c > 0$ such that every $p \in \mathcal{A}_n$ with $p(0) = 1$ has at most $c \log n$ *real* zeros? If not, what is the best possible upper bound for the number of *real* zeros of polynomials $p \in \mathcal{A}_n$? What is the best possible upper bound for the number of *distinct real* zeros of polynomials $p \in \mathcal{A}_n$?

Odlyzko asked the next question after observing computationally that no $p \in \mathcal{A}_n$ with $n \leq 25$ had a repeat root of modulus greater than one.

Problem 14.12. Prove or disprove that a polynomial $p \in \mathcal{A}_n$ has all its repeated zeros at 0 or on the unit circle.

One can show, not completely trivially, that there are polynomials $p \in \mathcal{F}_n$ with repeated zeros in $(0, 1)$ up to multiplicity 4.

Problem 14.13. Can the multiplicity of a zero of a $p \in \cup_{n=1}^{\infty} \mathcal{F}_n$ in

$$\{z \in \mathbb{C} : 0 < |z| < 1\}$$

be arbitrarily large?

A negative answer to the above question would resolve an old conjecture of Lehmer concerning Mahler's measure. (See [3].)

Boyd [20] shows that there is an absolute constant c such that every $p \in \mathcal{L}_n$ can have at most $c \log^2 n / \log \log n$ zeros at 1. Is is easy to give polynomials $p \in \mathcal{L}_n$ with $c \log n$ zeros at 1.

Problem 14.14. Prove or disprove that there is an absolute constant c such that every polynomial $p \in \mathcal{L}_n$ can have at most $c \log n$ zeros at 1. Prove or disprove that there is an absolute constant c such that every polynomial $p \in \mathcal{L}_n$ can have at most $c \log n$ zeros in $[-1, 1]$.

Problem 14.15. Can Boyd's result be extended to the class $\mathcal{K}_n$? In other words, is there an absolute constant c such that every $p \in \mathcal{K}_n$ can have at most $c \log^2 n / \log \log n$ zeros at 1? (It is easy to give polynomials $p \in \mathcal{L}_n$ with $c \log n$ zeros at 1.)

Problem 14.16. Prove or disprove that every polynomial $p \in \mathcal{L}_n$ has at least one zero in the annulus

$$\left\{ z \in \mathbb{C} : 1 - \frac{c}{n} < |z| < 1 + \frac{c}{n} \right\},$$

where $c > 0$ is an absolute constant.

Problem 14.17. Prove or disprove that $N(p_n) \to \infty$, where $N(p_n)$ denotes the number of real zeros of

$$p_n(t) := \sum_{k=0}^{n} a_{k,n} \cos kt, \qquad a_{k,n} = \pm 1, \;\; k = 0, 1, 2, \ldots, n,$$

in the period $[0, 2\pi)$.

The next question is a version of an old and hard unsolved problem known as the Tarry-Escott Problem.

Problem 14.18. Let $N \in \mathbb{N}$ be fixed. Let $a(N)$ be the smallest value of k for which there is a polynomial $p \in \cup_{n=1}^{\infty} \mathcal{F}_n$ with exactly k nonzero terms in it and with a zero at 1 with multiplicity at least N. Prove or disprove that $a(N) = 2N$.

To prove that $a(N) \geq 2N$ is simple. The fact that $a(N) \leq 2N$ is known for $N = 1, 2, \ldots, 10$, but the problem is open for every $N \geq 11$. The best known upper bound for $a(N)$ in general seems to be $a(N) \leq cN^2 \log N$ with an absolute constant $c > 0$. See [17]. Even improving this (like dropping the factor $\log N$) would be a significant achievement.

Problem 14.19. It would be interesting to see if $\|S_n\|_{p,\partial D}^{comp} \leq c$ is possible for any $0 < p < 2$ with an absolute constant c, where S_n is the polynomial truncation operator defined in Section 13.

In light of Theorems 6.3 and 6.4, we ask the following questions.

Problem 14.20. Does

$$\lim_{n \to \infty} \frac{\log \left(\inf_{0 \neq p \in \mathcal{F}_n} \|p\|_{[0,1]} \right)}{\sqrt{n}}$$

exist? If it does, what is it? Does

$$\lim_{n \to \infty} \frac{\log \left(\inf_{0 \neq p \in \mathcal{A}_n} \|p(-x)\|_{[0,1]} \right)}{\log^2(n+1)}$$

exist? If it does, what is it?

References

1. F. Amoroso, Sur le diamètre transfini entier d'un intervalle réel, Ann. Inst. Fourier, Grenoble **40** (1990), 885–911.

2. B. Aparicio, New bounds on the minimal Diophantine deviation from zero on $[0, 1]$ and $[0, 1/4]$, Actus Sextas Jour. Mat. Hisp.-Lusitanas (1979), 289–291.

3. F. Beaucoup, P. Borwein, D. Boyd, and C. Pinner, Multiple roots of $[-1, 1]$ power series, J. London Math. Soc. (2) **57** (1998), 135–147.

4. J. Beck, Flat polynomials on the unit circle – note on a problem of Littlewood, Bull. London Math. Soc. **23** (1991), 269–277.

5. A. T. Bharucha-Reid and M. Sambandham, *Random Polynomials*, Academic Press, Orlando, 1986.

6. A. Bloch and G. Pólya, On the roots of certain algebraic equations, Proc. London Math. Soc. **33** (1932), 102–114.

7. E. Bombieri and J. Vaaler, Polynomials with low height and prescribed vanishing, in *Analytic Number Theory and Diophantine Problems, (Stillwater, OK, 1984)*, Birkhäuser, Boston, 1987, 53–73.

8. P. Borwein and T. Erdélyi, *Polynomials and Polynomial Inequalities*, Springer-Verlag, New York, 1995.

9. P. Borwein and T. Erdélyi, The integer Chebyshev Problem, Math. Comp. **65** (1996), 661–681.

10. P. Borwein and T. Erdélyi, Questions about polymomials with $\{-1, 0, 1\}$ coefficients, Constr. Approx. **12** (1996), 439–442.

11. P. Borwein and T. Erdélyi, Littlewood-type problems on subarcs of the unit circle, Indiana Univ. Math. J. **46** (1997), 1323–1346.

12. P. Borwein and T. Erdélyi, Markov and Bernstein type inequalities for polynomials with restricted coefficients, Ramanujan J. **1** (1997), 309–323.

13. P. Borwein and T. Erdélyi, On the zeros of polynomials with restricted coefficients, Illinois J. Math. (1997), 667–675.

14. P. Borwein and T. Erdélyi, Markov-Bernstein type inequalities under Littlewood-type coefficient constraints, Indag. Math. N.S. (2) **11** (2000), 159–172.

15. P. Borwein and T. Erdélyi, Trigonometric polynomials with many real zeros and a Littlewood-type problem, Proc. Amer. Math. Soc. **129** (2001), 725–730.

16. P. Borwein and T. Erdélyi, Lower bounds for the merit factors of trigonometric polynomials, manuscript.

17. P. Borwein, T. Erdélyi, and G. Kós, Littlewood-type problems on [0, 1], Proc. London Math. Soc. (3) **79** (1999), 22–46.

18. P. Borwein and C. Ingalls, The Prouhet, Tarry, Escott problem, Ens. Math. **40** (1994), 3–27.

19. J. Bourgain, Sur le minimum d'une somme de cosinus, Acta Arith. **45** (1986), 381–389.

20. D. Boyd, On a problem of Byrnes concerning polynomials with restricted coefficients, Math. Comput. **66** (1977), 1697–1703.

21. J. S. Byrnes, On polynomials with coefficients of modulus one, Bull. London Math. Soc. **9** (1977), 171–176.

22. J. S. Byrnes and B. Saffari, Unimodular polynomials: a wrong theorem with very fortunate consequences, Bull. London Math. Soc., to appear.

23. J. S. Byrnes and D. J. Newman, Null steering employing polynomials with restricted coefficients, IEEE Trans. Antennas and Propagation **36** (1988), 301–303.

24. F. W. Carrol, D. Eustice, and T. Figiel, The minimum modulus of polynomials with coefficients of modulus one, J. London Math. Soc. **16** (1977), 76–82.

25. J. Clunie, On the minimum modulus of a polynomial on the unit circle, Quart. J. Math. **10** (1959), 95–98.

26. P.J. Cohen, On a conjecture of Littlewood and idempotent measures, Amer. J. Math. **82** (1960), 191–212.

27. R.A. DeVore and G.G. Lorentz, *Constructive Approximation*, Springer-Verlag, Berlin, 1993.

28. P. Erdős, Some unsolved problems, Michigan Math. J. **4** (1957), 291–300.

29. P. Erdős, An inequality for the maximum of trigonometric polynomials, Annales Polonica Math. **12** (1962), 151–154.

30. P. Erdős, Some old and new problems in approximation theory: research problems, Constr. Approx. **11** (1995), 419–421.

31. T. Erdélyi, The resolution of Saffari's Phase Problem, C. R. Acad. Sci. Paris Sér. I Math. (10) **331** (2000), 803–808.

32. T. Erdélyi, The Phase Problem of ultraflat unimodular polynomials: the resolution of the conjecture of Saffari, Math. Annalen to appear.

33. T. Erdélyi, On the zeros of polynomials with Littlewood-type coefficient constraints, Michigan Math. J. **49** (2001), 97–111.

34. T. Erdélyi, How far is a sequence of ultraflat unimodular polynomials from being conjugate reciprocal, Michigan Math. J. **49** (2001), 259–264.

35. T. Erdélyi, Proof af Saffari's near orthogonality conjecture on ultra-flatsequences of unimodular polynomials, C. R. Acad. Sci. Paris Sér. I Math. to appear.

36. T. Erdélyi, On the real part of ultraflat sequences of unimodular polynomials, manuscript.

37. T. Erdélyi, The norm of the polynomial truncation operator on the unit circle and on $[-1, 1]$, Colloquium Mathematicum to appear.

38. P. Erdős and P. Turán , On the distribution of roots of polynomials, Annals of Math. **57** (1950), 105–119.

39. Le Baron O. Ferguson, *Approximation by Polynomials with Integral Coefficients*, Amer. Math. Soc., Rhode Island, 1980.

40. G. T. Fielding, The expected value of the integral around the unit circle of a certain class of polynomials, Bull. London Math. Soc. **2** (1970), 301–306.

41. L. K. Hua, *Introduction to Number Theory*, Springer-Verlag, Berlin, Heidelberg, New York, 1982.

42. M. Kac, On the average number of real roots of a random algebraic equation, II, Proc. London Math. Soc. **50** (1948), 390–408.

43. J-P. Kahane, Sur les polynômes à coefficients unimodulaires, Bull. London Math. Soc, **12** (1980), 321–342.

44. J-P. Kahane, *Some Random Series of Functions*, Cambridge Studies in Advanced Mathematics **5**, Cambridge, 1985, Second Edition.

45. S. Konjagin, On a problem of Littlewood, Izv. A. N. SSSR, Ser. Mat. (2) **45** (1981), 243–265.

46. S. Konjagin, On a question of Pichorides, C. R. Acad. Sci. Paris Sér I Math. (4) **324** (1997), 385–388.

47. T.W. Körner, On a polynomial of J.S. Byrnes, Bull. London Math. Soc. **12** (1980), 219–224.

48. J. E. Littlewood, On the mean value of certain trigonometric polynomials, J. London Math. Soc. **36** (1961), 307–334.

49. J. E. Littlewood, On the mean value of certain trigonometric polynomials (II), J. London Math. Soc. **39** (1964), 511–552.

50. J. E. Littlewood, The real zeros and value distributions of real trigonometrical polynomials, J. London Math. Soc., **41** (1966), 336–342.

51. J. E. Littlewood, On polynomials $\sum \pm z^m$ and $\sum e^{\alpha_m i} z^m$, $z = e^{\theta i}$, J. London Math. Soc., **41** (1966), 367–376.

52. J. E. Littlewood, *Some Problems in Real and Complex Analysis*, Heath Mathematical Monographs, Lexington, Massachusetts, 1968.

53. J. E. Littlewood and A.C. Offord, On the number of real roots of a random algebraic equation (II), Proc. Cam. Phil. Soc. **35** 1939, 133–148.

54. L. Lovász, J. Spencer, and K. Vesztergombi, Discrepancy of set systems and matrices, European J. of Comin. **7** (1986), 151–160.

55. K. Mahler, On two extremal properties of polynomials, Illinois J. Math., **7** (1963), 681–701.

56. O. C. McGehee, L. Pigno and B. Smith, Hardy's inequality and the L^1-norm of exponential sums, Ann. Math. **113** (1981), 613–618.

57. Milovanović, G.V., D. S. Mitrinović, and Th. M. Rassias, *Topics in Polynomials: Extremal Problems, Inequalities, Zeros*, World Scientific, Singapore, 1994.

58. F. Nazarov, Local estimates of exponential polynomials and their applications to the inequalities of uncertainty principle type, St. Petersburg Math. J. **5** (1994), 663–717.

59. D. J. Newman and J. S. Byrnes, The L^4 norm of a polynomial with coefficients ± 1, MAA Monthly **97** (1990), 42–45.

60. D. J. Newman and A. Giroux, Properties on the unit circle of polynomials with unimodular coefficients, Proc. Amer. Math. Soc. **109** (1) (1990), 113–116.

61. A. Odlyzko and B. Poonen, Zeros of polynomials with 0,1 coefficients, Ens. Math. **39** (1993), 317–348.

62. S. K. Pichorides, Notes on trigonometric polynomials, in *Conference on Harmonic Analysis in Honor of Antoni Zygmund*, Vol. I, II (Chicago, Ill., 1981), Wadsworth Math. Ser., Wadsworth, Belmont, Calif., 1983, 84–94.

63. G. Pólya and G. Szegő, *Problems and Theorems in Analysis*, Volume I, Springer-Verlag, New York, 1972.

64. H. Queffelec and B. Saffari, Unimodular polynomials and Bernstein's inequalities, C. R. Acad. Sci. Paris Sér. I Math. (3) **321** (1995), 313–318.

65. H. Queffelec and B. Saffari, On Bernstein's inequality and Kahane's ultraflat polynomials, J. Fourier Anal. Appl. **2** (1996), 519–582.

66. B. Saffari, The phase behavior of ultraflat unimodular polynomials, in *Probabilistic and Stochastic Methods in Analysis, with Applications* (Il Ciocco, 1991), Kluwer Acad. Publ., Dordrecht, 1992, 555–572.

67. R. Salem and A. Zygmund, Some properties of trigonometric series whose terms have random signs, Acta Math. **91** (1954), 254–301.

68. E. Schmidt, Über algebraische Gleichungen vom Pólya-Bloch-Typos, Sitz. Preuss. Akad. Wiss., Phys.-Math. Kl. (1932), 321.

69. I. Schur, Untersuchungen über algebraische Gleichungen, Sitz. Preuss. Akad. Wiss., Phys.-Math. Kl. (1933), 403–428.

70. B. Solomyak, On the random series $\sum \pm \lambda^n$ (an Erdős problem), Annals of Math. **142** (1995), 611–625.

71. G. Szegő, Bemerkungen zu einem Satz von E. Schmidt über algebraische Gleichungen, Sitz. Preuss. Akad. Wiss., Phys.-Math. Kl. (1934), 86–98.

72. P. Turán, *On a New Method of Analysis and its Applications*, Wiley, New York, 1984.

Tamás Erdélyi
Department of Mathematics
Texas A&M University
College Station, Texas 77843
USA
terdelyi@math.tamu.edu

Parameterization of Manifold Triangulations

Michael S. Floater, Kai Hormann, and Martin Reimers

Abstract. This paper proposes a straightforward method of parameterizing manifold triangulations where the parameter domain is a coarser triangulation of the same topology. The method partitions the given triangulation into triangular patches bounded by geodesic curves and parameterizes each patch individually. We apply the global parameterization to remeshing and wavelet decomposition.

§1. Introduction

This paper deals with the parameterization of manifold triangulations of arbitrary topology. By a triangle T we mean the convex hull of three non-collinear points $v_1, v_2, v_3 \in \mathbb{R}^3$, *i.e.*, $T = [v_1, v_2, v_3]$. We call a set of triangles $\mathcal{T} = \{T_1, \ldots, T_m\}$ a manifold triangulation if

1) $T_i \cap T_j$ is either empty, a common vertex, or a common edge, $i \neq j$, and

2) $\Omega_\mathcal{T} = \bigcup_{i=1}^{m} T_i$ is an orientable 2-manifold.

In this paper we propose a method for constructing a parameterization of a given manifold triangulation $\mathcal{T}$ over a coarse manifold triangulation $\mathcal{D}$. By **parameterization** we mean a homeomorphism,

$$\phi : \Omega_\mathcal{D} \to \Omega_\mathcal{T},$$

as illustrated in Fig. 1. Thus $\Omega_\mathcal{D}$ will be the parameter domain of $\Omega_\mathcal{T}$ and ϕ will be bijective, continuous, and have a continuous inverse.

The basic idea is to partition $\Omega_\mathcal{T}$ into triangular surface patches S_i, each bounded by geodesic curves, such that there is a one-to-one correspondence between the triangles T_i of $\mathcal{D}$ and the patches S_i. We then construct local, piecewise linear parameterizations

$$\phi_i : T_i \to S_i,$$

Approximation Theory X: Abstract and Classical Analysis   197
Charles K. Chui, Larry L. Schumaker, and Joachim Stöckler (eds.), pp. 197–209.
Copyright ⓒ 2002 by Vanderbilt University Press, Nashville, TN.
ISBN 0-8265-1415-4.

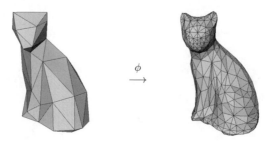

Fig. 1. Parameterization.

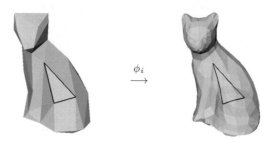

Fig. 2. Local parameterization.

as illustrated in Fig. 2, which match pairwise at the vertices and edges of $\mathcal{D}$ so that the combined mapping

$$\phi(x) = \phi_i(x), \qquad x \in T_i,$$

is also continuous and one-to-one.

The first method for parameterizing manifold triangulations was that of Eck et al. [5], who proposed partitioning by growing Voronoi-like regions. Local parameterization was done through discrete harmonic maps. Later, Lee et al. [16] proposed a method in which the parameterization is carried out iteratively in parallel with mesh simplification. We believe our method is conceptually simpler and unlike in [5] and [16], it guarantees a one-to-one parameterization under only mild conditions on the parameter domain.

We complete the paper by giving an application of the method to a wavelet decomposition of the surface $\Omega_\mathcal{T}$. We subdivide the parameter domain dyadically and sample the surface $\Omega_\mathcal{T}$ at the vertices, yielding an approximate surface $\Omega_\mathcal{S}$ which can be decomposed using some chosen wavelet scheme.

After completing this paper, we became aware of the recent publication [21] by Praun, Sweldens and Schröder where a set of triangulations are parameterized over a common parameter domain, and where geodesic curves play an important role.

§2. Finding the Parameter Domain

Before we construct a parameterization, we need to determine an appropriate parameter domain, which in this setting will be a manifold triangulation $\mathcal{D}$ of the same topology as $\mathcal{T}$, but with typically much fewer triangles. Thus we regard $\mathcal{T}$ as a fine triangulation and $\mathcal{D}$ as a coarse one. Moreover, for our method, we also need a correspondence between the vertices of $\mathcal{D}$ and some suitable points on $\Omega_{\mathcal{T}}$. We will make clear what we mean by 'suitable' later.

If we are not given a parameter domain and correspondence in advance, we propose using one of the many known algorithms for *mesh decimation* to find one. For a survey of methods, see [11,15]. One major class of these algorithms modifies a given triangulation by iteratively applying a decimation operator such as *vertex removal* or *half-edge collapse* that decreases the number of triangles by two in each step without changing the topology of the mesh surface. All these algorithms result in a coarser triangulation $\mathcal{D}$ that approximates the shape of the initial mesh $\mathcal{T}$. Moreover, these algorithms usually provide a correspondence between the vertices v_i of $\mathcal{D}$ and points on $\Omega_{\mathcal{T}}$. In fact, many of these algorithms never move vertices at each step, only delete them, in which case the vertices of the final coarse triangulation are also vertices of the original triangulation $\mathcal{T}$. In our numerical examples, we have used the method of Campagna [1].

§3. Partitioning

The first step in the parameterization method is to partition the surface $\Omega_{\mathcal{T}}$ into triangular regions S_i corresponding to the triangles T_i of $\mathcal{D}$.

Recall that for each vertex v_j of $\mathcal{D}$ there is a suitable corresponding point x_j in $\Omega_{\mathcal{T}}$. Then for each edge $[v_j, v_k]$ of $\mathcal{D}$ we construct a shortest geodesic path $\Gamma(x_j, x_k) \subset \Omega_{\mathcal{T}}$ between the two points x_j and x_k of $\Omega_{\mathcal{T}}$. We suppose that the parameter domain and vertex correspondence are such that these geodesic paths partition the surface $\Omega_{\mathcal{T}}$ into triangular patches corresponding to the triangles of $\mathcal{D}$. Thus to each triangle $T_i = [v_j, v_k, v_\ell]$ there should be a triangular region $S_i \subset \Omega_{\mathcal{T}}$ bounded by the three curves $\Gamma(x_j, x_k)$, $\Gamma(x_k, x_\ell)$, $\Gamma(x_\ell, x_j)$. In this case the patches $\{S_i\}$ constitute a surface triangulation of the surface $\Omega_{\mathcal{T}}$ corresponding to the triangulation $\mathcal{D}$. In fact, it is known in graph theory [10] that the curves will induce such a surface triangulation if

1) the only intersections between the curves is at common endpoints and

2) the cyclic ordering of the curves around a common vertex is the same as the cyclic ordering of the corresponding edges in $\mathcal{D}$.

We thus require a parameter domain $\Omega_{\mathcal{D}}$ and a vertex correspondence with the property that conditions (1) and (2) hold. We have found in our

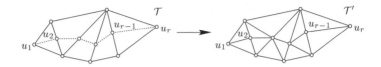

Fig. 3. Refining $\mathcal{T}$ in order to embed a geodesic curve.

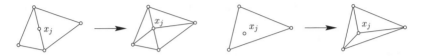

Fig. 4. Pre-process of inserting a point x_j into $\mathcal{T}$.

numerical examples that these conditions have always been met, where we have used mesh decimation to generate $\mathcal{D}$, under one of the usual stopping criteria. In fact, theoretically we could ensure that conditions (1) and (2) hold by simply making them part of the stopping criterion. In practice, however, we did not find this necessary. In [21], however, the geodesic paths typically do not form a valid partition due to the fact that a whole set of triangulations are being parameterized over the same domain. The solution in [21] is to modify the geodesic curves so as to prevent patch boundary crossings by suitably modifying the algorithm of [14].

Since the shortest path across a triangle is a straight line, we see that geodesics on triangulations are polygonal curves with nodes lying on the edges of the triangulation. Thus, a minimal geodesic corresponding to the edge $[v_j, v_k]$ of $\mathcal{D}$ is a union of line segments,

$$\Gamma(x_j, x_k) = [u_1, u_2] \cup [u_2, u_3] \cup \cdots \cup [u_{r-1}, u_r], \tag{1}$$

where $u_1 = x_j$, $u_r = x_k$, and the points u_p are located on the edges of $\mathcal{T}$. We can therefore insert all segments of these geodesic curves into the triangulation $\mathcal{T}$, yielding a refined triangulation $\mathcal{T}'$. All points u_p will become vertices in $\mathcal{T}'$, and the segments $[u_p, u_{p+1}]$ will become edges of $\mathcal{T}'$. The geometry of $\mathcal{T}'$ is the same as that of $\mathcal{T}$, i.e., $\Omega_{\mathcal{T}'} = \Omega_{\mathcal{T}}$. Fig. 3 illustrates the refinement when the endpoints x_j and x_k of the curve are themselves vertices of $\mathcal{T}$, which is the case in our implementation. In this case, each segment $[u_p, u_{p+1}]$ will split the triangle in $\mathcal{T}$ containing it into one triangle and one quadrilateral. When making the refinement $\mathcal{T}'$, we simply split the quadrilateral into two triangles by choosing either one of the two diagonals.

If one of the points x_j was not a vertex of $\mathcal{T}$, we would apply a pre-process before applying the above procedure. In the pre-process, we would insert x_j into $\mathcal{T}$ by adding appropriate edges emanating from it, depending on whether x_j lies on an edge of $\mathcal{T}$ or is in the interior of some triangle of $\mathcal{T}$; see Fig. 4.

The problem of computing geodesics on polygonal surfaces is a classic one and has received considerable attention, see *e.g.* [2,12,14,17,18]. We have used Chen and Han's shortest path algorithm [2]; see also [12]. The method is exact and based on unfolding triangles. However, as the method is computationally quite expensive, we compute the geodesic $\Gamma(x_j, x_k)$ by restricting the search to a small subtriangulation $\mathcal{T}_{jk}$ of $\mathcal{T}$. The shape of $\mathcal{T}_{jk}$ is chosen to be an 'ellipse' with x_j and x_k its 'focal points'. We take $\mathcal{T}_{jk}$ to be the set of all triangles in $\mathcal{T}$ containing a vertex v of $\mathcal{T}$ such that

$$d_a(v, x_j) + d_a(v, x_k) \leq K d_a(x_j, x_k),$$

where $d_a(x, y)$ denotes an *approximation* to the geodesic distance $d(x, y)$ between two points x and y of $\Omega_{\mathcal{T}}$, and $K > 1$ is some chosen constant. The idea here is that we can use a much faster algorithm to compute the approximative distance $d_a(x, y)$. We have used the method of Kimmel and Sethian [14] to compute $d_a(x, y)$ and subsequently find $\mathcal{T}_{jk}$. Since the latter method gives such a close approximation to the true geodesic distance, we have been able to set $K = 1.06$ and then in almost all cases we have run, the geodesic path $\Gamma(x_j, x_k)$ has been contained in $\Omega_{\mathcal{T}_{jk}}$. In the event that the path is not contained in $\Omega_{\mathcal{T}_{jk}}$, we simply increase K. Eventually, for large enough K we are guaranteed to locate $\Gamma(x_j, x_k)$.

The point about the refinement is that each patch $S_i \subset \Omega_{\mathcal{T}}$ is now the union of triangles in $\mathcal{T}'$, *i.e.*,

$$S_i = \Omega_{\mathcal{T}_i},$$

where $\mathcal{T}_i$ is a subtriangulation of $\mathcal{T}'$. This will enable us to apply a standard parameterization method to parameterize each patch S_i.

§4. Parameterization

For each triangle $T_i = [v_j, v_k, v_\ell]$ of $\mathcal{D}$, we now have a corresponding triangular patch $S_i \subset \Omega_{\mathcal{T}}$, the union of the triangles in a subtriangulation $\mathcal{T}_i$ of $\mathcal{T}'$. It remains to construct a parameterization $\phi_i : T_i \to S_i$. We do this by taking ϕ_i to be the inverse of a piecewise linear mapping $\psi_i : S_i \to T_i$. In other words ψ_i will be continuous over S_i and linear over each triangle in $\mathcal{T}_i$. By linear (or affine), we mean that for any triangle $T = [v_1, v_2, v_3]$ in $\mathcal{T}_i$,

$$\psi_i(\lambda_1 v_1 + \lambda_2 v_2 + \lambda_3 v_3) = \lambda_1 \psi_i(v_1) + \lambda_2 \psi_i(v_2) + \lambda_3 \psi_i(v_3), \qquad (2)$$

for any real numbers $\lambda_1, \lambda_2, \lambda_3 \geq 0$ which sum to one. Thus ψ_i will be completely determined by the parameter points $\psi_i(v) \in \mathbb{R}^3$ for vertices v of $\mathcal{T}_i$.

We will use the method of [6]. First of all, due to construction, corresponding to the vertices v_j, v_k, v_ℓ of T_i, there must be three corresponding 'corner' vertices x_j, x_k, x_ℓ in the boundary of $\mathcal{T}_i$. We therefore set $\psi_i(x_r) = v_r$, for $r \in \{j, k, \ell\}$. Next we determine ψ_i at the remaining boundary vertices of $\mathcal{T}_i$. The boundary of $\mathcal{T}_i$ consists of three geodesic curves, each with endpoints in $\{x_j, x_k, x_\ell\}$. Consider the edge $[v_j, v_k]$ and the corresponding geodesic curve $\Gamma(x_j, x_k)$ in (1). We take ψ_i at the vertices u_p to be a chord length parameterization, that is we demand that

$$\psi_i(u_p) = (1 - \lambda_p)\psi_i(u_1) + \lambda_p\psi_i(u_r),$$

where

$$\lambda_p = \frac{L_p}{L_r} \quad \text{with} \quad L_p = \sum_{q=1}^{p-1} \|u_{q+1} - u_q\|,$$

Finally, we determine ψ_i at the interior vertices of $\mathcal{T}_i$ by solving a sparse linear system of equations. For each interior vertex v of $\mathcal{T}_i$, we demand that

$$\psi_i(v) = \sum_{w \in N_v} \lambda_{vw}\psi_i(w),$$

where the weights λ_{vw} are strictly positive and sum to one. Here N_v denotes the set of neighbouring vertices of v. It was shown in [6] that this linear system always has a unique solution. We take the weights λ_{vw} to be the shape-preserving weights of [6], which were shown to have a reproduction property and tend to minimize distortion.

Since the equations force each point $\psi_i(v)$ to be a convex combination of its neighbouring points $\psi_i(w)$, ψ_i is a so-called **convex combination map**. As to the question of whether ψ_i is injective, we have the following.

Proposition 1. *The mapping* $\psi_i : S_i \to T_i$ *is one-to-one.*

Proof: Due to Theorem 6.1 of [7], it is sufficient to show that no dividing edge of $\mathcal{T}_i$ is mapped into the boundary of T_i. By a dividing edge we understand an edge $[v, w]$ of $\mathcal{T}_i$ which is an interior edge of $\mathcal{T}_i$ and whose endpoints v and w are boundary vertices of $\mathcal{T}_i$. Thus we simply need to show that for such an edge, the two mapped points $\psi_i(v)$ and $\psi_i(w)$ do not belong to the same edge of T_i.

Assume for the sake of contradiction that the points $\psi_i(v)$ and $\psi_i(w)$ belong to the same edge of T_i. Due to our construction this implies that v and w lie on the same geodesic curve. Since the edge $[v, w]$ is obviously the shortest geodesic curve connecting v and w, it must belong to the boundary of $\mathcal{T}_i$, contradicting our assumption that $[v, w]$ is a dividing edge. $\square$

We complete this section by showing that the method has linear precision in the sense that the parameterization ϕ is locally linear if $\Omega_{\mathcal{T}}$ is locally planar.

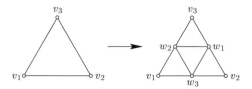

Fig. 5. Dyadic refinement.

Proposition 2. *Suppose that for some triangle $T_i = [v_j, v_k, v_\ell]$, the triangle $[x_j, x_k, x_\ell]$ is contained in Ω_T. Then we have $S_i = [x_j, x_k, x_\ell]$ and the mapping $\phi_i : T_i \to S_i$ is linear in the sense of (2).*

Proof: If $[x_j, x_k, x_\ell] \subset \Omega_T$, then the line segments $[x_j, x_k], [x_k, x_\ell], [x_\ell, x_j]$ are the geodesic boundary curves of S_i since they are trivially shortest paths, and thus $S_i = [x_j, x_k, x_\ell]$. Due to chord length parameterization of the boundary curves, the whole boundary of S_i is mapped affinely into the boundary of T_i. Since moreover we use shape-preserving parameterization for interior vertices, Proposition 6 in [6] shows that ψ_i, and therefore ϕ_i, is linear. $\square$

In fact, due to construction, the whole parameterization ϕ depends continuously on the vertices of T. This implies that when Ω_T is locally close to being planar, the parameterization ϕ will locally be close to being linear.

§5. Application to Wavelet Decomposition

The parameterization ϕ of the manifold surface Ω_T allows us to approximate Ω_T by a new manifold surface which can be decomposed using a wavelet scheme.

To explain how this works, first set $\mathcal{D}^0 = \mathcal{D}$ and consider its dyadic refinement $\mathcal{D}^1$. By dyadic refinement we mean that we divide each triangle $T = [v_1, v_2, v_3]$ in $\mathcal{D}^0$ into the four congruent subtriangles,

$$[v_1, w_2, w_3], \quad [w_1, v_2, w_3], \quad [w_1, w_2, v_3], \quad [w_1, w_2, w_3],$$

where

$$w_1 = \frac{v_2 + v_3}{2}, \qquad w_2 = \frac{v_3 + v_1}{2}, \qquad w_3 = \frac{v_1 + v_2}{2},$$

are the midpoints of the edges of T; see Fig. 5. This refinement is also referred to as the one-to-four split, and the set of all such subtriangles forms a triangulation $\mathcal{D}^1$. Similarly, we can refine $\mathcal{D}^1$ to form $\mathcal{D}^2$, and so on and we have $\Omega_{\mathcal{D}_j} = \Omega_\mathcal{D}$ for all $j = 0, 1, \ldots$.

Next let S^j be the linear space of all functions $f^j : \Omega_\mathcal{D} \to \mathbb{R}^3$ which are continuous on $\Omega_\mathcal{D}$ and linear on each triangle of $\mathcal{D}^j$. These linear spaces are clearly nested,

$$S^0 \subset S^1 \subset S^2 \subset \cdots,$$

and the dimension of S^j is the number of vertices in $\mathcal{D}^j$.

The nested spaces S^j allow us to set up a multiresolution framework which can be used to decompose a given element f^j of S^j. For each $j = 1, 2, \ldots$, we choose W^{j-1} to be some subspace of S^j such that

$$S^{j-1} \oplus W^{j-1} = S^j,$$

where $\oplus$ denotes in general a direct sum. We call W^{j-1} a wavelet space and we can decompose any space S^j into its coarsest subspace S^0 and a sequence of wavelet spaces,

$$S^j = S^{j-1} \oplus W^{j-1} = \cdots = S^0 \oplus W^0 \oplus \cdots \oplus W^{j-1}. \tag{3}$$

Within this framework, we can decompose a given function $f^j \in S^j$ into *levels of detail*. Equation (3) implies that there exist unique functions $f^{j-1} \in S^{j-1}$ and $g^{j-1} \in W^{j-1}$ such that

$$f^j = f^{j-1} + g^{j-1},$$

and we regard f^{j-1} as an approximation to f^j at a lower resolution or level of detail. The function g^{j-1} is the error introduced when replacing the original function f^j by its approximation f^{j-1}. We can continue this decomposition until $j = 0$. Then we will have $f^0 \in S^0$ and $g^i \in W^i$, $i = 0, 1, \ldots, j-1$ with

$$f^j = f^0 + g^0 + g^1 + \cdots + g^{j-1},$$

and f^0 will be the coarsest possible approximation to f^j.

Now, we can use the parameterization $\phi : \Omega_{\mathcal{D}} \to \Omega_{\mathcal{T}}$ to decompose and approximate $\Omega_{\mathcal{T}}$. We simply let f^j be that element of S^j such that

$$f^j(v) = \phi(v),$$

for all vertices v of $\mathcal{D}^j$. For sufficiently large j, f^j will be a good approximation to ϕ in which case the Hausdorff distance between $\Omega_{\mathcal{T}}$ and the image $f^j(\Omega_{\mathcal{D}})$ will be small. Thus a wavelet decomposition of f^j will be an approximate decomposition of ϕ and hence the manifold $\Omega_{\mathcal{T}}$ itself.

Many wavelet spaces and bases have been proposed for decomposing piecewise linear spaces over triangulations, see for example [8,22]. In our implementation we have taken the wavelet space W^{j-1} to be orthogonal to S^{j-1} with respect to the weighted inner product

$$\langle f, g \rangle = \sum_{T \in \mathcal{D}} \frac{1}{\mathrm{a}(T)} \int_T f(x)g(x) \, dA, \qquad f, g \in C(\Omega_{\mathcal{D}})$$

where $\mathrm{a}(T)$ is the area of triangle T. For the wavelet basis we have applied the wavelets constructed in [8] which generalize to this manifold setting in a trivial way. As usual we used the hat (nodal) functions as bases for the nested spaces S^j themselves. We have implemented the filterbank algorithms developed in [9], adapted to this manifold setting.

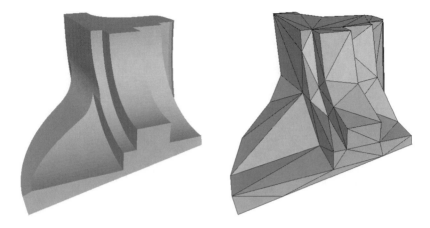

Fig. 6. Fine triangulation $\mathcal{T}$ with 12,946 triangles (left) and coarse triangulation $\mathcal{D}$ with 114 triangles (right).

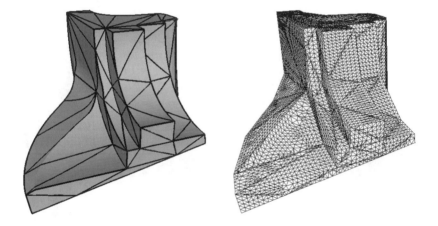

Fig. 7. Geodesic paths on $\Omega_{\mathcal{D}}$ (left) and parameterization of $\Omega_{\mathcal{T}}$ over $\Omega_{\mathcal{D}}$ (right).

§6. Numerical Example

Though the parameterization method applies to manifolds of arbitrary topology, we will simply illustrate with the manifold triangulation $\mathcal{T}$ on the left of Fig. 6, which is homeomorphic to a sphere. We first computed a coarser triangulation $\mathcal{D}$ by mesh decimation as described in [1], shown on the right of Fig. 6.

For the 171 edges in $\mathcal{D}$ we then computed the corresponding geodesic paths in $\Omega_{\mathcal{T}}$ as shown on the left of Fig. 7. These were then embedded into $\mathcal{T}$, yielding the refined triangulation $\mathcal{T}'$. Each triangular patch S_i of $\mathcal{T}'$ was then parameterized over the corresponding triangle T_i of $\mathcal{D}$, the

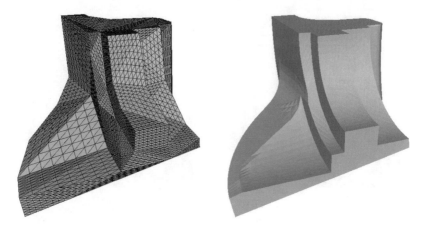

Fig. 8. Dyadic refinement $\mathcal{D}^3$ (left) and approximation $f^3(\Omega_\mathcal{D})$ of $\Omega_\mathcal{T}$ with 7,296 triangles (right).

Fig. 9. Wavelet reconstruction of $f^3(\Omega_\mathcal{D})$ using 5% (left) and 20% (right) of the wavelet coefficients.

result being shown on the right of Fig. 7. Each linear system was solved using the bi-conjugate gradient method.

Figures 8 and 9 illustrate the use of the parameterization $\phi : \Omega_\mathcal{D} \to \Omega_\mathcal{T}$ for wavelet decomposition. First, we dyadically refined $\mathcal{D}$ three times, yielding the triangulation $\mathcal{D}^3$, and took $f^3 \in S^3$ to be the piecewise linear interpolant to ϕ. Its image $f^3(\Omega_\mathcal{D})$ approximates the given triangulation $\Omega_\mathcal{T}$; see Fig. 8. We then decomposed f^3 into $f^0 \in S^0$ and $g^i \in W^i$, $i = 0, 1, 2$. Sorting the 3,591 wavelet coefficients by magnitude and using only the largest 5% or 20%, respectively, to reconstruct f^3 gives the triangulations shown in Fig. 9.

We have implemented the whole method in C++ and measured the following computation times on an SGI Octane with a 400 MHz R12000 processor.

mesh reduction	35 sec.
geodesic paths	16 sec.
parameterization	20 sec.
refinement	1 sec.
decomposition	2 sec.
reconstruction	1 sec.
total	75 sec.

These computation times appear to scale roughly linearly with respect to the number of triangles in the mesh. For the "horse" data set, which has 96,966 triangles, we found that the increase in computation times of all the algorithms was only slightly worse than linear. The total computation time for the horse was 741 seconds as opposed to 75 seconds for the fan-disk, giving a factor of roughly 10 while the ratio of number of triangles, 96,966 : 12,946, is roughly 8.

Acknowledgments. This work was supported in part by the European Union research project "Multiresolution in Geometric Modeling (MIN-GLE)" under grant HPRN-CT-1999-00117.

References

1. Campagna, S., L. Kobbelt, and H.-P. Seidel, Efficient generation of hierarchical triangle meshes, in *The Mathematics of Surfaces VIII*, R. Cripps (ed.), Information Geometers, 1998, 105–123.

2. Chen, J. and Y. Han, Shortest paths on a polyhedron, J. Comput. Geom. and Appl. **6** (1996), 127–144.

3. Dijkstra, E. W., A note on two problems in connexion with graphs, Numer. Math. **1** (1959), 269–271.

4. Duchamp, T., A. Certain, T. DeRose, and W. Stuetzle, Hierarchical computation of PL harmonic embeddings, Technical Report, University of Washington, 1997.

5. Eck, M., T. DeRose, T. Duchamp, H. Hoppe, M. Lounsbery, and W. Stuetzle, Multiresolution analysis of arbitrary meshes, Comp. Graphics (SIGGRAPH '95 Proc.) **29** (1995), 173–182.

6. Floater, M. S., Parameterization and smooth approximation of surface triangulations, Comput. Aided Geom. Design **14** (1997), 231–250.

7. Floater, M. S., One-to-one piecewise linear mappings over triangulations, Preprint, University of Oslo, June 2001.

8. Floater, M. S. and E. G. Quak, Piecewise linear prewavelets on arbitrary triangulations, Numer. Math. **82** (1999), 221–252.

9. Floater, M. S., E. G. Quak, and M. Reimers, Filter bank algorithms for piecewise linear prewavelets on arbitrary triangulations, J. Comput. Appl. Math. **119** (2000), 185–207.

10. Gross, J. L. and T. W. Tucker, *Topological Graph Theory*, John Wiley & Sons, 1987.

11. Heckbert, P. S. and M. Garland, Survey of polygonal surface simplification algorithms, Technical Report, Carnegie Mellon University, 1997.

12. Kaneva, B. and J. O'Rourke, An implementation of Chen & Han's shortest paths algorithm, in *Proc. of the 12th Canadian Conf. on Comput. Geom.*, 2000, 139–146.

13. Kapoor, S., Efficient computation of geodesic shortest paths, in *ACM Symposium on Theory of Computing*, 1999, 770–779.

14. Kimmel R. and J. A. Sethian, Computing geodesic paths on manifolds, in *Proc. of National Academy of Sciences* **95**, 1998, 8431–8435.

15. Kobbelt, L., S. Campagna, and H.-P. Seidel, A general framework for mesh decimation, in *Graphics Interface 98 Proc.*, 1998, 43–50.

16. Lee, A., W. Sweldens, P. Schröder, L. Cowsar, and D. Dobkin, Multiresolution adaptive parametrization of surfaces. Comp. Graphics (SIGGRAPH '98 Proc.) **32** (1998), 95–104.

17. Mitchell, J. S. B., D. M. Mount, and C. H. Papadimitriou, The discrete geodesic problem, SIAM J. Comput. **16** (1987), 647–668.

18. Pham-Trong, V., N. Szafran, and L. Baird, Pseudo-geodesics on three-dimensional surfaces and pseudo-geodesic meshes, Numer. Algorithms **6** (2001), 305–315.

19. Pinkall, U. and K. Polthier, Computing discrete minimal surfaces and their conjugates, Experimental Mathematics **2** (1993), 15–36.

20. Polthier K. and M. Schmies, Straightest geodesics on polyhedral surfaces, in *Mathematical Visualisation*, C. Hege and K. Polthier (eds.), Springer, 1998, 135–150.

21. Praun E., W. Sweldens, and P. Schröder, Consistent mesh parameterizations, Comp. Graphics (SIGGRAPH '01 Proc.) (2001), 179–184.

22. Schröder, P. and W. Sweldens, Spherical wavelets: Efficiently representing functions on the sphere, Comp. Graphics (SIGGRAPH '95 Proc.) **29** (1995), 161–172.

Michael S. Floater
SINTEF
Post Box 124, Blindern
N-0314 Oslo, Norway
mif@math.sintef.no

Kai Hormann
Computer Graphics Group
University of Erlangen-Nürnberg
Am Weichselgarten 9
D-91058 Erlangen, Germany
hormann@cs.fau.de

Martin Reimers
Institutt for informatikk
Universitetet i Oslo
Post Box 1080, Blindern
N-0316 Oslo, Norway
martinre@ifi.uio.no

Limit Theorems for Polynomial Approximations with Hermite and Freud Weights

Michael I. Ganzburg

Abstract. The purpose of this paper is to study certain limit relations between the errors of best approximation by polynomials and entire functions of exponential type in the weighted $L_p(\mathbb{R})$-metric with an exponential weight. Some applications are discussed. In particular, we show that the error of best polynomial approximation of $|x|^\lambda$ with a Freud weight can be asymptotically expressed via the Bernstein constant.

§1. Introduction

Weighted polynomial approximation on the real line $\mathbb{R}$ with an exponential weight $W(x) = \exp(-Q(x))$ has attracted much attention since the 1970s. The basic results of approximation theory such as Bernstein-Markov-Nikolskii type inequalities and Jackson-Bernstein theorems, have been proved for even smooth Q of polynomial growth (Freud weights) and faster than polynomial growth (Erdős weights), see [4,5,6,7,16,17,18,20] for references and discussions.

In this paper we study certain limit relations between weighted polynomial and harmonic approximations and apply them to some problems of approximation theory. In particular, we show that the error of best polynomial approximation of $f_\lambda(x) := |x|^\lambda$ with a Freud weight can be asymptotically expressed via the Bernstein constant.

Limit theorems for polynomial approximation on $[-1, 1]$ were discovered by Bernstein [3] and developed by Raitsin [19] and the author [9,10,11,12,13]. They have proved to be useful in Jackson theorems,

Approximation Theory X: Abstract and Classical Analysis
Charles K. Chui, Larry L. Schumaker, and Joachim Stöckler (eds.), pp. 211–221.
Copyright Ⓒ 2002 by Vanderbilt University Press, Nashville, TN.
ISBN 0-8265-1415-4.

Bernstein-Markov inequalities, exact constants of approximation on certain classes, approximation of individual functions and other areas [1,2,9,11,12,13,21]. In particular, if f is a function of polynomial growth on $\mathbb{R}$ and $A_\sigma(f)_p < \infty$, then [10,11,13,19]:

$$\lim_{n \to \infty} (n/\sigma)^{1/p} E_n(f((n/\sigma)\cdot), L_p[-1,1]) = A_\sigma(f)_p, \quad 0 < p < \infty, \quad (1)$$

$$\lim_{n \to \infty} E_n(f(d_n\cdot), L_\infty[-1,1]) = A_\sigma(f)_\infty, \quad (2)$$

where $d_n = n(1 - \gamma_n)/\sigma$, and $\{\gamma_n\}_{n=1}^\infty$ is a sequence of real numbers satisfying the following conditions: (a) $0 \le \gamma_n \le 1$, (b) $\lim_{n \to \infty} \gamma_n/n = 0$, (c) $\liminf_{n \to \infty} \gamma_n n^\delta > 0$ for some $\delta \in (0, 2/3)$. Here, $E_n(h, L_p[-1,1])$ is the error of best approximation of h in $L_p[-1,1]$ by polynomials of degree at most n, and $A_\sigma(f)_p$ the error of best approximation of f by entire functions of exponential type σ in $L_p(\mathbb{R})$.

Limit relations (1) and (2) immediately imply the following asymptotic:

$$\lim_{n \to \infty} n^\lambda E_n(f_\lambda, L_p[-1,1]) = B_{\lambda,p}, \quad 0 < p \le \infty, \quad (3)$$

where $B_{\lambda,p} = A_1(f_\lambda)_p$ is the Bernstein constant [2,13,14,22].

In this paper we prove a weighted analogue of (1) for approximation with the Hermite weight and $p \in [1, 2]$ (Theorem 1). The proof is based on an estimate of the error of best polynomial approximation of an entire function of exponential type (Theorem 2) and on a weighted version of a V. A. Markov-type inequality for the coefficients of an algebraic polynomial (Theorem 3). The proofs of Theorems 1, 2, and 3 are given in Sections 7, 5, and 6, respectively. Extension of these theorems to general Freud weights and some applications are discussed in Sections 3 and 4.

Notation. Let $\mathcal{P}_n$ be the set of all polynomials of degree at most n, B_σ the class of all entire functions of exponential type $\sigma > 0$, and $L_{p,W}$, $0 < p \le \infty$, a quasi-normed space with quasi-norm

$$\|f\|_{p,W} := \left(\int_\mathbb{R} (|f(x)|W(x))^p dx \right)^{1/p}, \qquad \|f\|_p := \|f\|_{p,1},$$

where W is a weight. For a locally integrable f on $\mathbb{R}$ we define

$$E_n(f, L_{p,W}) := \inf_{P \in \mathcal{P}_n} \|f - P\|_{p,W}, \qquad A_\sigma(f)_p := \inf_{g \in B_\sigma} \|f - g\|_p.$$

Throughout the paper C denotes a positive constant independent of $n, k, m, \sigma, t, x, \varepsilon, A, P_n, g, g_1$, and does not necessarily denote the same constant in different occurrences.

§2. The Hermite Weight

Let $H(x) := \exp(-x^2)$. Our result about the limit connection between the errors of best approximation by polynomials and entire functions of exponential type now can be stated as follows.

Theorem 1. *If f is a function of polynomial growth on $\mathbb{R}$ and $A_\sigma(f)_p < \infty$ with $\sigma > 0$, then*

$$\lim_{n \to \infty} (b_n/\sigma)^{1/p} E_n(f((b_n/\sigma)\cdot), L_{p,H}) = A_\sigma(f)_p, \tag{4}$$

where $b_n := 2\sqrt{n}$ and $p \in [1, 2]$.

The following two theorems are used in the proof of Theorem 1, but they are also interesting in themselves. The first of them is a refinement of Theorem 1 when f is an entire function of exponential type, and the second one is a V. A. Markov-type inequality with the Hermite weight.

Theorem 2. *If a function $g \in B_{qb_n}$, $0 < q < 1$, satisfies the inequality $|g(x)| \le An^M(1 + |x|)^N$ on $\mathbb{R}$, then for every $p \in [1, \infty]$,*

$$E_n(g, L_{p,H}) \le Ae^{-na}, \tag{5}$$

where $a > 0$ is independent of n and g.

Theorem 3. *For any polynomial $P_n(x) = \sum_{k=0}^{n} c_k x^k$ and for every $p \in (0, 2]$, the following inequalities hold:*

$$|c_k| \le \frac{C(b_n)^{k+1/p}}{k!} \|P_n\|_{L_{p,H}}, \qquad 0 \le k \le n. \tag{6}$$

§3. Freud Weights

Let $W := e^{-Q}$, where $Q : \mathbb{R} \to \mathbb{R}$ is even, continuous and Q'' is continuous in $(0, \infty)$, $Q(0) = 0$, $Q' > 0$, $Q'' > 0$ in $(0, \infty)$, and for some $A, B, C > 0$,

$$A \le xQ''(x)/Q'(x) \le B, \quad x^2|Q'''(x)|/Q'(x) \le C, \qquad x \in (0, \infty).$$

We define the Mhaskar-Rachmanov-Saff number $a_n = a_n(Q)$ to be the positive root of the equation

$$n = (2/\pi) \int_0^1 a_n t Q'(a_n t) dt / \sqrt{1 - t^2}.$$

Next we define the number

$$b_n = b_n(Q) := (2/\pi) \int_0^1 Q'(a_n t) \sqrt{1 - t^2} dt/t + n/a_n.$$

In particular, the weight $W(x) = \exp(-|x|^\alpha)$ with $\alpha > 1$ satisfies these conditions and $b_n = \alpha a_n/(\alpha - 1)$.

It is possible to show that relations (4), (5) and a version of inequality (6) with H replaced by W hold for all $p \in [1, \infty)$. For $p = \infty$ the following analogue of (2) and (4) is valid:

$$\lim_{n \to \infty} E_n(f(d_n\cdot), L_{\infty,W}) = A_\sigma(f)_\infty,$$

where $d_n = b_n(1 - \gamma_n)/\sigma$, and $\{\gamma_n\}_{n=1}^\infty$ is a sequence of real numbers satisfying conditions like (a), (b), and (c).

The proofs of all these results are fairly complicated, and will be given in the paper [15].

§4. Applications and Extensions

The limit theorems for polynomial approximation with Freud weights can be applied to exact constants of weighted approximation on certain classes, Jackson's theorems, Bernstein-Markov inequalities, approximation of functions like $|x|^\lambda \ln^k |x|$, and some other problems; see [15] for details.

In particular, for $f_\lambda(x) = |x|^\lambda$, where $\lambda > -1/p$ and $1 \le p \le \infty$, the following asymptotic holds:

$$\lim_{n \to \infty} b_n^{\lambda+1/p} E_n(f_\lambda, L_{p,W}) = B_{\lambda,p}, \tag{7}$$

where $B_{\lambda,p}$ is defined by (3). Note that for the Hermite weight and $p \in [1, 2]$, relation (7) follows immediately from Theorem 1.

We remark that a version of Theorem 1 holds also for Erdős weights, and it is possible to extend (4) to all locally integrable functions on $\mathbb{R}$.

§5. Proof of Theorem 2

We first prove (5) for $g(x) = \exp(itx)$, where $|t| \le 2q_1\sqrt{n}$ and $q_1 \in (0, 1)$. Using the generating function relation for the Hermite polynomials H_k [8]

$$\exp(2uz - z^2) = \sum_{k=0}^\infty H_k(u)z^k/k!,$$

and choosing $z = it/2^{3/2}, u = \sqrt{2}x$, we obtain

$$\exp(ixt) = \exp(-t^2/8) \sum_{k=0}^\infty \frac{(it)^k}{2^{3k/2}k!} = H_k(\sqrt{2}x). \tag{8}$$

It follows from the inequalities

$$\|H_k(\sqrt{2}\cdot)\|_{p,H} \le Ck^{1/(2p)} \sup_{x\in\mathbb{R}} \exp(-x^2)|H_k(\sqrt{2}x)| \le ck^{1/(2p)}\sqrt{k!2^k} \tag{9}$$

that for each fixed $t \in \mathbb{R}$ the series (8) converges in $L_{p,H}$ for $1 \leq p \leq \infty$. Note that the first estimate in (9) is a consequence of the Nikolskii-type inequality [20], while the second one was proved by Cramér [8].

Next putting

$$P_n(t, x) := \exp(-t^2/8) \sum_{k=0}^{n} \frac{(it)^k}{2^{3k/2} k!} = H_k(\sqrt{2}x), \tag{10}$$

we obtain from (8)–(10) and Stirling's formula

$$|| \exp(it\cdot) - P_n(t,\cdot)||_{p,H} \leq e^{-t^2/8} \sum_{k=n+1}^{\infty} \frac{|t|^k}{2^{3k/2} k!} = ||H_k(\sqrt{2}\cdot)||_{p,H}$$

$$\leq C e^{-t^2/8} \sum_{k=n+1}^{\infty} \frac{k^{1/(2p)} |t|^k e^{k/2}}{2^k k^{k/2 + 1/4}} = C \sum_{k=n+1}^{\infty} k^{1/(2p)} \exp\left(\frac{k}{2} \varphi\left(\frac{t^2}{4k}\right)\right), \tag{11}$$

where $\varphi(y) := \ln y - y + 1$. Since φ increases in $(0,1)$ and $\varphi(q_1) < 0$, it follows from (11) that for $|t| \leq 2q_1\sqrt{n}$,

$$E_n(\exp(it\cdot), L_{p,H}) \leq || \exp(it\cdot) - P_n(t,\cdot)||_{p,H}$$

$$\leq C \sum_{k=n+1}^{\infty} k^{1/(2p)} \exp(k\varphi(q_1)/2) \leq C n^{1/(2p)} \exp(n\varphi(q_1)/2). \tag{12}$$

Now let $g_1 \in L_2 \cap B_{q_1 b_n}$, where $b_n = 2\sqrt{n}$ and $0 < q_1 < 1$. Then by the Paley-Wiener theorem, $g_1(x) = \int_{-q_1 b_n}^{q_1 b_n} h(t) e^{ixt} dt$ with $h \in L_2[-q_1 b_n, q_1 b_n]$. It easily follows from (10) that $Q_n(x) = \int_{-q_1 b_n}^{q_1 b_n} h(t) P_n(t,x) dt$ is a polynomial of degree n. Hence using (12), we have

$$E_n(g_1, L_{p,H}) \leq ||g_1 - Q_n||_{p,H} \leq \int_{-q_1 b_n}^{q_1 b_n} || \exp(it\cdot) - P_n(t,\cdot)||_{p,H} |h(t)| dt$$

$$\leq C n^{1/(2p)+1/4} \exp(n\varphi(q_1)/2) ||g_1||_2. \tag{13}$$

Finally, let $g \in B_{qb_n}$ satisfy the condition $|g(x)| \leq An^M (1 + |x|)^N$ for all $x \in \mathbb{R}$, where $b_n = 2\sqrt{n}$, $0 < q < 1$ and $M, N \geq 0$. Then the function $g_1(x) := g(x)(\sin(\varepsilon x)/(\varepsilon x))^{N+2}$ with $\varepsilon \in (0,1)$, belongs to $B_{q_1 b_n} \cap L_2$, where $0 < q_1 := q + \varepsilon(N+2)/b_n < (1+q)/2$, if $0 < \varepsilon < (1-q)b_n/(2N+2)$, and

$$||g_1||_2 \leq CAn^M \varepsilon^{-N-1/2}. \tag{14}$$

Furthermore using the inequality $1 - \sin y/y \leq y^2/6, y \geq 0$, we have

$$\|g - g_1\|_{p,H} \le A n^M \left(\int_{\mathbb{R}} (1 + |x|)^{pN} \left(1 - \left(\frac{\sin(\varepsilon x)}{\varepsilon x} \right)^{N+2} \right)^p e^{-px^2} dx \right)^{1/p}$$

$$\le C A n^M \varepsilon^2 \left(\int_{\mathbb{R}} (1 + |x|)^{p(N+2)} e^{-px^2/2} dx \right)^{1/p} = C A n^M \varepsilon^2. \tag{15}$$

Then, by (13)–(15),

$$E_n(g, L_{p,H}) \le E_n(g_1, L_{p,H}) + \|g - g_1\|_{p,H}$$
$$\le C A n^M (\varepsilon^{-N-1/2} n^{1/(2p)+1/4} \exp(n\varphi(q_1)/2) + \varepsilon^2). \tag{16}$$

Minimizing the right-hand side of (16) over all $\varepsilon \in (0, (1-q)b_n/(N+2))$, we arrive at inequality (5). $\square$

§6. Proof of Theorem 3

We first prove (6) for $p = 2$. For every polynomial $P_n(x) = \sum_{k=0}^n c_k x^k = \sum_{k=0}^n c_k(P_n) x^k$, we consider the Hermite expansion (see [8])

$$P_n(x) = \sum_{s=0}^n A_s \alpha_s H_s(\sqrt{2}x),$$

where

$$\alpha_s := \|H_s(\sqrt{2}\cdot)\|_{2,H}^{-1} = (2/\pi)^{1/4} 2^{-s/2} (s!)^{-1/2}, \qquad 0 \le s \le n. \tag{17}$$

Furthermore,

$$\|P_n\|_{2,H} = \left(\sum_{s=0}^n A_s^2 \right)^{1/2}; \qquad c_k = \sum_{s=k}^n A_s \alpha_s B_{s,k}, \quad 0 \le k \le n, \tag{18}$$

where

$$B_{s,k} = \begin{cases} \frac{(-1)^{d-m} 2^{3m} (2d)!}{(d-m)!(2m)!}, & \text{if } s = 2d \text{ is even} \\ & \text{and } k = 2m \text{ is even,} \\ \frac{(-1)^{d-m} 2^{3m-3/2} (2d-1)!}{(d-m)!(2m-1)!}, & \text{if } s = 2d - 1 \text{ is odd} \\ & \text{and } k = 2m - 1 \text{ is odd,} \\ 0, & \text{otherwise,} \end{cases} \tag{19}$$

are the coefficients of the Hermite polynomials $H_s(\sqrt{2}x) = \sum_{k=0}^s B_{s,k} x^k$ (see [8]). It easily follows from (18) that

$$I_{n,k} := \sup_{\|P_n\|_{2,H} \le 1} |c_k(P_n)| = \left(\sum_{s=k}^n (\alpha_s B_{s,k})^2 \right)^{1/2}, \qquad 0 \le k \le n. \tag{20}$$

Then using (17)–(20) and taking into account Stirling's formula, we obtain for an even $k = 2m$,

$$
I_{n,2m}^2 = \sum_{d=m}^{[n/2]} (\alpha_{2d} B_{2d,2m})^2 = (2/\pi)^{1/2} \frac{2^{6m}}{(2m)!^2} \sum_{d=m}^{[n/2]} \frac{(2d)!}{(d-m)!^2 2^{2d}}
$$

$$
= (2/\pi)^{1/2} \frac{2^{4m}}{(2m)!^2} \left(\sum_{r=1}^{[n/2]-m} \frac{(2m+2r)!}{r!^2 2^{2r}} + (2m)! \right)
$$

$$
\leq C \frac{2^{4m}}{(2m)!^2} \left(\sum_{r=1}^{[n/2]-m} \frac{(2m+2r)^{2m+1/2}}{r} + (2m)! \right)
$$

$$
= C \frac{2^{4m}}{(2m)!^2} \left(\sum_{r=1}^{m} + \sum_{r=m+1}^{[n/2]=-m} + (2m)! \right)
$$

$$
\leq C \frac{2^{4m}}{(2m)!^2} \left(\sum_{l=1}^{4m} l^{2m+1/2} + \sum_{l=1}^{n} l^{2m-1/2} + (2m)! \right)
$$

$$
\leq C \frac{2^{4m}}{(2m)!^2} \left(\frac{(4m)^{2m+3/2}}{2m+3/2} = + n^{2m+1/2} + n^{2m} \right) \leq C \frac{2^{4m}}{(2m)!^2} n^{2m+1/2}.
$$

Hence for an even k,

$$
I_{n,k} \leq C \frac{(2\sqrt{n})^{k+1/2}}{k!}. \tag{21}
$$

Similarly for an odd $k = 2m - 1$,

$$
I_{n,2m-1}^2 = \sum_{d=m}^{\langle n \rangle} (\alpha_{2d-1} B_{2d-1,2m-1})^2 = C \frac{2^{6m}}{(2m-1)!^2} \sum_{d=m}^{\langle n \rangle} \frac{(2d-1)!}{(d-m)!^2 2^{2d}}
$$

$$
\leq C \frac{2^{4m}}{(2m-1)!^2} \left(\sum_{r=1}^{\langle n \rangle - m} \frac{(2m+2r-1)^{2m-1/2}}{r} + (2m-1)! \right)
$$

$$
\leq C \frac{2^{4m}}{(2m-1)!^2} \left(\sum_{l=1}^{4m-2} l^{2m-1/2} + \sum_{l=1}^{n} l^{2m-3/2} + (2m-1)! \right)
$$

$$
\leq C \frac{2^{4m}}{(2m-1)!^2} n^{2m-1/2}.
$$

Therefore (21) holds for all $k, 0 \leq k \leq n$. Thus (6) is proved for $p = 2$. Theorem 3 now follows from the Nikolskii-type inequality [20]

$$
\|P_n\|_{2,H} \leq C(\sqrt{n})^{1/p-1/2} \|P_n\|_{p,H}, \qquad 0 < p < 2. \quad \square
$$

§7. Proof of Theorem 1

We first prove the inequality

$$\limsup_{n\to\infty}(b_n/\sigma)^{1/p}E_n(f((b_n/\sigma)\cdot),L_{p,H}) \le A_\sigma(f)_p, \quad 1 \le p < \infty. \quad (22)$$

It is known [1,13,21] that if $A_\sigma(f)_p < \infty$ for $1 \le p < \infty$ and $\sigma > 0$, then

$$\lim_{\tau\to\sigma} A_\tau(f)_p = A_\sigma(f)_p. \quad (23)$$

Then for a small enough $\varepsilon \in (0,\sigma)$, $A_{\sigma-\varepsilon}(f)_p < \infty$, and there exists $g_\sigma \in B_{\sigma-\varepsilon}$ such that $A_{\sigma-\varepsilon}(f)_p = \|f - g_\sigma\|_p$, see [13,21]. Moreover, if $|f(x)| \le A(1 + |x|)^N$, then $|g_\sigma(x)| \le CA(1 + |x|)^N$, see [11,13]. Next,

$$(b_n/\sigma)^{1/p}E_n(f((b_n/\sigma)\cdot),L_{p,H})$$
$$\le (b_n/\sigma)^{1/p}(\|f((b_n/\sigma)\cdot) - g_\sigma((b_n/\sigma)\cdot)\|_{p,H} + E_n(g_\sigma((b_n/\sigma)\cdot),L_{p,H}))$$
$$\le A_{\sigma-\varepsilon}(f)_p + (b_n/\sigma)^{1/p}E_n(g_\sigma((b_n/\sigma)\cdot),L_{p,H}). \quad (24)$$

Since $g(\cdot) := (b_n/\sigma)^{1/p}g_\sigma((b_n/\sigma)\cdot) \in B_{qb_n}$, where $q = (\sigma - \varepsilon)/\sigma$, and $|g(x)| \le CAn^{(N+1/p)/2}(1 + |x|)^N$, it follows from (5) that

$$\lim_{n\to\infty} E_n(g,L_{p,H}) = 0.$$

Hence (24) yields the inequality

$$\limsup_{n\to\infty}(b_n/\sigma)^{1/p}E_n(f((b_n/\sigma)\cdot),L_{p,H}) \le A_{\sigma-\varepsilon}(f)_p. \quad (25)$$

Letting $\varepsilon \to 0$ in (25) and taking account of (23), we obtain (22).

To prove the inequality

$$\liminf_{n\to\infty}(b_n/\sigma)^{1/p}E_n(f((b_n/\sigma)\cdot),L_{p,H}) \ge A_\sigma(f)_p, \quad 0 < p \le 2, \quad (26)$$

we assume without loss of generality that there exists the finite limit

$$\lim_{n\to\infty} (b_n/\sigma)^{1/p}E_n(f((b_n/\sigma)\cdot),L_{p,H}).$$

First let $f \in L_p$, and let $P_n(x) := \sum_{k=0}^{n} c_k x^k$ be the polynomial of best approximation to $f_n(x) := (b_n/\sigma)^{1/p}f((b_n/\sigma)x)$ in $L_{p,H}$. Then for $n = 1, 2, \ldots$,

$$\|P_n\|_{p,H} \le 2\|f_n\|_{p,H} \le 2\|f\|_p. \quad (27)$$

Next, by Theorem 3 and (27),

$$|c_k| \le C\frac{b_n^{k+1/p}}{k!}\|f\|_{p,H}, \quad 0 \le k \le n.$$

Hence the coefficients of the polynomial

$$Q_n(x) := (b_n/\sigma)^{-1/p} P_n((\sigma/b_n)x) = \sum_{k=0}^{n} a_{k,n} x^k$$

satisfy the condition $|a_{k,n}| \leq C/k!$ for $n = 1, 2, \ldots$ and $0 \leq k \leq n$. This shows that the set $\{Q_n\}_{n=1}^{\infty}$ is a precompact set in B_σ, i.e., there exist a sequence $\{n_s\}_{s=1}^{\infty}$ and $g_0 \in B_\sigma$ such that

$$\lim_{s \to \infty} Q_{n_s}(z) = g_0(z) \tag{28}$$

uniformly in any interval $[-R, R]$ (see [1,21]).

Further for fixed numbers $R > 0$ and $\varepsilon \in (0, 1)$ we define $s_0 := \inf\{s : n_s \geq R^2\sigma^2/(4|\ln(1-\varepsilon)|)\}$. Then for $s > s_0$ we have

$$(b_n/\sigma)^{1/p} E_{n_s}(f((b_{n_s}/\sigma)\cdot), L_{p,H}) = \left(\int_{\mathbb{R}} |f(y) - Q_n(y)|^p e^{-p(\sigma y/b_{n_s})^2} dy \right)^{1/p}$$

$$\geq (1 - \varepsilon) \left(\int_{-R}^{R} |f(y) - Q_n(y)|^p dy \right)^{1/p}. \tag{29}$$

Now by (28) and (29),

$$\lim_{n \to \infty} (b_n/\sigma)^{1/p} E_n(f((b_n/\sigma)\cdot), L_{p,H})$$

$$\geq (1 - \varepsilon) \left(\int_{-R}^{R} |f(y) - g_0(y)|^p dy \right)^{1/p}.$$

Letting $\varepsilon \to 0$ and $R \to \infty$ and using (23), we obtain (26) for $f \in L_p$.

If f is a function of polynomial growth on $\mathbb{R}$ and $A_\sigma(f)_p < \infty$, then $f_1 = f - g \in L_p$ for some $g \in B_{\sigma-\varepsilon}$ and $\varepsilon \in (0, \sigma)$. Since , by Theorem 2,

$$\lim_{n \to \infty} (b_n/\sigma)^{1/p} E_n(g((b_n/\sigma)\cdot), L_{p,H}) = 0,$$

we have

$$\liminf_{n \to \infty} (b_n/\sigma)^{1/p} E_n(f((b_n/\sigma)\cdot), L_{p,H})$$

$$\geq \liminf_{n \to \infty} (b_n/\sigma)^{1/p} f_1((b_n/\sigma)\cdot), L_{p,H})$$

$$\geq A_\sigma(f_1)_p = A_\sigma(f)_p.$$

Hence (26) follows. Finally, inequalities (22) and (26) yield (4). $\square$

Acknowledgments. The author thanks Doron Lubinsky and Steven Damelin for inspiring discussions.

References

1. Akhiezer, N. I., *Lectures on Approximation Theory*, 2nd edition, Nauka, Moscow, 1965.

2. Bernstein, S. N., On the best approximation of $|x|^p$ by means of polynomials of extremely high degree, Izv. Akad Nauk SSSR **2** (1938), 169–180 (Russian).

3. Bernstein, S. N., On the best approximation of continuous functions on the whole real axis by means of entire functions of given degree 5, Dokl. Akad Nauk SSSR **54** (1946), 479–482 (Russian).

4. Damelin, S. B., Converse and smoothness theorems for Erdős weights, J. Approx. Theory **93** (1998), 349–398.

5. Damelin, S. B., Smoothness theorems for Erdős weights II, J. Approx. Theory **97** (1999), 220–239.

6. Damelin, S. B. and D. S. Lubinsky, Jackson theorems for Erdős weights in $L_p(0 < p \leq \infty)$, J. Approx. Theory **94** (1998), 333–382.

7. Ditzian, Z. and D. S. Lubinsky, Jackson and smoothness theorems for Freud weights in $L_p(0 < p \leq \infty)$, Constr. Approx. **13** (1997), 99–152.

8. Erdelyi, A., W. Magnus, F. Oberhettinger and F. G. Tricomi, *Higher Transcendental Functions*, vol. II, McGraw-Hill, New York, 1953.

9. Ganzburg, M. I., Multidimensional limit theorems of the theory of best polynomial approximation, Siberian Math. J. **23** (1983), 316–331.

10. Ganzburg, M. I., Limit theorems for the best polynomial approximations in the L_∞ metric, Ukrainian Math. J. **43**(3) (1991), 299–305.

11. Ganzburg, M. I., Limit theorems in approximation theory, Anal. Math. **18** (1992), 37–57.

12. Ganzburg, M., Limit relations in approximation theory and their applications, in *Approximation Theory VIII, Vol. 1: Approximation and Interpolation*, Charles K. Chui and Larry L. Schumaker (eds.), World Scientific Publishing Co., Inc., Singapore, 1995, 223–232.

13. Ganzburg, M. I., Limit theorems and best constants of approximation theory, in *Handbook on Analytic-Computational Methods in Applied Mathematics*, G. A. Anastassiou (ed.), Chapman & Hall/CRC, Boca Raton, FL, 2000, 507–569.

14. Ganzburg, M. I., New limit relations for polynomial approximations, in *Trends in Approximation Theory*, Kirill Kopotun, Tom Lyche, and Mike Neamtu (eds), Vanderbilt University Press, Nashville, 2001, 113–122.

15. Ganzburg, M. I., Limit theorems for polynomial approximation with an exponential weight, in preparation.

16. Lubinsky, D. S., An update on orthogonal polynomials and weighted approximation on the real line, Acta Applic. Math. **3** (1993), 121–164.

17. Lubinsky, D. S., Ideas of weighted polynomial approximation on $(-\infty, \infty)$, in *Approximation Theory VIII, Vol. 1: Approximation and Interpolation*, Charles K. Chui and Larry L. Schumaker (eds.), World Scientific Publishing Co., Inc., Singapore, 1995, 371–396.

18. Mhaskar, H. N., *Introduction to the Theory of Weighted Polynomial Approximation*, World Scientific, River Edge, NJ, 1996.

19. Raitsin, R. A., S. N. Bernstein's limit theorem for the best approximation in the mean and some of its applications, Izv. VUZ Mat. **12**(10) (1968), 81–86 (Russian).

20. Saff, E. B. and V. Totik, *Logarithmic Potentials with External Fields*, Springer-Verlag, Berlin, 1997.

21. Timan, A. F., *Theory of Approximation of Functions of a Real Variable*, Pergamon Press, New York, 1963.

22. Varga, R. S. and A. J. Carpenter, On the Bernstein conjecture in approximation theory, Constr. Approx. **1** (1985), 333–348.

Michael I. Ganzburg
Dept. of Mathematics
Hampton University
Hampton, VA 23668
michael.ganzburg@hamptonu.edu

Complex Haar Spaces Generated by Shifts Applied to a Single Function

Walter Hengartner and Gerhard Opfer

Abstract. Some of the known Haar spaces are generated by shifts of a single function G. There are examples of two types. In one case the spaces generated are real spaces defined on compact intervals, and in the other case, the generated spaces are also Haar spaces on compact subsets of the complex plane $\mathbb{C}$. Under the assumption that G is analytic we are able to characterize those functions G which generate Haar spaces in the plane $\mathbb{C}$.

§1. Background

We denote by $\mathbb{N}$ the set of all positive integers, by $\mathbb{Z}$ the integers, by $\mathbb{R}$ the real and by $\mathbb{C}$ the complex numbers. The letter $\mathbb{K}$ stands for either $\mathbb{R}$ or $\mathbb{C}$ and Π_n denotes the set of all polynomials with degree at most n, where, in general, complex coefficients are permitted. We will distinguish between the equality sign $=$ and the sign $:=$ (possibly also $=:$) where the latter stands for a defining relation. What is defined appears on the side where the colon is. Let D be any compact Hausdorff space and $X := C(D)$ the set of all continuous functions defined on D with values in $\mathbb{K}$. With the norm $||f|| := \max_{t \in D} |f(t)|$ for all $f \in X$, usually called the uniform norm, the space X becomes a Banach space over $\mathbb{K}$, or since functions can even be multiplied pointwise, a Banach algebra. The main underlying theorem is the following.

Theorem 1. *For a fixed $n \in \mathbb{N}$ let $H \subset X$ be an n-dimensional subspace of X. The following four statements are equivalent:*

i) *For any selection of n pairwise distinct points $t_j \in D$ and any set of n numbers $\eta_j \in \mathbb{K}$ the interpolation problem*

$$h(t_j) = \eta_j, \quad j = 1, 2, \ldots, n$$

Approximation Theory X: Abstract and Classical Analysis
Charles K. Chui, Larry L. Schumaker, and Joachim Stöckler (eds.), pp. 223–238.

has a unique solution $h \in H$.

ii) Any $h \in H \backslash \{0\}$ has at most $n - 1$ zeros in D.

iii) Let $H := \langle h_1, h_2, \ldots, h_n \rangle$, i.e. H is the linear hull of the linearly independent functions $h_j \in X$, $j = 1, 2, \ldots, n$. Then the $(n \times n)$ matrix

$$\mathbf{M} := (h_j(t_k)), \quad j, k = 1, 2, \ldots, n$$

is non-singular for any choice of pairwise distinct points $t_j \in D$, $j = 1, 2, \ldots, n$.

iv) Any $f \in X$ possesses a unique best approximation $\hat{h} \in H$, i.e., $\|f - \hat{h}\| < \|f - h\|$ for all $h \in H \backslash \{\hat{h}\}$.

Proof: See MEINARDUS, [12, 1967, p. 16–17]. $\square$

Definition 2. A finite dimensional subspace H of X with dimension n is called a **Haar space** if it possesses one of the properties given in Theorem 1.

Unfortunately, Haar spaces do not generally exist. This is governed by the following two theorems.

Theorem 3. Let $\mathbb{K} = \mathbb{R}$. An n-dimensional subspace $H \subset X$ with $n \geq 2$ can only be a Haar space if D is topologically equivalent to a closed subset of the unit circle $C := \{z \in \mathbb{C} : |z| = 1\}$ with at least n points.

Proof: See MAIRHUBER, [11, 1956], CURTIS, [5, 1958]. $\square$

This result reduces Haar spaces in the real case essentially to the following cases: D is a compact interval in $\mathbb{R}$, D is a complete circle, which reduces H to p-periodic functions on $[0, p[\subset \mathbb{R}$, D consists of finitely many points (at least n).

Theorem 4. Let $\mathbb{K} = \mathbb{C}$ and D locally connected. An n-dimensional subspace $H \subset X$ with $n \geq 2$ can only be a Haar space if D is topologically equivalent to a closed subset of $\mathbb{C}$ and contains at least n points.

Proof: See HENDERSON & UMMEL, [8, 1973]. $\square$

Historically the first theorem restricting the domain of definition D in the case of complex Haar spaces $H \subset X$ to subsets of $\mathbb{C}$ was given by SCHOENBERG & YANG, [16, 1961]. The consequence of Theorem 3 and Theorem 4 is that Haar spaces for $D \subset \mathbb{K}^n$ with $n \geq 2$ do not in general exist.

§2. Examples

In the literature there are lists of Haar spaces which are generally not very long. Standard sources are KARLIN & STUDDEN [9, 1966] and DUNHAM [6, 1974]. One finds two examples where the Haar spaces are generated by shifts applied to a single function f. These examples are $f(t) := 1/t$ and $f(t) := \exp(-t^2)$. In the end of § 3 we will mention another example. The main question of this paper is whether one can characterize the functions with this property. This problem was also raised for the real case by CHENEY & LIGHT [4, 2000, p. 76]. Radial basis splines (CHENEY & LIGHT [4, 2000, Ch. 15 & 16]) evolved also from shifts applied to a single function. Let us first inspect the two mentioned examples.

Example 5. Let $f = 1/t$ for all $t \in \mathbb{C}\backslash\{0\}$ and let D be any compact set in $\mathbb{C}$ and $S := \complement D$ with respect to $\mathbb{C}$, where $\complement$ stands for the complement. Let $s_j \in S$, $j \in \mathbb{N}$ any sequence of pairwise distinct points and $h_j(t) := f(t - s_j)$, $j \in \mathbb{N}$. Then for all n, the spaces $H_n := \langle h_1, h_2, \ldots, h_n \rangle$ are Haar spaces with $H_n \subset H_{n+1}$. This is easy to see: A typical element of H_n is of the form

$$\eta_n := \sum_{j=1}^{n} a_j h_j, \quad a_j \in \mathbb{C}.$$

This expression can be given the form

$$\eta_n(t) = \sum_{j=1}^{n} a_j h_j(t) = \frac{1}{\prod_{j=1}^{n}(t - s_j)} \sum_{j=1}^{n} a_j \prod_{k \neq j}(t - s_k) =: \frac{p(t)}{q(t)},$$

where $p \in \Pi_{n-1}, q \in \Pi_n$. Since q has no zeros in D, η_n is well defined and η_n has at most $n - 1$ zeros in D, since p has at most $n - 1$ zeros in $\mathbb{C}$, provided p is non-trivial.

In this example $f(t) := 1/t$ universally generates Haar spaces in the real and in the complex case.

Example 6. Let $f(t) := \exp(-t^2)$, $t \in \mathbb{K}$. We consider spaces spanned by $h_j(t) := f(t - s_j)$ with arbitrary but pairwise distinct shifts $s_j \in \mathbb{K}$, $j \in \mathbb{N}$. We have $h_j(t) = \exp(-s_j^2)\exp(-t^2)\exp(2s_j t)$. Thus, for all $n \in \mathbb{N}$ we have

$$H_n := \langle h_1, h_2, \ldots, h_n \rangle = \exp(-t^2)\langle \exp(2s_1 t), \exp(2s_2 t), \ldots, \exp(2s_n t) \rangle.$$

Since e^z has no zeros for all $z \in \mathbb{C}$, the problem is reduced to the investigation of the space

$$\tilde{H}_n := \langle \exp(s_1 t), \exp(s_2 t), \ldots, \exp(s_n t) \rangle.$$

If $\mathbb{K} = \mathbb{R}$ the space $\tilde{H}_n$ belongs in the catalogue of well known Haar spaces, KARLIN & STUDDEN, [9, 1966, Example 1, p. 9-10]. If, however, $\mathbb{K} = \mathbb{C}$ let us consider $\tilde{H}_2$. According to Theorem 1, part iii, $\tilde{H}_2$ is a Haar space if and only if $\det \mathbf{M} \neq 0$, where $\mathbf{M} := (\exp(s_j t_k))$, $j, k = 1, 2$ for arbitrary

$t_1 \neq t_2$. Now, $\det \mathbf{M} = \exp(s_1t_1 + s_2t_2) - \exp(s_1t_2 + s_2t_1)$. This expression is zero if and only if $s_1t_1 + s_2t_2 = s_1t_2 + s_2t_1 + 2k\pi\mathrm{i}$ for some $k \in \mathbb{Z}$. This is equivalent to $(t_1 - t_2)(s_1 - s_2) = 2k\pi\mathrm{i}$. Now it is easy to see, that for $k \neq 0$ there are solutions with $t_1 \neq t_2$ for any given $s_1 \neq s_2$. One example with $k \neq 0$ is $s_1 = \pi\mathrm{i}/2, s_2 = -\pi\mathrm{i}/2$; $t_1 = k, t_2 = -k$. In the complex case the spaces $\tilde{H}_n$ are no longer Haar spaces for $n \geq 2$.

In this example $f(t) := \exp(-t^2)$ does not universally generate Haar spaces in the real and in the complex case.

§3. The Complex Analytic Case

We will denote by $\mathbb{D}$ the open, unit disk in $\mathbb{C}$ and correspondingly $\overline{\mathbb{D}}$ will denote the closed, unit disk.

Definition 7. Let $n \in \mathbb{N}$ be a fixed natural number. A function G defined on $\mathbb{C}\backslash\{0\}$ with values in $\mathbb{C}$ will be called an n-dimensional Haar space generator if for each set of n pairwise distinct points $s_1, s_2, \ldots, s_n \in \mathbb{C}\backslash\overline{\mathbb{D}}$, the functions h_j defined by $h_j(z) := G(z - s_j)$, $j = 1, 2, \ldots, n$ for $z \in \overline{\mathbb{D}}$ span an n-dimensional Haar space.

Example 8. Let $G(z) := z^{m-1}$ with $m \geq 1$ fixed. Then G is an m-dimensional Haar space generator but not an $(m + 1)$-dimensional Haar space generator. We leave open the case whether it is an ℓ-dimensional Haar space generator for $2 \leq \ell < m$.

Definition 9. A function G defined on $\mathbb{C}\backslash\{0\}$ with values in $\mathbb{C}$ will be called a universal Haar space generator if for each $n \in \mathbb{N}$, it is an n-dimensional Haar space generator. A universal Haar space generator will be abbreviated by UHG.

Theorem 10. *Let G be analytic on $\mathbb{C}\backslash\{0\}$. Then G is a UHG if and only if G is of the form*

$$G(z) := \frac{\mathrm{e}^{az+b}}{z}, \quad a \in \mathbb{C}, \, b \in \mathbb{C}. \tag{1}$$

Proof: a) Sufficiency of the condition: It is essentially the same proof as in Example 5. Fix a and $b \in \mathbb{C}$. Then the n functions generated by shifts from G have the form

$$h_j(z) := A\frac{\mathrm{e}^{a(z-s_j)}}{z - s_j}, \quad 1 \leq j \leq n, \quad A \neq 0.$$

They are linearly independent since the poles at s_j cannot be removed by non-trivial linear combinations $\eta_n := \sum_{j=1}^{n} \mu_j h_j$. Now we have $\eta_n(z) = \sum_{j=1}^{n} \mu_j h_j(z) = \mathrm{e}^{az}\frac{p(z)}{q(z)}$, where p is a polynomial of degree at most $n - 1$

and q a polynomial of degree n, and where the zeros of q are not in $\overline{\mathbb{D}}$. Therefore, each non-trivial η_n has at most $n-1$ zeros in $\mathbb{C}$ and hence in $\overline{\mathbb{D}}$. We conclude that G is a UHG.

b) Necessity of the condition: Suppose G is an analytic UHG. The proof that G is of the form (1) is separated into several Lemmata.

Lemma 11. G *does not vanish on* $\mathbb{C}\backslash\{0\}$.

Proof: Let $n = 1$, and suppose that G vanishes at $z_0 \neq 0$. If $|z_0| > 1$, then $\tilde{G}(z) := G(z + z_0) = G(z - (-z_0))$ vanishes at the origin where $-z_0 \notin \overline{\mathbb{D}}$. If $z_0 \in \overline{\mathbb{D}}\backslash\{0\}$, put $s = -\frac{2+|z_0|}{2|z_0|}z_0$. Then $s \notin \overline{\mathbb{D}}$ and $\tilde{G}(z) := G(z - s)$ vanishes at $z := \frac{-2+|z_0|}{2}\frac{z_0}{|z_0|}$ which is in $\mathbb{D}$. Both cases are hence excluded by the hypothesis that G is a UHG, and in particular a 1-dimensional Haar space generator. $\square$

Lemma 12. G *has the form*

$$G(z) := z^m e^{\phi(z)}, \quad z \in \mathbb{C}, \quad m \in \mathbb{Z}, \quad \text{where } \phi \text{ is an entire function.} \quad (2)$$

Proof: Define the function

$$F(z, t, s) := \frac{G(z - t)}{G(z - s)}, \quad |z| \le 1, |t| > 1, |s| > 1, s \neq t. \quad (3)$$

This definition is permissible since by Lemma 11, G has no zeros in $\mathbb{C}\backslash\{0\}$. We are studying the number of solutions of $F(z, t, s) = \mu$ with a constant $\mu \in \mathbb{C}\backslash\{0\}$ since this is equivalent to studying the number of zeros of the linear combination $G(z - t) - \mu G(z - s)$ of two shifts, where $t \neq s$ and $t, s \notin \overline{\mathbb{D}}$. The function G has by definition an isolated singularity at the origin. There are only three possibilities: (a) This singularity is removable, (b) this singularity is a pole, or (c) this singularity is an essential singularity. We shall now show, that the case (c) is not possible. Suppose that the origin is an essential isolated singularity for G. Choose $t > 1$ and $s > t + 10$ on the positive real axis such that $G(t - s)$ is well defined and, according to Lemma 11, $G(t - s) \neq 0$. Then, by Picard's Theorem, (AHLFORS [1, 1966, p. 297]) there is a $\mu \in \mathbb{C}$ such that $F(z, t, s) = \mu$ admits infinitely many solutions in z in every neighbourhood of $z = t$. Furthermore, there is a cone $\{z : |\arg(z - t) - \beta| < \frac{\pi}{4}\}$ ending at t which contains infinitely many zeros of $F(z, t, s) - \mu$. Put $t_1 := -(1 + \epsilon)e^{i\beta}$, $z_1 := z - t + t_1$ and $s_1 := s - t + t_1$. We may choose $\epsilon > 0$ sufficiently small such that $F(z_1, t_1, s_1) - \mu$ vanishes at several points in $\mathbb{D}$. Observe that t_1 and s_1 are outside $\overline{\mathbb{D}}$. Hence, G is not a 2-dimensional Haar space generator. Thus, G has an isolated singularity at the origin, which is either removable or a pole, and in addition G has no zeros in $\mathbb{C}\backslash\{0\}$. In other words, we have shown that G has to be of the form defined in (2). Furthermore, all F defined by (3) are meromorphic on $\mathbb{C}$ having at most one pole and $F(0, t, s) \neq 0$, $F(0, t, s) \neq \infty$. $\square$

Lemma 13. *Let $\mu \in \mathbb{C}\backslash\{0\}$ be fixed and let F be defined as in (3). Then $F - \mu$ has at most one zero in the disk $\Delta(p) := \{z : |z - p| \leq 1\}$ provided $|p| > |t| + |s|$. Furthermore, this zero is simple.*

Proof: Fix $|s| > 1, |t| > 1$, and $|p| > |t| + |s|$ and set $z_1 := z - p$, $t_1 := t - p$, $s_1 := s - p$. Then $F(z, t, s) - \mu$ contains at most one simple zero on the disk $\Delta(p)$. Indeed, $F(z, t, s) - \mu$ vanishes at a point $z^* \in \Delta(p)$ if and only if (setting $z_1^* := z^* - p$)

$$F(z_1^*, t_1, s_1) - \mu = F(z^* - p, t - p, s - p) - \mu = 0. \tag{4}$$

We have $|z_1^*| \leq 1, |t_1| = |p - t| \geq |p| - |t| > |t| + |s| - |t| = |s| > 1$ and analogously $|s_1| = |p - s| > 1$ which implies the above statement. $\square$

Lemma 14. *F defined in (3) is of the form*

$$F(z, t, s) = \frac{G(z - t)}{G(z - s)} = \left[\frac{z - t}{z - s}\right]^m e^{a(t,s)z^2 + b(t,s)z + c(t,s)}. \tag{5}$$

Proof: Consider the partition of $\mathbb{C}$ by the squares

$$R_{j,k} := \left\{(x, y) \in \mathbb{R}^2 : \frac{j}{2} \leq x \leq \frac{j+1}{2}, \quad \frac{k}{2} \leq y \leq \frac{k+1}{2}, \quad j, k \in \mathbb{Z}\right\} \tag{6}$$

and denote by $n(r, F)$ the number of zeros of the function $F(z, t, s) - \mu$ in the disk $\{z : |z| \leq r\}$ for a fixed $\mu \in \mathbb{C}\backslash\{0\}$. Fix an $r_0 \in \mathbb{N}$ with $r_0 > |t| + |s|$. Each closed disk $\{z : |z| \leq r\}$, $r \in \mathbb{N}$, is covered by $16r^2$ squares of the form (6). Moreover, $|n(r_0, F)| \leq M < \infty$ uniformly in μ. This follows directly from the argument principle for meromorphic functions, AHLFORS [1, 1966, p. 151]. Hence, by Lemma 13 we conclude that

$$n(r, F) = n(r_0, F) + [n(r, F) - n(r_0, F)] \leq n(r_0, F) + 16r^2 = O(r^2), \tag{7}$$

as $r \to \infty$. Observe that (7) holds for all $r > r_0$ and all $\mu \in \mathbb{C}$. Hence, by the first and second fundamental theorem of R. Nevanlinna (NEVANLINNA [14, 1953, p. 168/256]), we conclude that F is meromorphic on $\mathbb{C}$ of maximal order two, which implies that $\exp(\phi(z - t) - \phi(z - s))$ is an entire function of maximal order two. The same conclusion may also be obtained by Cartan's theorem (CARTAN, [2, 1928; 3, 1929]). Therefore, $\phi(z - t) - \phi(z - s)$ must be a polynomial of degree $d \leq 2$. In other words, F has to be of the form given in (5). $\square$

For more details of this proof, see Section 4: Appendix.

Lemma 15. G *is of the form*

$$G(z) = z^m e^{Az^3 + Bz^2 + Cz + D}, \quad m \in \mathbb{Z}, \quad A, B, C, D \in \mathbb{C}. \tag{8}$$

Proof: We evaluate the constants $a(t, s)$, $b(t, s)$, $c(t, s)$ occurring in (5) of Lemma 14. Let us define a polynomial (in z with parameters t, s) by

$$p(z; t, s) := \phi(z - t) - \phi(z - s) = a(t, s)z^2 + b(t, s)z + c(t, s).$$

This form is motivated by the proof of the previous Lemma 14. The polynomial p has the simple property $p(z; s, t) = -p(z; t, s)$, which implies the relations $a(t, s) = -a(s, t)$, $b(t, s) = -b(s, t)$ and $c(t, s) = -c(s, t)$ for all t, s. If we compute the derivatives of p with respect to z, t, s we obtain

$$\phi'(z - t) - \phi'(z - s) = 2a(t, s)z + b(t, s),$$
$$-\phi'(z - t) = a_t(t, s)z^2 + b_t(t, s)z + c_t(t, s),$$
$$\phi'(z - s) = a_s(t, s)z^2 + b_s(t, s)z + c_s(t, s).$$

If we compare the negative of the first equation with the sum of the second and third equation we obtain (omitting the arguments)

$$a_t + a_s = 0, \ b_t + b_s = -2a, \ c_t + c_s = -b.$$

Computing the second derivatives with respect to z, t, s we obtain

$$\phi''(z - t) - \phi''(z - s) = 2a(t, s),$$
$$\phi''(z - t) = a_{tt}(t, s)z^2 + b_{tt}(t, s)z + c_{tt}(t, s),$$
$$-\phi''(z - s) = a_{ss}(t, s)z^2 + b_{ss}(t, s)z + c_{ss}(t, s).$$

Comparing the sum of the last two equations with the first equation yields

$$a_{tt} + a_{ss} = 0, \ b_{tt} + b_{ss} = 0, \ c_{tt} + c_{ss} = 2a.$$

Combining these equations we obtain the partial differential equations:

$$a_{tt} = 0, \ a_{ss} = 0, \ b_{tt} = -2a_t, \ b_{ss} = -2a_s, \ c_{tt} = -b_t, \ c_{ss} = -b_s.$$

Taking the above mentioned symmetry into account, we find the solutions to be

$$a(t, s) = \alpha(t - s), \ b(t, s) = -\alpha(t^2 - s^2) + \beta(t - s),$$

$$c(t, s) = \frac{\alpha}{3}(t^3 - s^3) - \frac{\beta}{2}(t^2 - s^2) + \gamma(t - s),$$

with arbitrary factors α, β, γ. In (5) put $z := 0$ and fix $s := s_0$. Then we obtain

$$F(0, -t, -s_0) = \frac{G(t)}{G(s_0)} = \left[\frac{t}{s_0}\right]^m e^{-\frac{\alpha}{3}(t^3 - s_0^3) - \frac{\beta}{2}(t^2 - s_0^2) - \gamma(t - s_0)},$$

from whence we conclude that G is of the form (8). $\square$

We summarize our results so far.

Proposition 16. *Let G be simultaneously an analytic one- and two-dimensional Haar space generator. Then G has to be of the form given in (8).*

Proof: This is a consequence of Lemma 11 to Lemma 15 where only Haar space generators up to dimension two were used. $\square$

Lemma 17. *In G of (8) we have $A = 0$.*

Proof: Suppose that the analytic function G of (8) has the property that $A \neq 0$. We shall show in this case that G is not even a 2-dimensional Haar space generator. Indeed, we have $F(z, s, t) - 1 = 0$ is equivalent to

$$(z - t)^m e^{A(z-t)^3 + B(z-t)^2 + C(z-t) + D} = (z - s)^m e^{A(z-s)^3 + B(z-s)^2 + C(z-s) + D}$$

or for $k \in \mathbb{Z}$ to

$$A(t^2 + ts + s^2) - (3Az + B)(t + s) + (3Az^2 + 2Bz + C) = \frac{m \log\left[\frac{z-t}{z-s}\right] + 2mk\pi i}{t - s}.$$

We choose $s := s(t)$ in such a way that $A(t^2 + ts + s^2) - B(t + s)$ (regarded as a quadratic polynomial in s) vanishes. This yields (as one possible choice)

$$s(t) := -\frac{At - B}{2A}\left[1 - \sqrt{1 - \frac{4At}{At - B}}\right]. \tag{9}$$

Observe that $s(t)$ is close to $e^{2\pi i/3}t$ if t is very large. We have $F(z, s, t) - 1 = 0$ if and only if

$$z - \frac{3Az^2 + 2Bz + C}{3A(t + s(t))} + \frac{m \log\left[\frac{z-t}{z-s(t)}\right] + 2mk\pi i}{3A(t^2 - s^2(t))} = 0, \quad k \in \mathbb{Z}. \tag{10}$$

Define

$$f_k(z) := \frac{3Az^2 + 2Bz + C}{3A(t + s(t))} - \frac{m \log\left[\frac{z-t}{z-s(t)}\right] + 2mk\pi i}{3A(t^2 - s^2(t))}, \quad k \in \mathbb{Z}.$$

Then, for large t, we have $t + s \approx t(1 + e^{2\pi i/3})$ and $t^2 - s^2 \approx t^2(1 - e^{-2\pi i/3})$. Furthermore, $\log\left[\frac{z-t}{z-s(t)}\right]$ will be close to $-2\pi i/3$. Now, choose $k_0 \in \mathbb{N}$. Then for t very large, the functions $f_k, 0 \leq k \leq k_0$, are analytic in the closed unit disk $\overline{\mathbb{D}}$ and $0 < |f_k(z)| \leq \frac{1}{2}$ on $\overline{\mathbb{D}}$. Applying Rouché's theorem (AHLFORS [1, 1966, p. 152]) to $f(z) := z$ and f_k we conclude that for each $k, 0 \leq |k| \leq k_0$, the function $f - f_k$ vanishes exactly once in $\mathbb{D}$. Therefore, we obtain several z of the form (10) which belong to the unit disk. This contradicts the fact that G is a 2-dimensional and hence UHG. Therefore, A has to be zero in (8). $\square$

Lemma 18. *In G of (8) we have $B = 0$ and hence $\phi(z) := Cz + D$.*

Proof: We proceed the same way as in Lemma 17 assuming that $A = 0$. Suppose to the contrary, that $B \neq 0$ in (8). Then $F(z, s, t) - 1 = 0$ is equivalent to

$$(z - t)^m e^{B(z-t)^2 + C(z-t) + D} = (z - s)^m e^{B(z-s)^2 + C(z-s) + D}$$

or to

$$B(s + t) - (2Bz + C) + \frac{m \log\left[\frac{z-t}{z-s}\right] + 2mk\pi i}{t - s} = 0, \quad k \in \mathbb{Z}. \tag{11}$$

Choose $s(t) := \frac{C}{B} - t$. Then (11) is equivalent to

$$z = \frac{m \log\left[\frac{z-t}{z-s}\right] + 2mk\pi i}{2B(t - s)} = \frac{m \log\left[\frac{z-t}{z+t-\frac{C}{B}}\right] + 2mk\pi i}{2B(2t - \frac{C}{B})}. \tag{12}$$

Now, choose $k_0 \in \mathbb{N}$. Then for t sufficiently large, the functions

$$f_k(z) := \frac{m \log\left[\frac{z-t}{z+t-\frac{C}{B}}\right] + 2mk\pi i}{2B(2t - \frac{C}{B})}, \quad 0 \leq |k| \leq k_0,$$

are analytic in the closed unit disk $\overline{\mathbb{D}}$ and $0 < |f_k(z)| \leq \frac{1}{2}$ on $\overline{\mathbb{D}}$. Applying Rouché's theorem (AHLFORS [1, 1966, p. 152]) to $f(z) := z$ and f_k we conclude that for each $k, 0 \leq |k| \leq k_0$, the function $f - f_k$ vanishes exactly once in $\mathbb{D}$. Therefore, we obtain several z of the form (12) which belong to the unit disk. This contradicts that G is a 2-dimensional and hence, a UHG. $\square$

Lemma 19. *The exponent m of the representation (2) is $m = -1$.*

Proof: So far, we have shown that an analytic UHG has to be of the form

$$G(z) = z^m e^{az+b}, \quad m \in \mathbb{Z}, \quad a, b \in \mathbb{C}. \tag{13}$$

Our aim is to show that $m = -1$. If $m \geq 0$, then for $n > m + 1$, the functions

$$v_j(z) := (z - s_j)^m e^{a(z-s_j)+b}, \quad 1 \leq j \leq n$$

form a linearly dependent set. Therefore, m has to be a negative integer.

If $m = -2$, the linear combination

$$h_3(z) := e^{at-b} \frac{1}{(z-t)^2} e^{a(z-t)+b}$$

$$+ e^{ae^{2\pi i/3}t - b - 4\pi i/3} \frac{1}{(z - e^{-2\pi i/3}t)^2} e^{a(z - e^{-2\pi i/3}t)+b}$$

$$+ e^{ae^{-2\pi i/3}t - b + 4\pi i/3} \frac{1}{(z - e^{2\pi i/3}t)^2} e^{a(z - e^{2\pi i/3}t)+b}$$

vanishes exactly at the points z where

$$\frac{1}{(z-t)^2} + \frac{1}{(e^{2\pi i/3}z - t)^2} + \frac{1}{(e^{-2\pi i/3}z - t)^2} = 0.$$

This is the case if and only if $z^3 = \frac{-t^3}{2}$. For $t := 1.1$ we have three zeros in the unit disk which shows that the case $m = -2$ is excluded.

Now let m be a negative integer, $m \leq -3$ and fix t on the real axis, $t > 1$. Then the linear combination

$$h_2(z) := e^{-(b+at)}(z+t)^m e^{a(z+t)+b} + (-1)^m e^{-b+at}(z-t)^m e^{a(z-t)+b} \quad (14)$$

vanishes if and only if $(t+z)^m + (t-z)^m = 0$ or, equivalently, if and only if

$$\psi(z) := \frac{t+z}{t-z} = (-1)^{1/m} = e^{\frac{-\pi i}{|m|} + \frac{2\pi i k}{|m|}}, \quad 0 \leq k \leq |m| - 1. \quad (15)$$

Observe, that ψ is a univalent (conformal) mapping from the unit disk $\mathbb{D}$ to the disk

$$\Omega := \left\{ w : \left| w - \frac{t^2+1}{t^2-1} \right| < \frac{2t}{t^2-1} \right\}. \quad (16)$$

Fix $t := 1.05$. Then Ω contains the two points $w_{1,2} := e^{\pm \pi i/|m|}$, $m \leq -3$. In other words, h_2 vanishes at the two points $z_1 := \psi^{-1}(e^{\pi i/|m|})$ and $z_2 := \psi^{-1}(e^{-\pi i/|m|}) = -z_1$ which belong to $\mathbb{D}$. Therefore, the cases $m \leq -3$ are also excluded. The only remaining case is $m = -1$. $\square$

Combining all Lemmata we have shown that an analytic UHG is of the form (1), which ends the proof of the main theorem. $\square$

We can prove much more. In particular, we obtain the following striking result.

Theorem 20. *Let G be an analytic n-dimensional Haar space generator for $n = 1, 2, 3$ and 4. Then G is a UHG. This result is best possible in the sense that 4 cannot be replaced by a smaller number.*

Proof: Suppose that G is an analytic n-dimensional Haar space generator for $n = 1, 2, 3$ and $n = 4$. From $n = 1$, we conclude that G does not vanish on $\mathbb{C}\backslash\{0\}$. The next section for $n = 2$ in the above proof, shows that G is of the form (13). The exclusion of the cases $m \leq -2$ and $m = 0, 1, 2$ are based on the same reasoning as before and uses only $n = 2, 3$ and $n = 4$. The only modification we have to add is for the cases $m \geq 3$. We can use the same arguments as for the cases $m \leq -3$. The linear combination h_2 already defined in (14) vanishes if and only if $(t + z)^m + (t - z)^m = 0$ or, equivalently, if and only if

$$\psi(z) := \frac{t + z}{t - z} = (-1)^{1/m} = e^{\frac{\pi i}{m} + \frac{2\pi i k}{m}}, \quad 0 \leq k \leq m - 1.$$

Then ψ is a univalent (conformal) mapping from the unit disk $\mathbb{D}$ to the disk Ω defined in (16). Fix $t := 1.05$. Then Ω contains the two points $w_{1,2} := e^{\pm \pi i/m}, m \geq 3$. In other words, h_2 vanishes at the two points $z_1 := \psi^{-1}(e^{\pi i/m})$ and $z_2 := \psi^{-1}(e^{-\pi i/m}) = -z_1$ which belong to $\mathbb{D}$. In all the arguments we used only the fact that G was an analytic n-dimensional Haar space generator with $n = 1, 2, 3$ and 4.

It remains to show that this result is best possible. Consider $G(z) = z^2$. We shall see that G is an analytic n-dimensional Haar space generator with $n = 1, 2, 3$ but not with $n = 4$ (compare Example 8).

Since $G(z) \neq 0$ for $z \neq 0$ we conclude that G is a 1-dimensional Haar space generator. Next, suppose that G is not a 2-dimensional Haar space generator. Then there must be a $|t| > 1$, an $|s| > 1$ and a $\mu \in \mathbb{C}$ such that there are at least two points z_1 and z_2 with

$$k(z) := \frac{t - z}{s - z} = \pm\sqrt{\mu}. \tag{17}$$

Since $w = 0$ and $w = \infty$ are not in the closure of the image $k(\mathbb{D})$ we conclude that $k(\mathbb{D})$ is a disk which cannot contain μ and $-\mu$. Therefore, G is a 2-dimensional Haar space generator. Next, observe that any linear combination of $(z-t)^2, (z-s)^2$ and $(z-u)^2$, where t, s and u are mutually distinct, cannot have more than two zeros in $\mathbb{D}$. Hence, G is a 3-dimensional Haar space generator. Finally, any four functions $(z - t_j)^2$, $j = 1, 2, 3, 4$ are linearly dependent and hence, G is not a 4-dimensional Haar space generator. $\square$

Theorem 20 may be given the following form.

Corollary 21. *Let G be analytic in $\mathbb{C}\backslash\{0\}$. For arbitrary pairwise distinct points $s_j \in \complement\overline{\mathbb{D}}$ and for arbitrary pairwise distinct $t_j \in \mathbb{D}$ we define the matrix*

$$\mathbf{M} := (m_{jk}) := \big(G(t_k - s_j)\big), \quad j, k = 1, 2, \ldots, n.$$

If $\mathbf{M}$ is non singular for $n = 1, 2, 3, 4$, then it is non singular for all $n \in \mathbb{N}$ and G has necessarily the form given in (1). The number $n = 4$ cannot be replaced by a smaller number.

Proof: Application of Theorem 1, part iii. to Theorem 20. $\square$

If in the definition (1) of G the constant a vanishes, the above matrix $\mathbf{M}$ is a so-called Cauchy matrix, see KNUTH [10, 1969, p. 36, 473].

Corollary 22. Let $m \geq 4$ and $g(z) := z^{m-1}$, $s_j \in \mathbb{C}\overline{\mathbb{D}}$, $j = 1, 2, \ldots, m+1$, $s_j \neq s_k$ for $j \neq k$. Define the spaces H_j on $\mathbb{D}$ by

$$H_j := \langle g(z - s_1), g(z - s_2), \ldots, g(z - s_j) \rangle, \quad j = 1, 2, \ldots, m + 1. \quad (18)$$

Then H_2 and H_{m+1} are not Haar spaces.

Proof: For H_2 we use the proof of Theorem 20 again, in which it is shown directly, that H_2 for $m \geq 4$ is not a Haar space. Since H_{m+1} has (at most) dimension m, it cannot be a Haar space. $\square$

It should be remarked that in the situation of Corollary 22 the spaces $H_1 := \langle (z - s_1)^m \rangle$ and $H_m (= \Pi_{m-1})$ are always Haar spaces. Since the sufficiency proof of Theorem 10 does not depend on any specific domain of definition, we have the following corollary.

Corollary 23. Let $D \subset \mathbb{C}$ be any compact set, $S := \mathbb{C}D$, G defined as in (1), and $n \in \mathbb{N}$. Then for any selection of n mutually disjoint points $s_j \in S$ the space $H_n := \langle h_1, h_2, \ldots, h_n \rangle$ is a Haar space with domain of definition D, where $h_j(z) := G(z - s_j)$, provided D contains at least n points.

Proof: A direct repitition of part a) of the proof of Theorem 10 suffices. $\square$

By Theorem 10 we also can identify many non-Haar spaces. One set of examples is produced by $g(z) := \frac{1}{z^m}$, $m \geq 2$ and the corresponding spaces generated by shifts. For $m = 2$ this was shown by RAHMAN & RUSCHEWEYH [15, 1983]. In the real case, the spaces $H_n := \langle g(z - s_1), g(z - s_2), \ldots, g(z - s_n) \rangle$ defined on $I := [-1, 1]$ are Haar spaces for all $m \geq 2$, provided, $|s_j| > 1$ and the points $s_1, s_2, \ldots, s_n$ are mutually distinct. Actually, in the mentioned literature and in the forthcoming book by MEINARDUS & WALZ [13, $\geq$ 2002, p. 104] the spaces spanned by

$$v_j(x) := \left(\frac{1}{1 - x_j x} \right)^m, \quad 0 < |x_j| < 1, \quad x \in [-1, 1] \quad (19)$$

were considered. But if we put $s_j := 1/x_j$, then

$$\widetilde{v}_j(x) := \left(\frac{1}{x - s_j} \right)^m, \quad |s_j| > 1, \quad x \in [-1, 1] \quad (20)$$

and v_j of (19) span the same spaces. We could even admit $s_1 := \infty$ which would lead to $\widetilde{v}_1 := 1$ corresponding to $x_1 = 0$. It is actually not so easy to show that v_j of (19) generate real Haar spaces on $[-1, 1]$, MEINARDUS & WALZ [13, ≥ 2002, Section 3.5].

Though the final result (Theorem 20) is at a first glance surprising, there is another result mentioned by HAYMAN [7, 1964, Theorem 2.6, p. 48], saying that two meromorphic functions f_1, f_2 in the complex plane coincide if $\{z : f_1(z) = a\} = \{z : f_2(z) = a\}$ for five different values of a, where five cannot be replaced by a smaller number.

§4. Appendix: Some Details with respect to Lemma 14

Let us recall that G analytic on $\mathbb{C}\backslash\{0\}$ has no zeros (Lemma 11) and that the origin is possibly a pole of G, but not an essential singularity (Lemma 12). In (3) we defined the function F, which played an important role at various places as follows:

$$F(z, t, s) := \frac{G(z - t)}{G(z - s)}, \quad |z| \leq 1, \ |t| > 1, \ |s| > 1, \ s \neq t. \qquad (3)$$

Since G may be regarded as a meromorphic function on all of $\mathbb{C}$ this applies to F, too. In addition, we have $F(0, t, s) \neq 0$, $F(0, t, s) \neq \infty$.

Now, let f be any meromorhic function on $\mathbb{C}$ with $f(0) \neq 0, f(0) \neq \infty$. We define the following quantities:

(1) $\log^+ \alpha := \max(0, \log \alpha)$ for all $\alpha \geq 0$, in particular, $\log^+ 0 = 0$,

(2) $M(r, f) := \max_{|z|=r} |f(z)|$, $r \geq 0$,

(3) $n(r, f) := \#\{z : f(z) = 0, |z| < r\}$ multiplicities counted, where $\#$ stands for *number of*,

(4) $N(r, f) := \int_0^r \frac{n(\sigma, f)}{\sigma} \, d\sigma$,

(5) $m(r, f) := \frac{1}{2\pi} \int_0^{2\pi} \log^+ |f(re^{it})| \, dt$,

(6) $T(r, f) := m(r, f) + N(r, \frac{1}{f})$, the so-called Nevanlinna characteristic,

(7) let h be positive on the non-negative real axis. Then

$\varrho(h) := \overline{\lim}_{r \to \infty} \frac{\log^+ h(r)}{\log r}$ is called the order of h. This number may be infinite, e. g. $h(r) := \exp(e^r)$, or finite, e. g. $h(r) = r^p, p \in \mathbb{Z}$ yields

$$\varrho(h) = \overline{\lim}_{r \to \infty} \frac{\log^+ r^p}{\log r} = \begin{cases} p & \text{for } p > 0, \\ 0 & \text{for } p \leq 0. \end{cases}$$

If the order is finite, then one can show that there are constants C, D such that

$\varrho(h) = \inf\{A : h(r) < Ce^{Ar} + D \text{ for all } r > 0\}$.

(8) $\varrho(f) := \varrho(T(r, f))$ is defined as the order of a meromorhic function f on $\mathbb{C}$.

   (8.1) If f is entire, then $\varrho(f) := \varrho(m(r, f))$, since in this case $n(r, \frac{1}{f}) = 0$, and therefore also $N(r, \frac{1}{f}) = 0$. In addition, one can show that in this case $T(r, f) \leq \log M(r, f) \leq \frac{R+r}{R-r} T(R, f)$, $R > r$.

(8.2) As an easy exercise we find $\varrho(z^p) = 0$ for all $p \in \mathbb{N}$.

(8.3) In the case $f(z) := z^{-p}$, $p \in \mathbb{N}$ we have $m(r, f) = 0$, $N(r, \frac{1}{f}) = N(r, z^p) = p \log r$ and thus, $\varrho(f) = \lim_{r \to \infty} \frac{\log^+ p \log r}{\log r} = 0$.

(8.4) For the entire function $f(z) := \exp(z^p)$ and $p \in \mathbb{N}$, we have $\varrho(f) := \varrho(m(r, f))$ and $m(r, f) = \frac{1}{2\pi} \int_0^{2\pi} \log^+ |\exp(r^p e^{ipt})| \, dt = \log^+ e^{r^p \cos pt} = r^p \cos^+ pt$, where $\cos^+ \alpha := \max(0, \cos \alpha)$. Now, $\int_0^{2\pi} \cos^+ pt \, dt = 2p \sin \frac{\pi}{2p}$, and it follows that $\varrho(\exp(z^p)) = \lim_{r \to \infty} \frac{\log^+ r^p \frac{p}{\pi} \sin \frac{\pi}{2p}}{\log r} = p$.

(9) The first fundamental theorem by R. Nevanlinna says:
$T(r, f) = T(r, \frac{1}{f-a}) + O(1)$ for all $a \in \mathbb{C}$.

If we apply the above results to $f(z) := F(z, t, s)$ we obtain:

(I) $n(r, F(z, t, s) - \mu) \leq M + 16r^2 \Rightarrow N(r, F(z, t, s) - \mu)$ is of order at most two for all μ,

(II) Cartan: $T(r, f) = \frac{1}{2\pi} \int_0^{2\pi} N(r, f(re^{it}) - e^{it}) \, dt + \log^+ |f(0)| \Rightarrow F(z, t, s)$ is of order at most two,

(III) the function

$$H(z, t, s) := \frac{(z-s)^m}{(z-t)^m} F(z, t, s) = \frac{(z-s)^m}{(z-t)^m} \frac{G(z-t)}{G(z-s)}$$

is an entire, nonvanishing function of order $\varrho \leq 2 \Rightarrow \varrho = 0$ or $\varrho = 1$ or $\varrho = 2 \Rightarrow F(z, t, s)$ is of the form (5).

More details can be found in the beginning two chapters of the book by HAYMAN [7, 1964].

Acknowledgments. We would like to thank Professor Günter Meinardus, University of Mannheim, who brought the paper by RAHMAN & RUSCHEWEYH to our attention and made available parts of a developing book by MEINARDUS & WALZ. We also would like to thank Professor Ron B. Guenther, Oregon State University, Corvallis, Oregon for carefully reading our manuscript.

References

1. Ahlfors, L. V., *Complex Analysis*, 2nd ed., McGraw-Hill, New York, 1966.

2. Cartan, H., Sur les systèmes de fonction holomorphes à variétés lacunaires et leurs applications. Thèse, Paris, 1928.

3. Cartan, H., Sur la fonction de croissance attachée à une fonction méromorphe de deux variables, et ses applications aux fonctions méromorphes d'une variable, C. R. Acad. Sci. Paris, **189** (1929), 521–523.

4. Cheney, E. W. and W. Light, *A Course in Approximation Theory*, Brooks/Cole, Pacific Grove, 2000.

5. Curtis, P. C., Jr., *n*-Parameter families and best approximation, Pacific J. Math., **IX** (1958), 1013–1027.

6. Dunham, C. B., Families satisfying the Haar condition, J. Approx. Theory, **12** (1974), 291–298.

7. Hayman, W. K., *Meromorphic Functions*, Clarendon Press, Oxford, 1964, 191 p.

8. Henderson, G. W., and B. R. Ummel, The nonexistence of complex Haar systems on nonplanar locally connected spaces, Proc. Amer. Math. Soc. **39** (1973), 640–641.

9. Karlin, S. and W. J. Studden, *Tchebycheff Systems: with Applications in Analysis and Statistics*, Interscience, New York, 1966, 586 p.

10. Knuth, D. E., *The Art of Computer Programming*, Vol. 1, Fundamental algorithms, Addison-Wesley, Reading, Mass., 1969.

11. Mairhuber, J. C., On Haar's theorem concerning Chebyshev approximation having unique solutions, Proc. Amer. Math. Soc. **7** (1956), 609–615.

12. Meinardus, G., *Approximation of Functions: Theory and Numerical Methods*, Springer, Berlin, Heidelberg, New York, 1967, 198 p. Translation by Larry L. Schumaker.

13. Meinardus, G. and G. Walz, *Approximation Theory*, to appear, Springer, Berlin, Heidelberg, New York, $\geq$ 2002, x p.

14. Nevanlinna, R., *Eindeutige analytische Funktionen*, 2. Aufl., Springer, Berlin, 1953.*

15. Rahman, Q. I. and St. Ruscheweyh, On the zeros of rational functions arising from certain determinants, in: Z. DITZIAN, A MEIR, S. D. RIEMENSCHNEIDER, A. SHARMA (eds.): Second Edmonton Conference on Approximation Theory [1982], AMS, Providence, RI, 1983, 333–339.

16. Schoenberg, I. J. and C. T. Yang, On the unicity of solutions of problems of best approximation, Ann. Mat. Pura Appl. **54** (1961), 1–12.

---

*Readers who prefer to read the English translation:

14′. R. Nevanlinna, *Analytic functions*, Springer, Berlin, 1970, 373 p. should definitely read the remarks by the reviewer, A. J. Lohwater, MR **43**, 1972, #5003, p. 917/918.

Walter Hengartner
Université de Laval
Département de Mathématique
Québec G1K 7P4, Canada
walheng@mat.ulaval.ca

Gerhard Opfer
University of Hamburg
Department of Mathematics
Bundesstr. 55
20146 Hamburg, Germany
opfer@math.uni-hamburg.de

Optimal Location with Polyhedral Constraints

R. Huotari, M. P. Prophet, and J. Ribando

Abstract. A Chebyshev center of a set A in a normed space is a location that minimizes the maximum distance to the set A. In many applications this center may be regarded as an **optimal location**. In this paper we demonstrate a finite algorithm for determining the optimal location subject to the constraint of a polyhedral set. This algorithm utilizes the procedure given in [6] for computing linearly constrained optimal locations.

§1. Introduction

If a subset A of a normed linear space has a smallest possible bounding sphere, the center of such a sphere is called a **Chebyshev center** of A. This notion has generated broad and sustained interest dating back to at least the 17th century, with recognition by Fermat and Sylvester. More modern results include significant theoretical advancements by A.L. Garkavi [5] and Amir and Ziegler [1,2]. Chebyshev-center characterizations of Hilbert space are given in [2,6]. Characterizations of the constrained Chebyshev center, and of its continuity, appear in [7,8,9,13]. In current applications it is common to refer to the Chebyshev center as an **optimal location**. Since the present paper is intended to be of use to those who apply the theory, we will adopt this mode of reference. The use of this terminology is found in industrial applications of the problem, which involve the determination of location of a new facility that will minimize maximum response time in servicing customers. See [4] for an overview of problems involving optimal location and [10] for several specific examples of solutions to optimal location problems. In general, **location theory** is the study of a set of optimal

Approximation Theory X: Abstract and Classical Analysis   239
Charles K. Chui, Larry L. Schumaker, and Joachim Stöckler (eds.), pp. 239–246.
Copyright © 2002 by Vanderbilt University Press, Nashville, TN.
ISBN 0-8265-1415-4.

locations associated with a particular optimization problem. See [12] for a large bibliographic overview of modern location theory.

In the following, $(X, \| \cdot \|)$ will denote a real inner-product space and A a finite subset of X. When we refer to an **algorithm**, we will always mean a procedure terminating with an exact answer after the execution of a finite number of steps. We are interested in algorithms that determine an optimal location of A relative to a constraint set. For example, if the constraint set is X itself, then the optimal location (commonly known in this case as the **absolute Chebyshev center**) is a distinguished element of the convex hull of A. A clever algorithm given in [3] iteratively determines this point by moving from each iterate x_j toward the set of points of A most distant from x_j. Among the attractive features of the algorithm in [3] is the ease with which it can be implemented on a computer. In [6] and [7] there is an investigation into generalizing the algorithm of [3] to handle constraint sets that are finite dimensional subspaces. Specifically, [6] shows how to metrically symmetrize A with respect to a constraint subspace V and then apply an unconstrained algorithm (such as in [3]) to obtain the V-relative optimal location of A.

In the current paper we seek an algorithm to determine the optimal location of A relative to a finite dimensional polyhedral set K; that is, K is a convex set of finite dimension possessing a finite number of faces. Because of our stipulations on A and K, we may assume that $X = \mathbb{R}^n$ for some n. (The assumption that K has finite dimension is not necessary, but we believe that K will have this property in most applications.) A polyhedral constraint set such as K offers challenges different from those occurring with linear constraint sets. For example, there is no natural 'K-symmetrization' of A as in [6]. It is the case, of course, that the boundary of K consists of a finite number of j-faces, $j = 0, 1, \ldots, n$, (where j denotes the dimension of the face) and as such there is a 'brute force' method of searching for the optimal location that is described as follows: determine the optimal location of A constrained to the affine span of *every* j-face, noting whether or not this location belongs to K. If the optimal location of A is in the interior of K, then it will agree with the unconstrained optimal location. Otherwise, the optimal location belongs to ∂K, and in particular, to the interior of a j-face of K ($j = 0$ is possible). Thus the optimal location relative to K is the solution to a linearly-constrained problem. This approach is obviously quite inefficient.

The algorithm given in Section 3 below performs an organized search of faces, beginning with the facets (highest dimensional faces) of K. It relies on the determination of optimal locations of A constrained by affine sets and half-spaces. Both of these constraint sets can be handled by a slightly modified version of the linearly constrained algorithm of [6]. For the reader's convenience we include a brief description of the linearly constrained algorithm in Section 2 below.

The idea which inspired our polyhedrally constrained algorithm is as follows. Suppose F and G are two convex sets in a metric space M, $x \in M$, and y (respectively, z, w) is a nearest point to x from F (resp., G, $F \cap G$). The idea is simply: if $y \in G$ then $y = w$. The algorithm uses an enumeration $\{H_1, \ldots, H_\sigma\}$ of half-spaces of which K is the intersection. It computes the optimal location of A relative to H_1 then uses the above idea to compute the optimal location of A relative to $H_1 \cap H_2$. The algorithm continues in this fashion to compute the optimal location of A in $H_1 \cap H_2 \cap H_3$ and so on. If at step j a new face K_j is created but the new optimal location is neither in the interior of H_j nor in the interior of K_j, it is necessary to nest a call of the algorithm with K replaced by the lower-dimensional set K_j.

The precise definition of our algorithm's target is as follows. We say that x^* is an **optimal location** (or **absolute Chebyshev center**) of A if for every $x \in X$

$$\max\{\|a - x^*\| : a \in A\} \leq \max\{\|a - x\| : a \in A\}, \tag{1}$$

and by $\mathcal{Z}(A)$ we denote the set of all optimal locations of A. In our setting $\mathcal{Z}(A)$ is always a singleton. If, in addition, K is a convex subset of X, we say that $x^* \in K$ is a K-**constrained optimal location** (or K-**relative Chebyshev center**) of A if (1) holds for every $x \in K$. In this case we denote the set of all K-constrained optimal locations of A by $\mathcal{Z}_K(A)$.

§2. An Algorithm for Linear Constraints

The algorithm described in this section first appeared in [6] as a generalization of the unconstrained 'Botkin algorithm' appearing in [3]. Indeed, the work in [6] generalizes the Botkin algorithm to the case where minimizing candidates must lie in a given affine set. Suppose $A = \{a_1, \ldots a_n\} \subset X$ and V is an affine subset of X. Let $x^* := \mathcal{Z}_V(A)$. We now wish to 'symmetrize' A with respect to V. Specifically, for each $a \in A$ let v_a denote the metric projection of a onto V and define $a_V := a + 2(v_a - a)$. Our symmetrization is accomplished by defining $A_V := A \cup \{a_V \mid a \in A\}$. The V-symmetrized far points of a point $x \in X$ will be denoted by $E_V(x) := \{a \in A_V : \|x - a\| = \max_{w \in A_V} \|x - w\|\}$. The idea behind the algorithm is as follows: from a current approximation x_k, let y_k be the best approximation to x_k from the convex hull of $E_V(x_k)$. Move from x_k to y_k along a straight line; x_{k+1} is defined to be the first point along the line at which $E_V(x_{k+1}) \neq E_V(x_k)$. If y_k is reached before such a point is found, then y_k is returned as the Chebyshev center. All iterates belong to V.

The Linearly Constrained Algorithm (LCA)

0. Let $k = 0$. Choose $a_0 \in A$; let $x_0 = \mathcal{P}_V(a_0)$.

1. If $E(x_k) = A$, stop (with $x^* := x_k$).

2. Let $y_k := Z_V(E(x_k))$, where $Z_V(E(x_k))$ is determined by calling LCA with $A := E(x_k)$. If $y_k = x_k$, stop (with $x^* := x_k$).

3. Let $x_{\alpha_k} := x_k + \alpha_k(y_k - x_k)$. Choose $\alpha_k \geq 0$ such that

$$\max_{a \in E(x_k)} \|x_{\alpha_k} - a\| = \max_{w \in A - E(x_k)} \|x_{\alpha_k} - w\|.$$

If $\alpha \geq 1$ stop (with $x^* := y_k$).

4. Let $x_{k+1} := x_k + \alpha_k(y_k - x_k)$, $k := k + 1$. Go to Step 1.   □

In [6] it was shown that LCA converges to $\mathcal{Z}_V(A)$; i.e., LCA$(A) = \mathcal{Z}_V(A)$. The following corollary demonstrates that LCA can be applied to constraint sets other than affine.

Corollary. *If $K \subset \mathbb{R}^n$ is a closed halfspace or a line segment, then $\mathcal{Z}_K(A)$ can be calculated using LCA.*

Proof: If K is a closed half-space of an affine set E, then the K-constrained optimal location of A is the best of the following two locations: the E-constrained optimal location of A (obtained via LCA) or the G-constrained optimal location of A, where G is the E-relative boundary of K. Since G is an affine set, $\mathcal{Z}_G(A)$ is determined via LCA. For a segment of a line, it is only necessary to consider the entire line and the endpoint(s) of the segment. □

§3. An Algorithm for Polyhedral Constraints

In this section we generalize the constraint set to an arbitrary polyhedral convex set K. Assuming knowledge of a finite number of half-spaces of which K is the intersection, and the availability of the LCA algorithm, we describe a way to calculate the K-constrained optimal location. For $0 \leq \alpha \leq n$ let $\mathcal{A}_\alpha$ consist of all polyhedral convex subsets of $\mathbb{R}^n$ of dimension α, and let $\mathcal{A} := \bigcup \{\mathcal{A}_\alpha : 0 \leq \alpha \leq n\}$. For every $C \in \mathcal{A}$ let aff(C) denote the affine hull of C. Let $\mathcal{H}(C)$ denote a finite choice of closed half-spaces of aff(C) such that $C = \bigcap \{H : H \in \mathcal{H}(C)\}$. We will denote by relint(C) the interior of C relative to aff(C). By the corollary at the end of Section 2, we have the ability to compute, as needed, $P(B) := \mathcal{Z}_B(A)$ for any B that is a closed half-space or line segment. Let $K' := K$.

The Polyhedrally Constrained Algorithm (PCA)

1. Suppose $\mathcal{H}(K') = \{H_1, \ldots, H_\sigma\}$. Let $G_0 := \mathrm{aff}(K')$.

2. If $P(G_0) \in K'$, let $z^* := P(G_0)$. Go to Step 9.

3. If $\dim(K') = 1$, let $z^* := P(K')$. Go to Step 9.

4. Let $G_1 = H_1$. Let $i := 0$.

5. Let $i := i + 1$. Let $G_{i+1} := G_i \cap H_{i+1}$.

6. If $P(G_i) \in \mathrm{relint}(H_{i+1})$, then $P(G_{i+1}) := P(G_i)$ and go to Step 8.

7. Call PCA (with $K' := \partial H_{i+1} \cap G_i$). Let $P(G_{i+1}) := \mathrm{PCA}(\partial H_{i+1} \cap G_i)$.

8. Let $z^* := P(G_{i+1})$. If $i + 1 < \sigma$, go to Step 5.

9. $\mathrm{PCA}(K') := z^*$. Stop.   □

We now prove that the PCA converges. The existence of nested inductions in the following proof mirrors the fact that the algorithm calls itself.

Theorem. *If $0 \leq \alpha \leq n$ and $K \in \mathcal{A}_\alpha$, then the algorithm PCA converges to $\mathcal{Z}_K(A)$.*

Proof: We will denote by $\mathrm{PCA}(Y)$ the result of the algorithm PCA when the constraint set Y is used. We now proceed by induction on α. The result is true by definition if $\alpha = 0$ or $\alpha = 1$. Suppose $\beta > 1$ and the theorem is true for every element of $\bigcup \{\mathcal{A}_\alpha : 0 \leq \alpha < \beta\}$. Suppose $K \in \mathcal{A}_\beta$. Let $\{H_i : i = 1, \ldots, \sigma\}$ and $\{G_m : m = 0, \ldots, \sigma\}$ be defined as in the description of the algorithm.

We now claim that for every $m \in \{0, \ldots, \sigma\}$,

$$P(G_m) = z_m^* := \mathcal{Z}_{G_m}(A) \tag{2}$$

We will prove the claim using induction on m. Indeed, for $m = 0$ the statement is true by definition. Let $k > 0$ and assume $P(G_{k-1}) = z_{k-1}^*$. We now consider the two cases described in Steps 6 and 7.

Suppose $z_{k-1}^* \in \mathrm{relint}(H_k)$. Then $P(G_k) = z_{k-1}^*$ from Step 6. We claim $z_k^* = z_{k-1}^*$. Indeed, since $\{z_k^*, z_{k-1}^*\} \subset G_k$ we must have z_k^* no further from A than is z_{k-1}^*. On the other hand, $G_k \subset G_{k-1}$ and thus z_{k-1}^* must be no further from A than is z_k^*. Thus we must have $z_{k-1}^* = z_k^*$ and, for this case, (2) is established.

Consider now the case $z_{k-1}^* \notin \mathrm{relint}(H_k)$. Let $J := \partial H_k \cap G_{k-1}$. Then, from Step 7, $P(G_k) = \mathrm{PCA}(J)$ and, since $\dim(J) < \beta$ (recall $H_i \in \mathrm{aff}(K)$) it must be true that $\mathrm{PCA}(J) = \mathcal{Z}_J(A)$. Thus $P(G_k) = \mathcal{Z}_J(A)$. We claim

$$\mathcal{Z}_J(A) = z_k^*. \tag{3}$$

Recall that, by definition, $z_k^* \in G_k = H_k \cap G_{k-1}$. If $z_k^* \in \partial H_k$ then this implies $z_k^* \in J$ and (3) is valid. Let us now assume $z_k^* \in \mathrm{relint}(H_k)$. We are assuming $z_{k-1}^* \notin \mathrm{relint}(H_k)$; let us suppose $z_{k-1}^* \in \partial H_k$. Then $z_{k-1}^* \in J$ and thus $z_{k-1}^* = \mathcal{Z}_J(A)$. But $J \subset G_k \subset G_{k-1}$ and therefore $z_k^* = z_{k-1}^* = \mathcal{Z}_J(A)$, establishing (3) for $z_{k-1}^* \in \partial H_k$. The last case to consider is that of z_{k-1}^* belonging to $(H_k)^c$, the complement of H_k, together with $z_k^* \in \mathrm{relint}(H_k)$. But this situation is not possible. Indeed, if $z_k^* \in \mathrm{relint}(H_k)$, then the line segment joining z_k^* and z_{k-1}^* (which must belong to G_{k-1} by convexity) would nontrivially intersect a G_k-relative neighborhood of z_k^* (recall once more that this is taking place in $\mathrm{aff}(K)$). The distance to A must decrease as one moves from z_k^* toward z_{k-1}^*. And this implies there is a point of G_k closer to A than z_k^*, a contradiction. Thus in all possible cases we have established (3) and, since

$$P(G_k) = \mathrm{PCA}(J) = \mathcal{Z}_J(A),$$

we have demonstrated the validity of (1) for every $m \in \{0, \ldots, \sigma\}$. In particular, since $G_\sigma = K$, we have

$$\mathrm{PCA}(K) = P(G_\sigma) = z_\sigma^* = \mathcal{Z}_K(A),$$

and the proof of the theorem is complete. $\square$

§4. Final Comments

The Polyhedral Constrained Algorithm (PCA) described above has yet to be implemented on a computer. It would be interesting to investigate the efficiency of such an implementation. There are several ways one can describe the efficiency, or complexity, of a proposed algorithm. For example, in [3] the complexities of several algorithms that solve optimal location problems are described via numerical data (produced via experimentation); the authors offer total run times and total iteration counts for various point sets belonging to finite-dimensional spaces of various sizes. In [7] efficiency of an algorithm is measured by bounding the number of approximations produced before the optimal location is reached. In that paper it was shown that the efficiency of the Linearly Constrained Algorithm (LCA) is bounded by a polynomial in $|A|$. The complexity of the PCA is a function of the number of constraints defining the constraint set as well as $|A|$. We conjecture that PCA is polynomial in $|A|$ and the number of linear inequalities defining K.

As LCA is used in PCA, it may be possible to incorporate PCA into the task of computing $\mathcal{Z}_K(A)$ for an arbitrary closed, convex set K. For example, let us assume that the best-approximations to the points of A from K are known (or can be computed). Let x_i denote the current guess to $\mathcal{Z}_K(A)$. We move, as in LCA, as far as we can in K toward x_i's farthest points, making a turn whenever LCA indicates, (assuming you can do so while staying in K) stopping at x_{i+1}. Is x_{i+1} the absolute Chebyshev center, or is x_{i+1} in the projection onto K of the convex hull of x_{i+1}'s far points? If the answer to either is yes, then we're done; otherwise we continue. Let K_{i+1} be the convex hull of the projections onto K of x_{i+1}'s far points. Use the PCA to find the K_{i+1} relative Chebyshev center, x_{i+2}. Let $i := i + 2$ and repeat. We conjecture that this process converges to $\mathcal{Z}_K(A)$ after a finite number of iterations.

References

1. Amir, D., *Characterizations of Inner Product Spaces*, Birkhauser Verlag, Basel, 1986.

2. Amir, D., and Z. Ziegler, Relative Chebyshev centers in normed linear spaces, I, J. Approx. Theory **29** (1980), 235–252.

3. Botkin, N. D., and V. L. Turova-Botkina, An algorithm for finding the Chebyshev center of a convex polyhedron, Appl. Math. and Optimiz. **29** (1994), 211–222.

4. Durier, R., Optimal locations and inner products, J. Math. Anal. Appl **207** (1997), 220–239.

5. Garkavi, A. L., The best possible net and the best possible cross-section of a set in a normed space, Amer. Math. Soc. Transl. **39** (1964) 111–132.

6. Huotari, R., and M. Prophet, Constrained optimal location, Numer. Func. Anal. Optim., to appear.

7. Huotari, R., and M. Prophet, On a constrained optimal location algorithm, J. Comput. Appl. Math., to appear.

8. Huotari, R., M. Prophet, and J. Shi, A farthest-point characterization of the relative Chebyshev center, Bull. Austral. Math. Soc. **54** (1996) 27–33.

9. Huotari, R., and S. Sahab, Strong unicity versus modulus of convexity, Bull. Austral. Math. Soc. **49** (1994), 305–310.

10. McAsey, M., and L. Mou, Existence and characterization of optimal locations, Journal of Global Optimizations **15** (1999), 85–104.

11. Megiddo, N., Linear-time algorithms for linear programming in R^3 and related problems, SIAM J. Comput **12** (1983), no. 4, 759–776.

12. Plastria, F., Continuous location problems, in *Facility Location: A Survey of Applications and Methods*, Z. Drezner, (ed.), Springer-Verlag, New York, 1995, 225–262.

13. Shi, S., and R. Huotari, Simultaneous approximation from convex sets, Computers Math. Applic. **30** (1995), 197–206.

Robert Huotari
Mathematics Department
Glendale Community College
Glendale, AZ 85302
r_huotari@yahoo.com

Michael P. Prophet and Jason Ribando
Department of Mathematics
University of Northern Iowa
Cedar Falls, IA 50614
jribando@math.uni.edu
prophet@math.uni.edu

Non-degenerate Rational Approximation

Franklin Kemp

Abstract. A best discrete minimax rational approximation exists and is unique in the "total degree" set of rationals that achieve a minimax error with $p+q = n-1$. Moreover, the minimax approximation's error equioscillates on at least $n+1$ points; so the approximation is non-degenerate. This evokes a "total degree" Remes one point exchange ascent algorithm. At each exchange it satisfies Chebyshev's equioscillation error condition on a reference of $n+1$ points for each of the n candidate rationals by transforming each condition to a symmetric eigenvalue problem via orthogonal polynomials on the reference. The eigenvalues are the equioscillation errors. Since the reference is the same for each rational, the same orthogonal polynomials apply with the fortunate result that the n symmetric matrices are the n leading principal submatrices of a grand $(n+1) \times (n+1)$ persymmetric matrix. We use inverse iteration with initial eigenvalue bounds, eigenvalue interlacing, and Rayleigh's quotient on this grand matrix to find all the equioscillation errors; hence, all the rationals are considered, but only those are kept whose denominator polynomials pass Sturm's test for zeros in order to ensure boundedness. The absolute value of the error is nondecreasing at each exchange; so the overall algorithm converges because the Remes one-point exchange algorithm converges for a single rational when it has a best rational approximation. Such an approximation always exists in the total degree set. Extensive test approximations are presented to validate the algorithm.

§1. Summary of Discrete Minimax Rational Approximation

Let $X \subset \mathbb{R}$ be a finite point set, I the smallest interval containing X, and $R(p,q)$ the set of all rationals $P/Q = \frac{a_0 + a_1 x + \cdots + a_p x^p}{b_0 + b_1 x + \cdots + b_q x^q}$ which are irreducible and whose denominator Q is nonzero on I. Chebyshev's theorem [11, p. 131] states that $P/Q \in R(p,q)$ is the minimax rational approximation of $y(x)$ on X iff

$$s\lambda = y - P/Q \qquad \text{(equioscillating error system)} \qquad (1)$$

Approximation Theory X: Abstract and Classical Analysis   247
Charles K. Chui, Larry L. Schumaker, and Joachim Stöckler (eds.), pp. 247–266.
Copyright © 2002 by Vanderbilt University Press, Nashville, TN.
ISBN 0-8265-1415-4.

holds on a **reference** of at least $n{+}1$ points $X_{ref} = \{x_0 < x_1 < \cdots < x_n\} \subset$
X. Here, the additional parameters are

$$|\lambda| = \max_X |y - P/Q|, \qquad \text{(maximum error amplitude)}$$
$$s(x_i) = (-1)^{n-i}, \ 0 \le i \le n, \qquad \text{(alternating sign function)}$$
$$d = \min(p - \deg P, q - \deg Q), \qquad \text{(deficiency of y w.r.t $R(p,q)$)}$$
$$n{+}1 = p{+}1{+}q{+}1{-}d. \qquad \text{(no. of max error alternations)}.$$

P/Q is **degenerate** if $d > 0$. The minimax polynomial approximation of
y exists and is unique [11, p. 33] while the minimax rational is unique
[11, p. 131] if it exists. Multiplying (1) by Q, we get

$$Q(y - s\lambda) - P = 0 \quad \text{on } X_{ref}, \qquad (2)$$

which is a linear homogeneous system of $n{+}1$ equations in the $n{+}1$ unknown
coefficients of P and Q with nonlinear parameter λ. It has a nontrivial
solution iff the determinant of its coefficient matrix, a polynomial of degree
$q{+}1$ in λ, is zero.

§2. Introduction

Discrete minimax rational approximation for a (p, q) pair can be viewed
as a combinatorial problem. One could solve the system (2) on every
reference to isolate the solution that satisfies $|\lambda| = \max_X |y - P/Q|$ with
$0 \notin Q(I)$. The following difficulties, however, are addressed in [4]: *(i)*
nonlinearity of the system (2), *(ii)* unbounded solutions of (2), *(iii)* large
number of references, *(iv)* references with $d > 0$ (**degeneracy**), and *(v)*
non-existence of the minimax approximant. The total degree algorithm
masters difficulties *(i)*, *(ii)* and *(iii)* and is unaffected by *(iv)* and *(v)*;
moreover, it finds the best rational approximant for a fixed number of
parameters, *i.e.*, the combined number of coefficients of P and Q.

Although we only discuss discrete approximation, the continuous case
can be handled by increasing the density of points, as Burke [2] has proved,
or by refining the best total degree approximant's reference via a many-
point exchange algorithm using continuous error curve extrema [3].

The paper contains three parts. The theory is developed in Sections 1–7, algorithms are presented in Sections 8–14, and numerical vali-
dation via 76 comparisons is given in Sections 15 and 16. In Section 3 we
show how degeneracy is eliminated in the set of rationals of total degree.
This dispatches difficulties *(iv)* and *(v)*. In Section 4 we recast difficulty
(i) by reworking A. Pelios' method of symmetrization and stabilization
via orthogonal polynomials of H. Werner's result that the equioscillation
errors are the eigenvalues of a generalized eigenvalue problem. Section 5
contains eigenvalue theory for the solution of multiple **symmetric eigenvalue**

problems which are all packaged in one grand persymmetric matrix. Sections 6 and 7 contain additional material, relating rational interpolation to rational approximation and presenting a link of any symmetric eigenvalue problem with a minimax rational approximation problem.

Sections 8–14 contain standard algorithms [3,5] for handling difficulties *(i)* and *(ii)*. After introducing notations (Section 8), we describe eigenvalue bounds and interlacing (Section 9), inverse iteration and Rayleigh quotient (Section 10), eigenvector orthogonality to solve *all* n symmetric eigenvalue problems of Section 5 for their $n(n + 1)/2$ eigenvalues (i.e. equioscillation errors) and eigenvectors (i.e. coefficients of Q) on each reference (Sections 11–12), Sturm's test for zeros to reject unbounded approximants in order to handle difficulty *(ii)* (Section 13), and a sketch of the proof of the convergence of the Remes one-point exchange ascent algorithm extended to the case of total degree (Section 14). The Remes algorithm is essential in order to reduce the number of references investigated easing difficulty *(iii)*.

§3. Elimination of Degeneracy

We present two basic results on degeneracy in this section.

Theorem 1. *(Deficiency)* *If P/Q is the minimax approximant of $y(x)$ out of $R(p, q)$ with deficiency d, then $P/Q \in R(p - d, q - d)$, but $P/Q \notin R(p - d - 1, q - d - 1)$.*

Proof: [11, p. 126] If $d = p - \deg(P) \le q - \deg(Q)$, then $\deg(P) = p - d$ and $\deg(Q) \le q - d$; hence, $P/Q \in R(p - d, q - d)$, but $P/Q \notin R(p - d - 1, q - d - 1)$ since $\deg(P) = p - d > p - d - 1$. A similar argument holds if $d = q - \deg(Q)$. $\square$

Powell [9, p. 117] gives an example of a minimax approximant of deficiency $d = 1$ for $R(2, 2)$ and shows how exchange algorithms can fail because they are usually designed to work with exactly $p + 1 + q + 1$ equioscillation errors. The deficiency occurs because there are no pole-free approximants with at least $p + 1 + q + 1$ equioscillation errors.

Theorem 2. *(Non-degeneracy) There exist pairs (p, q) such that the best minimax rational approximation of y with $p + q = n - 1$ exists, is unique, and is non-degenerate; hence, it is characterized by a reference set of at least $n + 1$ points.*

Proof: Approximation in $R(n - 1, 0)$ is polynomial. $\square$

Remark 1. There may be no best discrete rational approximation. Best approximation of $y(0) = 1, y(1) = y(2) = 0$ from $R(1, 0)$ is $\frac{3-2x}{4}$ with minimax error $\frac{1}{4}$, but there are approximations from $R(0, 1)$ like $\frac{1}{500x+1}$ with maximum absolute error arbitrarily small, but nonzero. In this case the total degree algorithm excludes $R(0,1)$ and returns the minimax approximation that exists in $R(1,0)$ which is clearly not best over all possible approximations from $R(1,0)$ and $R(0,1)$.

Remark 2. Fig. 1 illustrates for $n = 7$ the triangle of deficiency sets with respect to the sets of total degree $n - 1$ (i.e. the sets in row $d = 0$) derived from the deficiency theorem. Suppose, for example, that P/Q is minimax degenerate of deficiency $d = 1$ with respect to $R(2,4)$. Then $P/Q \in R(1,3)$ holds by the deficiency theorem, but, of course, P/Q is also in $R(3,3)$ and $R(1,5)$. If neither of these sets have minimax approximations, and P/Q's minimax error is less than all non-degenerates in row $d = 0$, it is conceivable that P/Q, a degenerate, is the best of all minimax approximations with $p + q \leq n - 1$. This type of circumstance has not been observed and, indeed, may never occur (which needs proof or counterexample); thus, it is likely that the best minimax approximation from row $d = 0$ is also the best overall degenerate and non-degenerate approximations in the triangle of deficiency sets.

$d = 3$				$R(0,0)$		
$d = 2$			$R(2,0)$	$R(1,1)$	$R(0,2)$	
$d = 1$		$R(4,0)$	$R(3,1)$	$R(2,2)$	$R(1,3)$	$R(0,4)$
$d = 0$	$R(6,0)$	$R(5,1)$	$R(4,2)$	$R(3,3)$	$R(2,4)$	$R(1,5)$ $R(0,6)$

Fig. 1. Deficiency Sets for the Case $p + q \leq 7 - 1$.

The non-degeneracy theorem enables us to construct an extension of the Remes one-point exchange algorithm which by-passes unbounded approximations and, significantly, makes exchanges only on references of exactly $n + 1$ points. It appears to answer Powell's query [9, p. 118] whether an exchange algorithm exists that can treat all cases successfully and handle degeneracies discussed by Ralston [10, p. 270] in approximating odd and even functions without making a change of variable like $x' = x^2$ for even functions to remove such degeneracies.

§4. Orthogonal Polynomials

We follow Pelios' development [8] replacing the monomials x^i of (2) with $n + 1$ orthogonal polynomials T_i under the weighted inner product over a reference $X_{ref} :$ $(f, g) = \sum_{X_{ref}} wfg$, with special weight function $w(x_i) = s(x_i) \prod_{j \neq i}^{n} (x_i - x_j)^{-1} > 0$. These $n + 1$ orthogonal polynomials satisfy the orthogonality conditions

$$(T_i, T_j) = \delta_{ij} \gamma_i, \tag{3}$$

and the three term recurrence

$$T_{i+1} = (x - \alpha_i)T_i - \beta_i T_{i-1}, \tag{4}$$

with initial conditions $T_0 = 1$ and $T_1 = (x-\alpha_0)T_0$, $\alpha_0 = (xT_0, T_0)/(T_0, T_0)$. Applying (3) to (4), we find $\gamma_i = (T_i, T_i)$, $\alpha_i = (xT_i, T_i)/\gamma_i$, $\beta_i = \gamma_i/\gamma_{i-1}$.

Because of the special weight function, the inner product equals the nth order divided difference Δ^n relative to X_{ref} (e.g. $\Delta^1 y = \frac{y_1 - y_0}{x_1 - x_0}$ etc.) which we compute recursively [9, p. 49]. Moreover, for polynomials P, Q, we have

$$(sP, Q) = \Delta^n(PQ) = \delta_{p+q,n}, \quad a_p b_q = 1, \quad 0 \le p + q \le n.$$

Orthogonal polynomials have many interesting properties which have been well-studied for the approximation of functions [9, pp. 136-149]. Some properties are the recurrence (4), $T_i = x^i + O(x^{i-1})$, and T_i having distinct zeros in (x_0, x_n) from which it follows that they interlace T_{i-1}'s zeros and $T_i(x_n) > 0$. For the special weight function at hand, Klopfenstein (see [8]) observed and proved the following remarkable additional properties which lead to compact calculations and reduced storage. We include the proof for sake of completeness.

Theorem 3. *The T_i satisfy:*
(1) $\gamma_{n-i}\gamma_i = 1$, $\alpha_{n-i} = \alpha_i$, $\beta_{n-i+1} = \beta_i$.
(2) $T_{n-i}/\sqrt{\gamma_{n-i}} = sT_i/\sqrt{\gamma_i}$.

Proof: [8] Since $(sT_0, T_n) = \Delta^n T_0 T_n = 1$ and $(sT_0, T_i) = \Delta^n T_0 T_i = 0$ for $i < n$, we conclude that sT_0 is parallel to T_n; i.e., $sT_0 = kT_n$ on X_{ref}. But $(sT_0, T_n) = (kT_n, T_n) = k(T_n, T_n) = k\gamma_n$ gives $k = \gamma_n^{-1}$. The identities $\gamma_0 = (T_0, T_0) = (sT_0, sT_0) = (kT_n, kT_n) = k^2 \gamma_n = 1/\gamma_n$ lead to $\gamma_0 \gamma_n = 1$. This implies that $\alpha_0 = (xT_0, T_0)/\gamma_0 = (xsT_0, sT_0)/\gamma_0 = (xT_n, T_n)/(\gamma_0 \gamma_n^2) = (xT_n, T_n)/\gamma_n = \alpha_n$, which is part (1) for 0 and n. Now we consider that $(sT_1, T_{n-1}) = \Delta^n T_1 T_{n-1} = 1$, $(sT_1, T_i) = \Delta^n T_1 T_i = 0$ for $i < n - 1$, and $(sT_1, T_n) = (sT_1, sT_0) = (T_1, T_0) = 0$. Hence, applying the same argument as before, we get $sT_1 = kT_{n-1}$ with $k = 1/\gamma_{n-1}$ and $\gamma_1 \gamma_{n-1} = 1$, which gives $\alpha_1 = \alpha_{n-1}$. It is clear how to continue inductively in order to prove part (1) of the theorem. Part (2) follows immediately from the above, since we found that $sT_i = T_{n-i}/\gamma_{n-i}$ and $\gamma_i \gamma_{n-1} = 1$. Moreover, $\beta_{n-i+1} = \beta_i$ follows from $\gamma_{n-i}\gamma_i = 1$. Note that if n is even, then $\gamma_{n/2}\gamma_{n/2} = 1$ implies $\gamma_{n/2} = 1$. $\square$

For the rest of the paper, the $n + 1$ polynomials T_i will be taken as an orthornormal set of polynomials, which means that each is scaled by $1/\sqrt{\gamma_i}$ so that $(T_i, T_j) = \delta_{ij}$.

§5. Multiple Symmetric Eigenvalue Problems

We follow Pelios' [8] symmetrization and stabilization of Werner's [13] generalized eigenvalue problem. Taking the inner product of (2) with $sT_i, 0 \leq i \leq q$, and letting $P = \sum_k a_k T_k$ and $Q = \sum_j b_j T_j$, we obtain

$$
\begin{aligned}
0 &= (sT_i, yQ) - (sT_i, s\lambda Q) - (sT_i, P) \\
&= (sT_i, y\sum_j b_j T_j) - \lambda(sT_i, s\sum_j b_j T_j) - (sT_i, \sum_k a_k T_k) \\
&= \sum_j b_j(sT_i, yT_j) - \lambda\sum_j b_j(T_i, T_j) - \sum_k a_k(T_{n-i}, T_k) \\
&= \sum_j b_j \Delta^n y T_i T_j - \lambda b_i,
\end{aligned}
$$

which in matrix notation is

$$ Sb = \lambda b, \tag{5} $$

where S is the $(q+1)\times(q+1)$ symmetric matrix with entries $s_{ij} = \Delta^n y T_i T_j$ and b is the $(q + 1) \times 1$ column vector of coefficients of Q represented in terms of the orthonormal polynomials T_i.

Taking the inner product of (2) with T_i, $0 \leq i \leq p$, we get

$$
\begin{aligned}
0 &= (T_i, yQ) - (T_i, s\lambda Q) - (T_i, P) \\
&= (T_i, y\sum_j b_j T_j) - \lambda(T_i, \sum_j sb_j T_j) - (T_i, \sum_k a_k T_k) \\
&= \sum_j b_j(T_i, yT_j) - \lambda\sum_j b_j(T_i, T_{n-j}) - \sum_k a_k(T_i, T_k) \\
&= \sum_j b_j(T_i, yT_j) - a_i \\
&= \sum_j b_j \Delta^n sy T_i T_j - a_i,
\end{aligned}
$$

which in matrix notation is

$$ Mb = a, \tag{6} $$

where M is the $(p + 1) \times (q + 1)$ matrix with entries $m_{ij} = \Delta^n sy T_i T_j$ and a is the $(p + 1) \times 1$ column vector of coefficients of P in terms of the orthonormal polynomials.

Once the eigenvalue problem (5) is solved for λ and the denominator coefficients b, the numerator coefficients a are found by multiplication of b in (6).

The equioscillation error λ of (5) is an eigenvalue of S, and b is its associated eigenvector. Because S is symmetric, there are $q+1$ real eigenvalues λ_i with associated orthogonal eigenvectors b_i. Each b_i produces a denominator polynomial Q_i.

At most one Q_i is nonzero on $[x_0, x_n]$, which can be seen as in the following proof of [8]. We find for $i \neq j$ that $(Q_i, Q_j) = (\sum b_{ik} T_k, \sum b_{jm} T_m) = \sum\sum b_{ik} b_{jm}(T_k, T_m) = \sum\sum b_{ik} b_{jm}\delta_{km} = \sum b_{ik} b_{jk} = 0$ since $b_i \perp b_j$; therefore, if Q_i and Q_j are nonzero on $[x_0, x_n]$, that is, Q_i and Q_j are of one sign on X_{ref}, then $(Q_i, Q_j) \neq 0$, contradicting $(Q_i, Q_j) = 0$.

To determine the exact number of Q_i's zeros in I, we evaluate a Sturm sequence [12] of polynomials at the end points of I. This is a theoretically guaranteed procedure for identifying a bounded Q_i^{-1}, only requiring evaluations of several low order polynomials at the end points of I.

The following simple theorem reveals a remarkable and useful connection among multiple eigenvalue problems associated with a reference X_{ref}.

Theorem 4. *If S_q, $0 \leq q \leq n-1$, denotes the $(q+1) \times (q+1)$ symmetric matrix of the eigenvalue problem associated with rational $R(n-1-q,q)$, then $S_0 \subset S_1 \subset \cdots \subset S_{n-1} \subset S_n$, where the notation "$\subset$" means that each matrix is the leading principal maximum submatrix of its successor and S_n is the $(n+1) \times (n+1)$ matrix using all $n+1$ orthonormal polynomials.*

Proof: The elements of any $S_q = \{\Delta^n y T_i T_j\}$, $0 \leq i,j \leq q$, depend only the first $q+1$ orthonormal polynomials up to degree q that appear in its denominator polynomial Q, so S_{q+1} has the same elements as S_q plus those involving T_{q+1}; hence, S_q is the leading $(q+1) \times (q+1)$ principal submatrix of S_{q+1}. $\square$

Solving the eigenvalue problem for each submatrix may provide a pole-free rational to compete with the other $n-1$ rationals for minimax approximation on a reference X_{ref}. Since the non-degeneracy result ensures there is a non-degenerate minimax rational on each reference, we solve n eigenvalue problems on each reference that appears at each stage of an exchange algorithm for a total of $n(n+1)/2$ eigenvalues and eigenvectors. This task simplifies because S_n is a persymmetric matrix, i.e. symmetric with respect to both diagonals. For notational simplicity in Theorem 5 and Corollary 6, we temporarily substitute S and M for S_n and M_n, respectively.

Theorem 5. *S is persymmetric with eigenvalues sy. M is S with columns reversed.*

Proof: Here $s_{ij} = \Delta^n y T_i T_j = \Delta^n y s T_i s T_j = \Delta^n y T_{n-i} T_{n-j} = s_{n-i,n-j}$. Moreover, it can be seen that $S = \{s_{ij}\} = \{\Delta^n y T_i T_j\} = \{(s y T_i, T_j)\} = $ T $\text{diag}(s(x_0)y(x_0), \cdots, s(x_n)y(x_n))$ T^t, where T $= \{\sqrt{w(x_j)}T_i(x_j)\}$ is an orthogonal matrix. For the matrix M we obtain $m_{ij} = \Delta^n s y T_i T_j = \Delta^n y T_i s T_j = \Delta^n y T_i T_{n-j}$. $\square$

The following are results from elementary linear algebra and the Courant minmax characterization of eigenvalues of real symmetric matrices, see [15, pp. 99–104].

Corollary 6. *The following statements are true:*
(1) *S is invertible iff $0 \notin y(X_{ref})$.*

(2) *The storage required for S is $(n+2)^2/4$ (n even) and $(n+1)(n+3)/4$ (n odd).*

(3) *The eigenvector associated with eigenvalue $s(x_j)y(x_j)$ is the jth column of the matrix T, say t_j, of Theorem 5.*

(4) *Spectral Decomposition of S:*

$$S = \sum_j s(x_j)y(x_j)t_j t_j^t = \sum_j \lambda_j t_{(j)} t_{(j)}^t,$$

where $\lambda_0 \le \lambda_1 \le \cdots \le \lambda_n$ are the $n+1$ ordered eigenvalues $\{sy\}$ of S and $t_{(0)}, t_{(1)}, \cdots, t_{(n)}$ their associated normalized eigenvectors.

(5) *Eigenvalue Minimax Characterization of S: For $0 \le i \le n$,*

$$\lambda_i(S) = \min_{\dim V = i+1} \max_{0 \ne b \in V \subset \mathbf{R}^{n+1}} b^t S b / b^t b = \max_{\dim V = n-i+1} \min_{0 \ne b \in V \subset \mathbf{R}^{n+1}} b^t S b / b^t b.$$

(6) *Eigenvalue Interlacing: S_{q-1}'s eigenvalues separate those of S_q, $1 \le q \le n$. In particular, the eigenvalues of S_{n-1} interlace the ordered sequence $\{sy\}$,*

(7) *the eigenvalues of S_i, $0 \le i \le n-1$, are bounded below by $\min\{sy\}$ and above by $\max\{sy\}$.*

§6. Rational Interpolation

There is no eigenvalue problem if the total degree of the rational is n, because that would be interpolation on $n+1$ points. We note the connection between S_n and **rational interpolation** (see [11, pp. 132–135] for existence issues) of y on X_{ref} by taking the inner product of (2) with $sT_i, 0 \le i \le n$, where $\lambda = 0$ and $p + q = n$ lead to

$$
\begin{aligned}
0 &= (sT_i, yQ) - (sT_i, P) \\
&= (sT_i, y\sum_j b_j T_j) - (sT_i, \sum_k a_k T_k) \\
&= \sum_j b_j(sT_i, yT_j) - \sum_k a_k(sT_i, T_k) \\
&= \sum_j b_j \Delta^n y T_i T_j - \sum_k a_k(T_{n-i}, T_k) \\
&= \sum_j b_j \Delta^n y T_i T_j - \sum_k a_k \delta_{n-i,k}.
\end{aligned}
$$

In matrix notation this is $Sb = a$, where S is the matrix whose columns are the first $q+1$ columns of S_n, b is the $(q+1) \times 1$ column vector $[b_0, b_1, \cdots, b_q]^t$, and a is the $(n+1) \times 1$ column vector $[0, \cdots, 0, a_p, \cdots, a_0]^t$. In particular, if $q = n$ and $p = 0$, then we have interpolation by the reciprocal polynomial $a_0 / \sum b_j T_j$ from $R(0, n)$ and $S = S_n$.

Theorem 7. *$R(n, 0)$ interpolates y on X_{ref}, while $R(0, n)$ interpolates y iff $0 \notin y(X_{ref})$.*

Proof: The first assertion is polynomial interpolation, while the second follows from $b = S_n^{-1}[0, \cdots, 0, 1]^t$ and Corollary 6 (1). $\square$

§7. Linear Transformation of a SEP to a RAP

In Section 5 we saw how a minimax rational approximation problem (RAP) is cast as a symmetric eigenvalue problem (SEP). Conversely, it is possible to transform a symmetric eigenvalue problem to a minimax rational approximation problem in which the eigenvalues are identified as equioscillation errors.

Theorem 8. *A symmetric eigenvalue problem can be linearly transformed to a minimax rational approximation problem.*

Proof: Let A be any $m \times m$ symmetric matrix with $m(m+1)/2$ independent entries a_{ij}, $0 \leq i \leq j \leq m - 1$. For, $m \geq 2$, set $n + 1 = m(m+1)/2$, choose any $X = X_{ref}$ of $n + 1$ points, and solve the $n + 1$ linear equations $\Delta^n y T_i T_j = a_{ij}, 0 \leq i \leq j \leq m - 1$, for $n + 1$ values of y. By construction, A is the leading $m \times m$ principal submatrix of *persymmetric* S_n associated with the total degree $= n - 1$ minimax rational approximation of y on X. This is minimax approximation of y on X by P/Q with $(p, q) = (n - m, m - 1)$. $\square$

§8. Notation for Eigenvalue Searching Algorithms

In order to describe the algorithms, we define several matrices. In Sections 9–11 S will denote the leading principal $(q + 1) \times (q + 1)$ symmetric (not persymmetric) submatrix of the $(n + 1) \times (n + 1)$ persymmetric matrix S_n exemplified in Figure 2 with M calculated from S as discussed in Section 5.

$$\begin{bmatrix} s_{00} & s_{01} & s_{02} & s_{03} \\ s_{01} & s_{11} & s_{12} & s_{02} \\ s_{02} & s_{12} & s_{11} & s_{01} \\ s_{03} & s_{02} & s_{01} & s_{00} \end{bmatrix}$$

Fig. 2. Persymmetric matrix S_n for $n = 3$.

An $(n+2) \times (n+1)$ matrix Λ of eigenvalues and bounds is constructed in the following way. Recall that we defined $\{\lambda\}$ as ordered $\{sy\}$. The first row contains only λ_0's. By Corollary 6 (7), λ_0 is a lower bound and λ_n is an upper bound on *all* eigenvalues. Beginning at row 3 column 1, n λ_n's run down the subdiagonal. By Corollary 6 (6) the last column beginning at row 2 contains $\lambda_0 \leq \lambda_1 \leq \cdots \leq \lambda_n$ which interlace the n eigenvalues (i.e. equioscillation errors) of the $R(0, n - 1)$ case. This arrangement of bounds makes a border around an upper triangular region which gets filled with the $n(n + 1)/2$ eigenvalues λ (equioscillation errors) from the n possible rational cases of total degree $n - 1$ with column q containing the $q + 1$ eigenvalues for the rationals of $\deg(Q) = q$, ordered increasingly down the column. The column index of Λ runs from 0 to n. Figure 3 displays an example of such a Λ for the case $n = 3$.

$$q = \quad \begin{matrix} 0 & 1 & 2 \end{matrix}$$

$$\begin{bmatrix} \lambda_0 & \lambda_0 & \lambda_0 & \lambda_0 \\ s_{00} & \lambda_{01} & \lambda_{02} & \lambda_0 \\ \lambda_3 & \lambda_{11} & \lambda_{12} & \lambda_1 \\ 0 & \lambda_3 & \lambda_{22} & \lambda_2 \\ 0 & 0 & \lambda_3 & \lambda_3 \end{bmatrix}$$

Fig. 3. Matrix Λ of eigenvalues and bounds for $n = 3$.

The columns of a $(n+2) \times (n(n+1)/2 + 2)$ matrix, say A, as shown in Figure 4 for $n = 3$, are filled with the $n + 1 = p + 1 + q + 1$ coefficients a followed by b of each approximation corresponding to each of the $n(n+1)/2$ eigenvalues plus the polynomial and reciprocal polynomial interpolations. The last row of A counts the number of zeros that each Q has in I. Thus, column indices 0 and $1 + n(n+1)/2$ contain the coefficients of the polynomial and reciprocal polynomial interpolations, respectively, with the coefficients of the n approximation cases $(p, q) = (n-1, 0), (n-2, 1), \cdots, (0, n-1)$ ordered in between. Specifically, the coefficients of each rational of denominator degree q have column indices between $1 + q(q+1)/2$ and $(q+1)(q+2)/2$. The bottom part of each column is the normalized eigenvector b associated with its eigenvalue.

(p,q) =	(3,0)	(2,0)	(1,1)	(1,1)	(0,2)	(0,2)	(0,2)	(0,3)
	a_0	a_{00}	a_{01}	a'_{01}	a_{02}	a'_{02}	a''_{02}	b_0
	a_1	a_{10}	a_{11}	a'_{11}	b_{20}	b'_{20}	b''_{20}	b_1
	a_2	a_{20}	b_{10}	b'_{10}	b_{21}	b'_{21}	b''_{21}	b_2
	a_3	b_{00}	b_{11}	b'_{11}	b_{22}	b'_{22}	b''_{22}	b_3
	0	0	?	?	?	?	?	?

Fig. 4. Matrix A of P and Q coefficients of orthogonal polynomials for $n = 3$.

When the best rational occurs with total degree that is less than $n-1$, we find its degree by checking its highest nonzero coefficients. If it has full degree $n - 1$, then we use $q = [\frac{-1+\sqrt{8j-7}}{2}]$, $1 \le j \le 1 + n(n+1)/2$, where column j of A contains the coefficients of the optimum rational. Here the square brackets denote the greatest integer function. This formula comes from solving $1 + q(q+1)/2 = j$ for q. Of course, p is obtained from $p = n - 1 - q$.

There are two eigenvalue searching algorithms which we dub "lohi" and "hilo". The lohi algorithm fills the upper triangular matrix inside Λ with eigenvalues between the lower and upper bounds λ_0, λ_n appearing in each column advancing through cases $(p, q) = (n - 1, 0), (n - 2, 1), \cdots, (0, n - 1)$ from left to right, column by column with eigenvalues ascending down each column from low to high. On the other hand, the hilo algorithm fills the upper triangular matrix inside Λ with eigenvalues starting with the column of eigenvalue bounds of the last column and advancing to the left column by column with eigenvalues decreasing up each column.

The **total degree Remes exchange** algorithm requires an $(n+1) \times n$ matrix, say E. As it ranges over references, each column of E corresponds to one of the n rational cases $(n-1, 0), (n-2, 1), \cdots, (0, n-1)$, respectively. In the discussion of the convergence of the total degree exchange algorithm below, we will see that the first element of such a column records the largest minimum equioscillation error met as long as its rational is bounded. The remaining n elements hold its associated reference index set. Some rational sets become inadmissable, in which case the first element where the equioscillation error would be stored is arbitrarily set to a very large value, and its reference index set where it became inadmissable is recorded. Furthermore, it may happen, as the method ranges over references, that some rationals are never encountered before the algorithm completes, so their columns are empty.

Figure 5 below is an example of an E matrix for the case of the total degree 6 approximation of y_{12} of Section 16. Reference indices actually run down columns, but are presented here in one line to shrink the display.

(p,q) =	(6,0)	(5,1)	(4,2)	(3,3)	(2,4)	(1,5)	(0,6)
	$.10460^0$	0	$.10186^0$	0	0	$.14831^0$	$.18019^0$
	0,1,3,5,10,15,17,19	0's	0,3,5,7,9,13,15,17	0's	0's	0,1,2,6,10,14,17,20	0,1,2,8,10,15,19,20

Fig. 5. Sample E matrix used in total degree exchanges.

§9. Bounds and Eigenvalue Interlacing

The lohi algorithm inserts eigenvalues in Λ in the following way. There is no searching in the $(p, q) = (n-1, 0)$ case, since the eigenvalue is s_{00} and already between λ_0 and λ_n. Coming to $(p, q) = (n-2, 1)$, we have two eigenvalues to find. By the **eigenvalue interlacing** property of symmetric matrices, one is between λ_0 and s_{00} and the other between s_{00} and λ_n. How we find eigenvalues between bounds is described in Sections 10 and 11 below. Once these two eigenvalues are located, we use them, the two outer bounds λ_0 and λ_n, and the eigenvalue interlacing property to locate the three eigenvalues of $(p, q) = (n-3, 2)$; thus, we successively exploit the interlacing and outer bounds locating all eigenvalues through case $(p, q) = (0, n-1)$ whose $n-1$ eigenvalues are interlaced between $\lambda_0 \leq \lambda_1 \leq \cdots \leq \lambda_n$.

The hilo algorithm runs through Λ in the opposite direction starting with $(p, q) = (0, n-1)$ knowing its n eigenvalues are interlaced between the bounds $\lambda_0 \leq \lambda_1 \leq \cdots \leq \lambda_n$. Once its n eigenvalues, $\lambda_{0,n-1} \leq \lambda_{1,n-1} \leq \cdots \leq \lambda_{n-1,n-1}$ are located, they become bounds for the $n-1$ eigenvalues of $(p,q) = (1, n-2)$ by the interlacing property, so as above, we carry on right to left successively ending at $(p,q) = (n-1,0)$ whose eigenvalue is already known, being s_{00}.

§10. Inverse Iteration and Rayleigh Quotient

To search for an eigenvalue associated with a (p,q) case known to be between a lower and upper bound, we guess a value on or between the bounds by binary chopping along with a guess at a value for its eigenvector. The guesses, say λ_0 and b_0, respectively, are fed into the matrix equation $b_i = (S - \lambda_i I)b_{i+1}$ which is solved for b_{i+1} which in turn is fed into $\lambda_{i+1} = b_{i+1}^t S b_{i+1}/b_{i+1}^t b_{i+1}$ updating both b_i and λ_i. These two equations are iterated until $|1 - \lambda_i/\lambda_{i+1}| < 10^{-11}$. From [5, pp. 308–309] we read that the iteration of the matrix equation is known as **inverse iteration** and the second defines **Rayleigh's quotient** which minimizes $\|(S - \lambda I)b\|_2$ at the true eigenvalue λ with its graph there like the bottom of a parabola. The error $\lambda - \lambda_i$ is roughly the square of the error in the eigenvector, which equals $(b - b_i)^t(b - b_i)$, producing cubic convergence to λ. When the iteration completes, b is used in (6) to give a and they both are stored in a column of A with index between $1 + q(q+1)/2$ and $(q+1)(q+2)/2$, the exact index corresponding to λ's algebraically ordered position in the $q+1$st column of Λ.

§11. Search Using Eigenvector Orthogonality

We did not say in Section 10 how to choose the initial guess for eigenvector b, b_0. What we do is suppose that if our initial guess of λ, λ_0, is good, then $(S - \lambda_0 I)b_0 = 0$ should be approximately true when b_0 is close to the true eigenvector. Since we don't want the trivial solution, we set the last component of b_0 to 1 and solve, dropping one row because we have just q unknowns, to get a b_0 as an estimate of b. This provides a guess when we don't have any eigenvectors at hand.

If we have already found i eigenvectors out of $q+1$ eigenvectors, say $b_0, \ldots, b_{i-1}$, which have been stored in A and are now searching for the $i+1$st eigenvalue λ_i of case (p,q), we know its eigenvector, say b_i, is orthogonal to the set of orthogonal eigenvectors $b_0, \ldots, b_{i-1}$ already found. But λ_i will be between our known bounds and as before, if we assume a guess at it, say λ_{i0}, is good, then $(S - \lambda_{i0} I)b_{i0} = 0$ should be approximately true when b_{i0} is close to b_i. In this case we again set the last component of b_{i0} to 1 and solve for b_{i0} by demanding that $b_{i0}^t b_j = 0$ for $0 \leq j \leq i-1$ and $(S' - \lambda_{i0} I)b_{i0} = 0$, where S' is the $(q-i) \times (q+1)$ submatrix of S. Of course, when q of the $q+1$ eigenvalues and eigenvectors have been found, we don't apply $(S - \lambda_{q+1,0}I)b_{q+1,0} = 0$ at all because we get q conditions from $b_{q+1,0}^t b_j = 0, 0 \leq j \leq q - 1$.

§12. Eigenvalue Search Refinements

Although we proceed with a guess of an eigenvalue for each (p, q) case using bounds in either the lohi or hilo algorithm, we may get an eigenvalue that falls outside our starting bounds in which case we don't discard it, but position it near where it belongs in the $q + 1$st column of Λ where eigenvalues must ultimately fall in algebraically ascending order down the column. We say "near" because eigenvalues can fall exactly on bounds in which case their position can only be determined when their neighbors are known. Of course, there is no positioning problem when eigenvalues are strictly within their bounds.

To handle the case of eigenvalues falling on bounds, we check to make sure, first of all, that if it is a repeated eigenvalue, its eigenvector is not repeated by testing for orthogonality. For example, there is no guarantee that an initial eigenvector guess which is orthogonal by construction to its antecedent brother eigenvectors will converge to an eigenvector also orthogonal to them. Secondly, we check to see how close an eigenvalue is to a bound after it comes out of inverse iteration. If it is within 10^{-12} of a bound, we set it equal to the bound to avoid any doubt whether it should be above or below the bound when it is positioned. This has the fortunate consequence that comparisons are made exactly without tolerances. Also, eigenvalues that are less than 10^{-12} in absolute value are set to zero.

Finally, if an eigenvalue and its eigenvector repeat, then the search between bounds must continue. We proceed by binary chopping the bounding interval. This usually suffices, but we may have to expand it to start the search with a different initial eigenvalue-eigenvector guess in order to produce a new eigenvalue and eigenvector.

§13. Sturm Test for Q's Zeros

Wilf [14, p. 235] gives the following formulation of Sturm's theorem for determining the exact number of zeros of a polynomial $Q = \sum d_i x^i$ in an interval $[a, b]$ as long as no zero is a multiple. Define $q + 1$ polynomials $C_q, C_{q-1}, \ldots, C_0$ of strictly decreasing degrees by $C_q = Q, C_{q-1} = Q'$, the derivative of Q, and

$$C_i = PC_{i-1} - C_{i-2}, \qquad 0 \leq i \leq q - 2, \tag{7}$$

where P is a polynomial such that $\deg(PC_{i-1}) = \deg(C_i)$. The coefficients of the sequence can be calculated recursively by equating like powers in (7) as follows:

$$
\begin{aligned}
c_{q,j} &= d_{q-j}, & 0 \le j \le q \\
c_{q-1,j} &= (j+1)d_{q-j-1}, & 0 \le j \le q-1 \\
\alpha_i &= c_{ii}/c_{i-1,i-1} \\
\beta_i &= (c_{i,i-1} - \alpha_i c_{i-1,i-2})/c_{i-1,i-1} \\
c_{i-2,i-j} &= \alpha_i c_{i-1,i-j-1} + \beta_i c_{i-1,i-j} - c_{i,i-j}
\end{aligned}
\left.\rule{0pt}{3.5em}\right\}
\begin{aligned}
j &= 2,3,\ldots,i; \\
i &= q,q-1,\ldots,2.
\end{aligned}
$$

Sturm's theorem [12, pp. 142–145] states that if $V(x)$ counts the number of sign changes in the sequence of $C_j(x)$'s, then the number of zeros in the interval $[a,b]$ is exactly $V(a)-V(b)$. Unfortunately, (7) is solvable only for monomials x^i, so we must represent $Q = \sum b_i T_i$ in terms of monomials

$$\sum b_i T_i = \sum d_i x^i. \tag{8}$$

However, the conversion can be done as a simple recursion. Equating like powers of (8), we get the upper triangular system of equations $d_i = \sum_{0 \le i \le j}^{q} \omega_{ij} b_j / \sqrt{\gamma_j}$, with $\omega_{ij} = \delta_{ij}$ and then recursively $\omega_{ij} = -(\alpha_{j-1}\omega_{ij-1} + \beta_{j-1}\omega_{ij-2}) + \omega_{i-1\,j-1}$, $0 \le i \le j-1; 1 \le j \le q$ where α_i and β_i are the coefficients of the three term recurrence of (4). We solve the system equivalently without directly using ω's by initializing two columns, l and r, where $l_i = 1, r_i = 0, d_i = b_i/\sqrt{\gamma_i}, 0 \le i \le q$, and then calculating recursively for $0 \le i \le j-1; 1 \le j \le q$:

$$
\begin{aligned}
t &= -(\alpha_{j-1}l_{i+1} + \beta_{j-1}r_{i+1}) + r_i \\
d_i &= d_i + (b_j/\sqrt{\gamma_j})t \\
r_{i+1} &= l_{i+1} \\
l_{i+1} &= t.
\end{aligned}
$$

§14. Convergence of the Total Degree Algorithm

Up to this point our development has been confined to a single reference X_{ref} of $n+1$ points. The total degree exchange algorithm works exactly like the one-point exchange algorithm when the degrees of P and Q are separately fixed, except now at each stage, the algorithm picks the best bounded rational from the set of up to n bounded rationals of total degree $n-1$ on all points X, so the algorithm is no longer confined to a single rational set. Just as in the one-point exchange, it creates a new reference by bringing in a point from X at which the current best rational has the greatest deviation and removing the point in X_{ref} which has the same sign as the greatest deviation and is closest to the new point. Closeness is determined by regarding the complete set of points of X to be a loop.

Theorem 9. *The total degree Remes one-point exchange algorithm converges to the minimax rational approximation of total degree $n - 1$.*

Proof: By Theorem 2, there is a non-degenerate minimax rational of total degree $n - 1$. This means every reference of $n + 1$ points has a bounded rational in *at least* one of the n sets $R(p, q)$ with $p + q = n - 1$ and $p, q \geq 0$ to which the Remes one-point exchange algorithm can be applied to converge to its minimax rational (see [9, p. 117], [10, p. 268], or [11, pp. 40-2,136-7]). If a bounded single rational in one of the sets eventually arrives at a reference on which it is no longer bounded, then its set cannot contain the best minimax non-degenerate rational, otherwise the Remes one-point exchange could continue with it; therefore, its set is removed from further consideration by recording its inadmissability in the E matrix.

Now the minimax error restricted to references strictly ascends for each new reference because it strictly ascends individually in each set of total degree by the Remes one-point exchange applied to that set. The maximum absolute error over X may or may not decrease, but cannot increase for if it did, it would mean the previous rational was better. Eventually, the minimax error on some reference will equal the maximum absolute error because the minimax reference error strictly increases at each stage while at worst the maximum absolute error remains constant. If a rational stays bounded from one of the n disjoint sets through all exchanges, then it remains a competitor for the minimax rational of total degree; therefore, we only need to show that the algorithm considers all such competitors.

On any reference all potential non-degenerate ultimately singly minimax rationals out of the n disjoint sets are available for selection, but only the one with the smallest equioscillation error bigger than the current minimum is selected. Thus it is possible for ultimately better non-degenerate rationals in other sets to be bypassed on a current reference because their present equioscillation error minimums exceed the current one. To avoid this straying, the total degree exchange algorithm updates the best minimums in each set and their associated reference index sets in the $(n+1) \times n$ matrix E. The E matrix is examined at each exchange to see if it has a better rational, i.e. one with a *smaller* equioscillation error in another one of the n disjoint sets than the current best in its set. When one in the E matrix is better, the algorithm switches to that rational and its reference in its set. In other words, we make the exchange with the smallest increase possible in the reference minimax error available at the exchange.

It can also happen that on a newly selected reference the minimax error exceeds the maximum absolute error which, of course, means this latest reference must be rejected. This is almost the same as the case of the current minimax error exceeding one in the E matrix, so the algorithm backtracks to the previous reference X_{ref} to find a lesser minimax error than the current one, but greater than the one just previously encountered and less than or equal to the current maximum absolute error on all the points and so the ascent process continues.

The only other possibility is that there can be sets that never appear in the E matrix. This happens when rationals in these sets are always found to be unbounded, or they always have minimum equioscillation errors on references that are examined by the algorithm that always exceed the current best rational. Apparently, good initial references can eliminate many exchanges because competitors get ranked quickly or are not met.
□

§15. Test Functions

Table 1 lists fifteen functions on equally spaced points and the minimax errors of their rational approximations by the differential correction algorithm [1,6] at several fixed numerator and denominator degrees. The $\pm$ appearing in the definition of y_{13} means the .1 alternates between $+.1$ and $-.1$ starting with $+.1$ as x advances in steps of .1 from 0 to 1 and is linear interpolation in between. As noted in [6], y_{12} and y_{13} were designed to give degenerate approximations because the points $\{0, \frac{1}{4}, \frac{1}{2}, \frac{3}{4}, 1\}$ and $\{0, .1, \cdots, 1\}$ are where $y_{12} - \frac{1}{1+x}$ and $y_{13} - \frac{1}{(\frac{11}{20}-x)^2}$ equioscillate, respectively. y_{14} has no best *strictly* rational approximation.

§16. Numerical Results

Table 2 contains signed minimax equioscillation errors and their reference index sets of the total degree Remes exchange algorithm coded in 230 lines of APL applied to the test functions in Table 1 for total degree up to 6. The sign on the equioscillation error is negative if the approximation is greater than the function on the first reference point. Comparing Table 2 with results of Table 1, we find, of course, that the total degree rational minimax equioscillation error is never less than a single rational minimax error. Of 75 possible comparisons, it is better in 50 cases (this number includes the 10 cases of y_{13} and y_{14} for which minimax approximations *exist* although other non-minimax approximations may exist with smaller error) and the same in the remaining 25 which are underlined in Table 1. Although the best approximation of total degree $n-1$ equioscillates on $n+1$ points, its actual total degree may be less; thus, for example, the best approximation of total degree 3 or 4 for odd function y_2 resides in R(3,0). The same applies to the odd function y_4. The minimax results for y_{12} reproduce the rigged equioscillation error .5 up to the 5 equioscillations; thereafter, the minimax error can equioscillate 6 or more times, so minimax approximations get better. Best approximations of y_{13} must be polynomials for $q \geq 2$ because minimax rational approximations of y_{13} must

y	I	X	$(1,1)$	$(0,2)$	$(2,2)$	$(1,3)$	$(4,2)$
$y_1 : e^x$	$[-1,1]$	51	$.20932^{-1}$	$.34791^{-1}$	$.86644^{-4}$	$.12392^{-3}$	$.21037^{-6}$
$y_2 : \sin x$	$[-3,3]$	21	$.62542^0$	$.99749^0$	$.30608^0$	$.30608^0$	$.66482^{-2}$
$y_3 : \sqrt{x}$	$[0,1]$	11	$.36243^{-1}$	$.18078^0$	$.77019^{-3}$	$.37281^{-2}$	$.10202^{-4}$
$y_4 : \mathrm{sgn}(\frac{1}{2}-x)$	$[0,1]$	21	$.81818^0$	$.10000^1$	$.26923^0$	$.26923^0$	$.70465^{-1}$
$y_5 : x$	$[0,1]$	26	$.58916^{-1}$	$.22594^0$	$.54260^{-1}$	$.48581^{-1}$	$.18717^{-1}$
$\quad \frac{1}{2}x + \frac{2}{5}$	$(1,2]$	25					
$y_6 : e^x$	$[0,1]$	11	$.30872^0$	$.20697^0$	$.86504^{-1}$	$.95354^{-1}$	$.30919^{-1}$
$\quad e + e^{-x} - e^{-1}$	$(1,2]$	10					
$y_7 : \ln(1+x)$	$[0,1]$	51	$.85978^{-3}$	$.92869^{-1}$	$.17028^{-5}$	$.74224^{-5}$	$.58255^{-8}$
$y_8 : erf\ x$	$[0,2]$	21	$.44085^{-1}$	$.19844^0$	$.13754^{-2}$	$.92931^{-3}$	$.44515^{-4}$
$y_9 : e^{-x^2}$	$[0,2]$	11	$.72164^{-1}$	$.69042^{-1}$	$.25586^{-2}$	$.41421^{-2}$	$.38213^{-4}$
$y_{10} : \Gamma(x)$	$[2,3]$	51	$.64376^{-2}$	$.64307^{-2}$	$.36395^{-4}$	$.55096^{-4}$	$.17428^{-6}$
$y_{11} : \Gamma(x)$	$[2,3]$	101	$.64420^{-2}$	$.64351^{-2}$	$.36432^{-4}$	$.55160^{-4}$	$.17660^{-6}$
$y_{12} : \frac{3}{2} - \frac{24x}{5}$	$[0,\frac{1}{4}]$	6	$.50000^0$	$.50000^0$	$.50000^0$	$.50000^0$	$.10186^0$
$\quad \frac{52x}{15} - \frac{17}{30}$	$(\frac{1}{4},\frac{1}{2}]$	5					
$\quad \frac{47}{14} - \frac{92x}{21}$	$(\frac{1}{2},\frac{3}{4}]$	5					
$\quad \frac{26x}{7} - \frac{19}{7}$	$(\frac{3}{4},1]$	5					
$y_{13} : \frac{1}{(\frac{11}{20}-x)^2} \pm .1$	$[0,1]$	11	$.19754^3$	$.10000^{0*}$	$.10000^{0*}$	$.10000^{0*}$	$.10000^{0*}$
$y_{14} : \delta(1-x)$	$[0,1]$	21	$.28934^{-7}$	$.31263^{-7}$	$.41851^{-7}$	$.16880^{-7}$	$.84017^{-7}$
$y_{15} : \cos(\frac{1}{2}x)$	$[0,\frac{6}{7}]$	7	$.30612^{-1}$	$.12451^0$	$.15114^{-3}$	$.77342^{-3}$	$.26476^{-6}$
	$[1,\frac{9}{5}]$	5					
	$[2,3]$	9					

Tab. 1. Differential correction algorithm minimax errors [1,6]. Here
* denotes the pole solution.

agree with $1/(\frac{11}{20} - x)^2$ which, although positive on $[0,1]$, is unbounded.
Uspensky [12, p. 139] indicates in the discussion of Sturm's sequence how
a multiple root of a polynomial can be detected by the occurrence of consecutive zeros in the sequence. Since we test terms of the Sturm sequence
for near zeros, i.e. values less than 10^{-6}, we detected the double zero of
Q's appearing for total degrees 2 to 6, thereby, shuning unbounded approximations of y_{13}. Since y_{14} has no discrete minimax strictly rational
approximation, our results produce minimax polynomial approximations
in contrast to the differential correction algorithm which returns rationals
whose error is near the well-defined zero error of a *non-existent* rational
approximation [1, p. 494]. This can lead to numerical inconsistencies as
shown in Table 1 for y_{14}, where the errors for $R(0,2) \subset R(2,2) \subset R(4,2)$
increase!

Finally, we also compare the total degree approximation of $\tan^{-1}(x)$
on $[0,1]$ to a Remes many-point exchange algorithm used in [3] that ap-

y_1	$(0,1)$	$(1,1)$	$(1,2)$	$(2,2)$	$(2,3)$	$(3,3)$
	$-.20906^{0}$	$.20932^{-1}$	$.16714^{-2}$	$-.86644^{-4}$	$-.42463^{-5}$	$.15425^{-6}$
	0,40,50	0,19,42,50	0,12,32,46,50	0,7,22,37,47,50	0,5,17,30,41,48,50	0,2,12,23,34,43,48,50
y_2	$(1,0)$	$(1,0)$	$(3,0)$	$(3,0)$	$(5,0)$	$(5,0)$
	$-.62542^{0}$	$.62542^{0}$	$-.86327^{-1}$	$-.86327^{-1}$	$.50693^{-2}$	$.50693^{-2}$
	6,14,20	0,6,14,20	0,2,7,13,18	0,2,7,13,18,20	0,1,4,8,12,16,19	0,1,4,8,12,16,19,20
y_3	$(1,0)$	$(1,1)$	$(2,1)$	$(2,2)$	$(3,2)$	$(3,3)$
	$-.12386^{0}$	$-.36243^{-1}$	$-.59233^{-2}$	$-.77019^{-3}$	$-.79865^{-4}$	$-.69725^{-5}$
	0,3,10	0,1,5,10	0,1,2,6,10	0,1,2,4,8,10	0,1,2,3,5,8,10	0,1,2,3,4,7,9,10
y_4	$(1,0)$	$(1,0)$	$(1,2)$	$(1,2)$	$(3,2)$	$(3,2)$
	$-.81818^{0}$	$-.81818^{0}$	$-.26923^{0}$	$.26923^{0}$	$-.70465^{-1}$	$-.70465^{-1}$
	0,9,11	0,9,11,20	7,9,11,13,20	0,7,9,11,13,20	0,5,8,9,11,12,15	0,5,8,9,11,12,15,20
y_5	$(1,0)$	$(1,1)$	$(3,0)$	$(1,3)$	$(3,2)$	$(3,3)$
	$-.15000^{0}$	$.58916^{-1}$	$.55075^{-1}$	$.48581^{-1}$	$-.40329^{-1}$	$.18206^{-1}$
	0,25,50	0,11,25,33	0,12,25,30,50	4,17,25,26,42,50	0,12,21,25,26,34,50	0,8,20,24,25,26,28,41
y_6	$(1,0)$	$(0,2)$	$(0,3)$	$(2,2)$	$(3,2)$	$(4,2)$
	$-.48771^{0}$	$-.20697^{0}$	$.10998^{0}$	$-.86504^{-1}$	$-.47661^{-1}$	$.30919^{-1}$
	0,10,20	0,10,15,20	0,7,10,13,20	0,4,8,10,13,20	0,5,9,10,12,16,20	0,3,7,9,10,11,14,20
y_7	$(1,0)$	$(1,1)$	$(2,1)$	$(2,2)$	$(3,2)$	$(3,3)$
	$-.29829^{-1}$	$-.85977^{-3}$	$-.49420^{-4}$	$-.17028^{-5}$	$-.87946^{-7}$	$-.32650^{-8}$
	0,22,50	0,10,34,50	0,5,21,41,50	0,3,14,29,44,50	0,2,10,21,34,45,50	0,2,7,15,26,38,47,50
y_8	$(1,0)$	$(2,0)$	$(1,2)$	$(1,3)$	$(3,2)$	$(4,2)$
	$-.17451^{0}$	$-.20524^{-1}$	$.44957^{-2}$	$-.92931^{-3}$	$-.46192^{-4}$	$.44515^{-4}$
	0,9,20	0,7,16,20	0,3,9,16,20	0,2,6,11,17,20	1,3,7,11,15,19,20	0,1,3,7,11,15,19,20
y_9	$(1,0)$	$(2,0)$	$(1,2)$	$(4,0)$	$(3,2)$	$(4,2)$
	$.10013^{0}$	$-.68551^{-1}$	$.93046^{-2}$	$.23775^{-2}$	$-.82000^{-3}$	$-.38213^{-4}$
	1,6,10	0,2,6,10	0,2,4,7,10	0,1,3,5,8,10	0,1,2,4,7,9,10	0,1,3,4,6,8,9,10
y_{10}	$(0,1)$	$(0,2)$	$(2,1)$	$(0,4)$	$(2,3)$	$(2,4)$
	$-.74658^{-2}$	$.64307^{-2}$	$.10356^{-3}$	$.51814^{-5}$	$.13544^{-5}$	$-.19923^{-7}$
	14,40,50	0,16,41,50	0,7,23,41,50	1,12,26,39,47,50	0,4,14,27,39,47,50	0,2,9,19,30,40,47,50
y_{11}	$(0,1)$	$(0,2)$	$(2,1)$	$(0,4)$	$(2,3)$	$(2,4)$
	$-.74658^{-2}$	$.64351^{-2}$	$.10382^{-3}$	$.51858^{-5}$	$.13555^{-5}$	$-.20097^{-7}$
	28,80,100	0,32,81,100	0,13,46,82,100	2,24,52,77,94,100	0,7,27,53,78,94,100	0,5,18,38,60,80,95,100
y_{12}	$(0,1)$	$(1,1)$	$(2,1)$	$(4,0)$	$(5,0)$	$(4,2)$
	$.50000^{0}$	$.50000^{0}$	$.50000^{0}$	$-.26845^{0}$	$-.26744^{0}$	$-.10186^{0}$
	0,5,20	0,5,10,15	0,5,10,15,20	0,2,5,10,15,18	0,2,5,10,15,20	0,3,5,7,9,13,15,17
y_{13}	$(1,0)$	$(2,0)$	$(3,0)$	$(4,0)$	$(5,0)$	$(6,0)$
	$-.19786^{3}$	$.16461^{3}$	$.16118^{3}$	$-.13062^{3}$	$-.12161^{3}$	$.88501^{2}$
	0,6,10	0,4,6,7	0,4,5,7,10	0,1,4,6,7,10	0,2,4,5,7,9,10	0,1,2,4,6,7,9,10
y_{14}	$(1,0)$	$(2,0)$	$(3,0)$	$(4,0)$	$(5,0)$	$(6,0)$
	$.47500^{0}$	$.40909^{0}$	$.32386^{0}$	$-.23909^{0}$	$.16761^{0}$	$-.11196^{0}$
	0,19,20	0,9,19,20	0,5,14,19,20	0,3,9,16,19,20	0,2,7,12,17,19,20	0,1,5,9,14,18,19,20
y_{15}	$(1,0)$	$(2,0)$	$(3,0)$	$(4,0)$	$(5,0)$	$(4,2)$
	$-.99249^{-1}$	$-.11633^{-1}$	$.11737^{-2}$	$.81868^{-4}$	$-.53476^{-5}$	$.26476^{-6}$
	0,9,20	0,6,14,20	0,3,9,16,20	0,2,7,12,18,20	0,1,5,9,14,18,20	0,1,4,8,11,16,19,20

Tab. 2. Total degree algorithm signed minimax errors and references.

proximates $\tan^{-1}(x)$ by a rational from $R(3,3)$. The result is that the best $R(3,3)$ approximation based on a continuous error extrema curve yields the minimax error of $.56399^{-6}$ which is greater than minimax error of $|-.32117^{-6}|$ on reference index set 0,3,10,19,30,40,47,50 of the best total degree 6 rational from $R(4,2)$ based on an X of 51 equally spaced points of $[0,1]$.

Acknowledgments. This paper is dedicated to George Bachman. The author thanks Joachim Stöckler for a correct formulation of Theorem 2.

References

1. Barrodale, I., M. J. D. Powell, and F. D. K. Roberts, The differential correction algorithm for rational l_∞ approximation, SIAM J. Numer. Anal. **9** (1972) 493–504.

2. Burke, M. E., Nonlinear best approximation on discrete sets, J. Approx. Theory **16** (1976) 133–141.

3. Curtis, A. and M. R. Osborne, The construction of minimax rational approximations to functions, Comp. J. **9** (1966) 286–293.

4. Dunham, C. B., Discrete Chebyshev approximation: alternation and the Remez algorithm, Z. Angew. Math. Mech. **58** (1979) 326–328.

5. Golub, G. H. and van Loan, C. F., *Matrix Computations*, John Hopkins Univ. Press, Baltimore, MD, 1983.

6. Kaufman, E. H. Jr., D. J. Leeming, and G. D. Taylor, A combined Remes-differential correction algorithm for rational approximation: experimental results, Comp. Math. Appl. **6** No. 2 (1980) 155–160.

7. Kemp, L. F., Rational approximation and symmetric eigenvalue bounds, *Approximation Theory V*, C. Chui, L. Schumaker, and J. Ward (eds.), Academic Press, New York, 1986, 415–417.

8. Pelios, A., Rational function approximation as a well-conditioned matrix eigenvalue problem, SIAM J. Numer. Anal. **4** No. 4 (1967) 542–547.

9. Powell, M. J. D., *Approximation Theory and Methods*, Cambridge Univ. Press, NY, 1981.

10. Ralston, A., Rational Chebyshev approximation, *Mathematical Methods for Digital Computers*, A. Ralston and H. Wilf (eds.), Vol. 2, John Wiley and Sons, NY, 1967, 264–284.

11. Rivlin, T. J., *An Introduction to the Approximation of Functions*, Dover, NY, 1981.

12. Uspensky, J. V., *Theory of Equations*, McGraw-Hill, New York, 1948.

13. Werner, H., A remark on the numerics of rational approximation and the rate of convergence of equally spaced interpolation of $|x|$, Circuit. Sys. Signal Process. **1** Nos. 3-4 (1982), 367–377. This reference is in English and contains the original result given in German of the equivalence of minimax rational approximation to a symmetric eigenvalue problem found in Arch. Rational Mech. Anal. **13** (1963) 330–347.

14. Wilf, H. S., The numerical solution of polynomial equations, *Mathematical Methods for Digital Computers*, A. Ralston and H. Wilf, eds., Vol. 1, John Wiley and Sons, NY, 1967, 233–241.

15. Wilkinson, J. H., *The Algebraic Eigenvalue Problem*, Oxford Univ. Press, NY, 1996.

Franklin Kemp
3004 Pataula Lane
Plano, TX 75074-8765
lfkemp@attbi.com

L_p Markov-Bernstein Type Inequalities on All Arcs of the Circle for Generalized Polynomials

C. K. Kobindarajah

Abstract. Let $0 < p < \infty$ and $0 \leq \alpha < \beta \leq 2\pi$. For generalized nonnegative algebraic polynomial P of (generalized) degree at most N and with restricted zeros, we establish L_p Markov-Bernstein type inequalities on arcs of the circle.

§1. Introduction and Results

The classical Bernstein's inequality for trigonometric polynomials

$$S_n(\theta) := \sum_{j=0}^{n} (c_j \cos j\theta + d_j \sin j\theta)$$

of degree at most n is

$$\|S_n'\|_{L_\infty[0,2\pi]} \leq n \|S_n\|_{L_\infty[0,2\pi]}.$$

In the 1950's V.S. Videnskii generalized the L_∞ inequality to the case where the interval over which the norm is taken is shorter than the period. He obtained that

$$\|S_n'\|_{L_\infty[-\omega,\omega]} \leq 2n^2 \cot \omega/2 \, \|S_n\|_{L_\infty[-\omega,\omega]}, \quad 0 < \omega < \pi, \text{ for } n \geq n_0(\omega). \tag{1}$$

As an L_p analogue of the above inequality D. S. Lubinsky [11] proved the following result.

Approximation Theory X: Abstract and Classical Analysis
Charles K. Chui, Larry L. Schumaker, and Joachim Stöckler (eds.), pp. 267–276.
Copyright Ⓒ 2002 by Vanderbilt University Press, Nashville, TN.
ISBN 0-8265-1415-4.

Theorem 1. Let $0 < p < \infty$ and $0 \le \alpha < \beta \le 2\pi$. Then for trigonometric polynomials S_n of degree at most n,

$$\int_\alpha^\beta |S_n'(\theta)|^p \left[\left| \sin\left(\frac{\theta - \alpha}{2}\right) \right| \left| \sin\left(\frac{\theta - \beta}{2}\right) \right| + \left(\frac{\beta - \alpha}{n}\right)^2 \right]^{p/2} d\theta$$

$$\le Cn^p \int_\alpha^\beta |S_n(\theta)|^p \, d\theta. \tag{2}$$

Here C is independent of α, β, n, and S_n.

In [8] we established an improvement of Theorem 1 as follows.

Theorem 2. Let $0 < p < \infty$ and $0 \le \alpha < \beta \le 2\pi$. Then for $n \ge 1$ and trigonometric polynomials S_n of degree at most n,

$$\int_\alpha^\beta |S_n'(\theta)|^p \left[\frac{\left| \sin\left(\frac{\theta-\alpha}{2}\right) \right| \left| \sin\left(\frac{\theta-\beta}{2}\right) \right| + \left(\frac{\beta-\alpha}{n}\right)^2}{\left(\cos\frac{\theta - \frac{\alpha+\beta}{2}}{2}\right)^2 + \left(\frac{1}{n}\right)^2} \right]^{p/2} d\theta$$

$$\le Cn^p \int_\alpha^\beta |S_n(\theta)|^p \, d\theta. \tag{3}$$

Here C is independent of α, β, n, S_n.

We deduced Theorem 2 from the following algebraic analogue.

Theorem 3. Let $0 < p < \infty$ and $0 \le \alpha < \beta \le 2\pi$. Let

$$\varepsilon_n(z) := \frac{1}{n} \left[\frac{\left| z - e^{i\alpha} \right| \left| z - e^{i\beta} \right| + \left(\frac{\beta-\alpha}{n}\right)^2}{\left| z + e^{i\frac{\alpha+\beta}{2}} \right|^2 + \left(\frac{1}{n}\right)^2} \right]^{1/2}. \tag{4}$$

Then for $n \ge 1$ and algebraic polynomials P of degree at most n,

$$\int_\alpha^\beta \left| (P'\varepsilon_n)\left(e^{i\theta}\right) \right|^p d\theta \le C \int_\alpha^\beta \left| P\left(e^{i\theta}\right) \right|^p d\theta. \tag{5}$$

Here C is independent of α, β, n, P.

In this paper, we prove that Theorem 3 still holds for generalized nonnegative algebraic polynomials with restricted zeros. This is stated in the next result.

Theorem 4. *Let*

$$P(z) := \omega \prod_{j=1}^{n} (z - z_j)^{r_j} \text{ with } r_j \geq 1, \ r_j \in \mathbb{R}, \text{ and } \omega \in \mathbb{C}$$

be a nonnegative generalized algebraic polynomial with generalized degree

$$N := \sum_{j=1}^{n} r_j.$$

Let $0 < p < \infty$ and $0 \leq \alpha < \beta \leq 2\pi$. Let

$$\varepsilon_N(z) := \frac{1}{N} \left[\frac{\left| z - e^{i\alpha} \right| \left| z - e^{i\beta} \right| + \left(\frac{\beta - \alpha}{N} \right)^2}{\left| z + e^{i\frac{\alpha+\beta}{2}} \right|^2 + \left(\frac{1}{N} \right)^2} \right]^{1/2},$$

and let

$$\triangle := \left\{ e^{i\theta} : \theta \in [\alpha, 2\pi - \alpha] \right\}.$$

Assume that for all j,

$$z_j \notin \bigcup_{z \in \triangle} \{ t : |t - z| \leq \varepsilon_N(z)/100 \} =: \gamma.$$

Then we have

$$\int_{\alpha}^{\beta} \left| (P' \varepsilon_N) \left(e^{i\theta} \right) \right|^p d\theta \leq C \int_{\alpha}^{\beta} \left| P \left(e^{i\theta} \right) \right|^p d\theta.$$

Here C is independent of α, β, N and P.

Note that our restriction on the zeros ensures that there is a single-valued branch of $P(z)$ in a neighbourhood of $\triangle$. Moreover, we may set
$P'(z) := P(z) \sum_{j=1}^{n} \frac{r_j}{z - z_j}$.

Our method of proof uses Carleson measures and Carleson's inequality, and also uses ideas from [8,11]. We shall prove Theorem 4 in the next section, based on a technical lemma whose statement we defer to the end of that section.

§2. The Proof of Theorem 4

Throughout, $C, C_0, C_1, C_2, \dots$ denote constants that are independent of α, β, ω, N and generalized nonnegative polynomials P of degree at most N or trigonometric polynomials S_n of degree at most n. They may however depend on p. The same symbol does not necessarily denote the same constant in different occurrences. We shall prove Theorem 4 in several steps:

Step 1. Reduction to the case $0 < \alpha < \pi$, $\beta = 2\pi - \alpha$: see [8,11].

Step 2. Pointwise estimates for $P'(z)$ when $p \geq 1$. As the zeros of P lie outside γ, we can choose a single-valued branch of $P(z)$ inside γ. Then, as in [8,11], by Cauchy's integral formula for derivatives and by Hölder's inequality we get

$$(|P'(z)|\, \varepsilon_N(z))^p \leq 100^p \frac{1}{2\pi} \int_{-\pi}^{\pi} \left| P\left(z + \frac{\varepsilon_N(z)}{100} e^{i\theta}\right) \right|^p d\theta. \tag{6}$$

Step 3. Pointwise estimates for $P'(z)$ when $p < 1$. Again, as P has no zeros inside γ we can choose a single-valued branch of P^p there with the properties

$$\frac{d}{dt} P(t)^p \big|_{t=z} = p P(z)^p \frac{P'(z)}{P(z)}$$

and

$$|P^p(t)| = |P(t)|^p.$$

Then by Cauchy's integral formula for derivatives and subharmonicity, we get

$$(|P'(z)|\, \varepsilon_N(z))^p \leq C_0 \cdot \frac{1}{2\pi} \int_{-\pi}^{\pi} \left| P\left(z + \frac{\varepsilon_N(z)}{100} e^{i\theta}\right) \right|^p d\theta, \tag{7}$$

where $C_0 := \max\left\{100^p, \left(\frac{100}{p}\right)^p\right\}$, see [8,11].

Step 4. Integration of the pointwise estimates. We obtain by integration of (7) that

$$\int_{\alpha}^{2\pi - \alpha} |(P'\varepsilon_N)(e^{i\theta})|^p d\theta \leq C_0 \int |P(z)|^p d\sigma, \tag{8}$$

where the measure σ is defined by

$$\int f \, d\sigma := \int_{\alpha}^{2\pi - \alpha} \left[\frac{1}{2\pi} \int_{-\pi}^{\pi} f\left(e^{is} + \frac{\varepsilon_N(e^{is})}{100} e^{i\theta} \right) d\theta \right] ds. \tag{9}$$

We now wish to pass from the right-hand side of (8) to an estimate over the whole unit circle. This passage would be permitted by a famous result of Carleson, provided P is analytic off the unit circle, and provided it has suitable behaviour at ∞. To take care of the fact that it does not have the correct behaviour at ∞, we need a conformal map.

Step 5. Conformal map Ψ of $\mathbb{C}\backslash\triangle$ onto $\{\omega : |\omega| > 1\}$: see [8,11]. Then (9) becomes

$$\int_{\alpha}^{2\pi-\alpha} \left|\left(P'\varepsilon_N\right)\left(e^{i\theta}\right)\right|^p d\theta$$
$$\leq C_1^p C_0 \left(\int \left|\frac{P}{\Psi^{N+\ell}}\right|^p (t)d\sigma^+(t) + \int \left|\frac{P}{\Psi^{N+\ell}}\right|^p \left(\frac{1}{t}\right) d\sigma^{\#}(t)\right), \tag{10}$$

where ℓ is the smallest integer $> 1/p$ and measures σ^* and $\sigma^{\#}$ are defined by

$$\sigma^+(S) := \sigma(S \cap \{z : |z| < 1\}),$$
$$\sigma^-(S) := \sigma(S \cap \{z : |z| > 1\}),$$
$$\sigma^{\#}(S) := \sigma^-(1/S) := \sigma^- \left(\left\{\frac{1}{t} : t \in S\right\}\right).$$

We next focus on handling the first integral on the right-hand side of (10).

Step 6. Estimating the integral involving σ^+. We know that if $|z_j| \leq 1$,

$$|z - z_j| = |1 - z\bar{z}_j| \text{ for } |z| = 1;$$
$$|z - z_j| \leq |1 - z\bar{z}_j| \text{ for } |z| \leq 1.$$

Then we split

$$P(z) = \omega \prod_{|z_j|\leq 1} (z - z_j)^{r_j} \prod_{|z_j|>1} (z - z_j)^{r_j},$$

and form

$$P^*(z) := \omega \prod_{|z_j|<1} (1 - z\bar{z}_j)^{r_j} \prod_{|z_j|>1} (z - z_j)^{r_j}, \quad |z| < 1.$$

Note that one can define a single-valued branch of $P^*(z)$, $|z| < 1$. We also have

$$|P(z)| = |P^*(z)|, \quad |z| = 1;$$
$$|P(z)| \leq |P^*(z)|, \quad |z| \leq 1. \tag{11}$$

Furthermore, for each factor $z - z_j$ in P, we define

$$b_j(z) := \begin{cases} (z - z_j) / \left(\frac{\Psi(z) - \Psi(z_j)}{1 - \overline{\Psi(z_j)}\Psi(z)} \right), & z \neq z_j \\ \left(1 - |\Psi(z_j)|^2 \right) / \Psi'(z_j), & z = z_j. \end{cases} \quad (12)$$

This is analytic in $\mathbb{C}\backslash\Delta$, does not have any zeros there, and moreover, since $|\Psi(z)| \to 1$ as $z \to \Delta$, we see that

$$|b_j(z)| = |z - z_j|, \, z \in \Delta; \quad |b_j(z)| \geq |z - z_j|, \, z \in \mathbb{C}\backslash\Delta.$$

(Recall that we extended Ψ to Δ as an exterior boundary value.) Define

$$P^{\#}(z) := \omega \left(\prod_{z_j \notin \Delta} b_j^{r_j}(z) \right).$$

Then we have from (11) that

$$|P| = |P^*| = |P^{\#}| \text{ on } \Delta,$$
$$|P| \leq |P^*| \leq |P^{\#}| \text{ in } \mathbb{C}\backslash\Delta,$$

and

$$\int \left| \frac{P}{\Psi^{N+\ell}} \right|^p (t) d\sigma^+(t) \leq \int \left| \frac{P^*}{\Psi^{N+\ell}} \right|^p (t) d\sigma^+(t). \quad (13)$$

Now from Carleson's inequality we get, with $N(\sigma^+)$ denoting the Carleson norm of σ^+,

$$\int \left| \frac{P}{\Psi^{N+\ell}} \right|^p d\sigma \leq \int \left| \frac{P^*}{\Psi^{N+\ell}} \right|^p d\sigma^+ \leq C_2 N(\sigma^+) \int_0^{2\pi} \left| \frac{P^*}{\Psi^{N+\ell}} (e^{i\theta}) \right|^p d\theta. \quad (14)$$

Recall that a positive Borel measure μ with support inside the unit ball is called a **Carleson measure** if there exists $A > 0$ such that for every $0 < h < 1$, and every sector

$$S := \left\{ re^{i\theta} : r \in [1 - h, 1]; \, |\theta - \theta_0| \leq h \right\},$$

we have

$$\mu(S) \leq Ah.$$

The smallest such A is called the **Carleson norm** of μ and is denoted by $N(\mu)$. One striking feature of such a measure is the inequality (Carleson's inequality)

$$\int |f|^p d\mu \leq C_p N(\mu) \int_0^{2\pi} \left| f(e^{i\theta}) \right|^p d\theta,$$

valid for every function f in the Hardy p space on the unit ball. Here C_p depends only on p. For more details see [4].

Step 7. Estimating the integral involving $\sigma^{\#}$. Applying Carleson's inequality to $h(t) := \left(P^* \left(\frac{1}{t} \right) / \Psi \left(\frac{1}{t} \right)^{N+\ell} \right)$, we get

$$\int \left| \frac{P^*}{\Psi^{N+\ell}} \right|^p \left(\frac{1}{t} \right) d\sigma^{\#}(t) \le C_2 N \left(\sigma^{\#} \right) \int_0^{2\pi} \left| \frac{P^*}{\Psi^{N+\ell}} \left(e^{-i\theta} \right) \right|^p d\theta$$

$$\Rightarrow \int \left| \frac{P}{\Psi^{N+\ell}} \right|^p \left(\frac{1}{t} \right) d\sigma^{\#}(t) \le C_2 N \left(\sigma^{\#} \right) \int_0^{2\pi} \left| \frac{P^*}{\Psi^{N+\ell}} \left(e^{-i\theta} \right) \right|^p d\theta.$$

Combined with (10) and (14), this gives

$$\int_\alpha^{2\pi-\alpha} \left| (P' \varepsilon_N) \left(e^{i\theta} \right) \right|^p d\theta \le$$

$$C_0 C_1^p C_2 \left(N \left(\sigma^+ \right) + N \left(\sigma^{\#} \right) \right) \int_0^{2\pi} \left| \frac{P^{\#}}{\Psi^{N+\ell}} \left(e^{i\theta} \right) \right|^p d\theta. \tag{15}$$

Step 8. Passing from the whole unit circle to $\triangle$ when $p > 1$. In [11] the author established an estimate of the form

$$\int_{\Gamma \backslash \triangle} |g(t)|^p \, |dt| \le C_3 \left(\int_\triangle |g_+(t)|^p \, |dt| + |g_-(t)|^p \, |dt| \right), \tag{16}$$

valid for all functions g analytic in $\mathbb{C} \backslash \triangle$, with limit 0 at ∞, and interior and exterior boundary values g_+ and g_- for which the right-hand side of (16) is finite. Applying (16) to $g := P^{\#}/\Psi^{N+\ell}$ we get

$$\int_{\Gamma \backslash \triangle} \left| \frac{P^{\#}}{\Psi^{N+\ell}} \right|^p \, |dt| \le C_3 \int_\triangle \left| P^{\#}(t) \right|^p \, |dt|$$

$$\Rightarrow \int_0^{2\pi} \left| \frac{P^{\#}}{\Psi^{N+\ell}} \left(e^{i\theta} \right) \right|^p d\theta \le (1 + C_3) \left(\int_\alpha^{2\pi-\alpha} \left| P \left(e^{i\theta} \right) \right|^p d\theta \right).$$

Here Γ denotes the whole unit circle, $|dt|$ denotes arclength on Γ, and C_3 is a constant depending only on p. Now (15) becomes

$$\int_\alpha^{2\pi-\alpha} \left| (P' \varepsilon_N) \left(e^{i\theta} \right) \right|^p d\theta$$

$$\le C_0 C_1^p C_2 (1 + C_3) \left(N \left(\sigma^+ \right) + N \left(\sigma^{\#} \right) \right) \int_\alpha^{2\pi-\alpha} \left| P \left(e^{i\theta} \right) \right|^p d\theta. \tag{17}$$

Step 9. Passing from the whole unit circle to $\triangle$ when $p \leq 1$.
Again we apply (16) with p replaced by $q := \ell p(> 1)$ and with

$$g := \left(P^\# / \Psi^N\right)^{p/q} \Psi^{-1} = (P^\# / \Psi^{N+\ell})^{p/q}. \tag{18}$$

The problem remains that, in general, g does not possess the required properties. To circumvent this, we proceed as follows: firstly, we may assume that P has full degree N. For, if P has degree less than N, we can add a term of the form $\delta (z - z_j)^N$, giving $P(z) + \delta (z - z_j)^N$, a polynomial of full degree N. Once our theorem is proved for such P, we can then let $\delta \to 0+$.

So assume that P has degree N. Then P / Ψ^N (and hence $P^\# / \Psi^N$) is analytic in $\mathbb{C} \backslash \triangle$ and has a finite non-zero limit at ∞, so is analytic at ∞. Now if all zeros of $P^\#$ lie on $\triangle$, then we may define a single-valued branch of g of (29) in $\overline{\mathbb{C}} \backslash \triangle$. As before, (16) with q replacing p gives

$$\int_{\Gamma \backslash \triangle} |g(t)|^q \, |dt| \leq C_3 \left(\int_\triangle |g_+(t)|^q \, |dt| + |g_-(t)|^q \, |dt| \right)$$

$$\Rightarrow \int_{\Gamma \backslash \triangle} \left| P^\# / \Psi^{N+\ell} \right|^p |dt| \leq 2C_3 \int_\triangle \left| P^\#(t) \right|^p |dt|$$

from which we obtain an estimate similar to (17). When $P^\#$ has zeros in $\mathbb{C} \backslash \triangle$, we adopt a standard procedure to "reflect" these out of $\mathbb{C} \backslash \triangle$. For each factor $z - z_j$ in P with $z_j \notin \triangle$, define $b_j(z)$ as in (12). We may now choose a branch of

$$g(z) := \left[d \left(\prod_{z_j \notin \triangle} b_j^{r_j}(z) \right) \Big/ \Psi(z)^N \right]^{p/q} \Big/ \Psi(z)$$

that is single-valued and analytic in $\mathbb{C} \backslash \triangle$, and has limit 0 at ∞. Then as $\Psi_\pm$ have absolute value 1 on $\triangle$, so that $|g_\pm|^q = |P|^p$ on $\triangle$, we deduce from (16) that

$$\int_{\Gamma \backslash \triangle} \left| P^\#(t) / \Psi(t)^{N+\ell} \right|^p |dt| \leq \int_{\Gamma \backslash \triangle} |g(t)|^q \, |dt|$$

$$\leq C_3 \int_\triangle \left(|g_+(t)|^q + |g_-(t)|^q \right) |dt| = 2C_3 \int_\triangle \left| P^\#(t) \right|^p |dt|.$$

Again we obtain an estimate similar to (17). Thus, we have

$$\int_\alpha^{2\pi - \alpha} \left| (P' \varepsilon_N) \left(e^{i\theta} \right) \right|^p d\theta$$

$$\leq C_0 C_1^p C_2 (1 + C_3) \left(N \left(\sigma^+ \right) + N \left(\sigma^\# \right) \right) \int_\alpha^{2\pi - \alpha} \left| P \left(e^{i\theta} \right) \right|^p d\theta,$$

for a unit circle as $|P(z)| = |P^*(z)|$, $|z| = 1$. Then by Lemma 5, this becomes

$$\int_\alpha^{2\pi-\alpha} \left|(P'\varepsilon_N)\left(e^{i\theta}\right)\right|^p d\theta \leq C_5 \int_\alpha^{2\pi-\alpha} \left|P\left(e^{i\theta}\right)\right|^p d\theta, \qquad (19)$$

where C_5 is a constant that depends only on the numerical constants C_j, $1 \leq j \leq 4$ that arise from

(a) the bound on the conformal map Ψ;

(b) Carleson's inequality;

(c) the norm of the Hilbert transform as an operator on $L_p(\Gamma)$ and the choice of ℓ;

(d) the upper bound on the Carleson norms of σ^+ and $\sigma^\#$ as before.

Lemma 5. *For some C_1 independent of N, α, β and P, $N(\sigma^+)+N(\sigma^\#) \leq C_1$.*

Acknowledgments. I would like to thank my supervisor Professor D. S. Lubinsky for his valuable suggestions and help.

References

1. Borwein, P. B., and T. Erdélyi, *Polynomials and Polynomial Inequalities*, Springer, 1995.

2. Carleson, L., Interpolation by bounded analytic functions and the Corona problem, Annals of Mathematics **76** (1962), 547–559.

3. DeVore, R. A., and G. G. Lorentz, *Constructive Approximation*, Springer-Verlag, Berlin, 1993.

4. Garnett, J. B., *Bounded Analytic Functions*, Academic Press, 1981.

5. Golinskii, L., Akhiezer's orthogonal polynomials and Bernstein-Szego method for a circular arc, J. Approx. Theory **95** (1998), 229–263.

6. Golinskii, L., D. S. Lubinsky and P. Nevai, Large sieve estimates on arcs of the circle, Journal of Number Theory, to appear.

7. Kobindarajah, C. K., Markov-Bernstein Inequalities, Ph.D. Thesis, Witwaterstand University, Johannesburg, South Africa, 2001.

8. Kobindarajah, C. K., and D. S. Lubinsky, L_p Markov-Bernstein inequalities on all arcs of the circle, J. Approx. Theory, to appear.

9. Levin, A. L., and D. S. Lubinsky, L_p Markov-Bernstein inequalities for Freud weights, J. Approx. Theory **77** (1994), 229–248.

10. Levin, A. L., and D. S. Lubinsky, Orthogonal polynomials associated with exponential weights, to appear.

11. Lubinsky, D. S., L_p Markov-Bernstein inequalities on arcs of the circle, J. Approx. Theory **108** (2001), 1–17.

C. K. Kobindarajah
Department of Mathematics
University of Witwatersrand
Wits 2050, Johannesburg
SOUTH AFRICA
036kobi@cosmos.wits.ac.za

and
Department of Mathematics
Eastern University
Chenkalady
SRI LANKA
kshant@yahoo.com

Positive Definite Kernel Interpolation on Manifolds: Convergence Rates

Jeremy Levesley and David L. Ragozin

Abstract. Local error estimates are presented for interpolation on Riemannian manifolds by translates of smooth kernels. For a function in the native space of a $\mathcal{C}^{2r}$ positive definite kernel k, the uniform error over a compact set V for interpolation by translates of k, the k-spline interpolant, goes to zero like ρ^r. Here ρ is the density of the interpolating set in the compact set V with respect to a fixed metric locally boundedly equivalent to the arclength metric. The local Euclidean space version is proved by combining the scale invariance of the Lebesgue constants for $2r$'th degree polynomial interpolation with the approximation order of $2r$'th degree Taylor expansions. The lift to an arbitrary compact V follows easily from norm minimization properties of the interpolant.

§1. Introduction

Our basic concern in this paper is to provide quantitative error estimates for the (uniform) approximation of a continuous function f on a Riemannian manifold X by *finite* linear combinations $s = \sum_{y \in Y} a_y k(\cdot, y)$ of a fixed kernel function k on $X \times X$. The most commonly considered examples of kernels have the form $k(x, y) = \phi(\|x - y\|)$, for $x, y \in \mathbb{R}^d$, so each function $k(\cdot, y)$ is usually referred to as a **translate** of k (or ϕ). Also, any linear combination, s, of translates is usually called a (**variational**) k-spline with knot set $Y = \{y : a_y \neq 0\}$.

We shall only allow kernel functions for which the **interpolating spline** $s_Y(f)$ always exists for an arbitrary finite set Y and function f. Recall that $s_Y(f)$ is defined by

$$s_Y(f) = \sum_{y \in Y} a_y k(\cdot, y), \quad \text{where } s_Y(f)(y) = f(y) \text{ for each } y \in Y. \quad (1)$$

Approximation Theory X: Abstract and Classical Analysis
Charles K. Chui, Larry L. Schumaker, and Joachim Stöckler (eds.), pp. 277–285.
Copyright Ⓒ 2002 by Vanderbilt University Press, Nashville, TN.
ISBN 0-8265-1415-4.

This interpolation problem can be solved for arbitrary finite Y only when every matrix $k_Y = [k(y, y')]$ is nonsingular. To assure this, we assume that k_Y is strictly positive definite for every choice of Y. Thus, our interpolants will be built from translates of a *real, symmetric strictly positive definite continuous* kernel k. For such k, we shall give a relatively simple proof of (local) error estimates for $|f(x) - s_Y(f)(x)|$.

Associated with any positive definite k and any subset $U \subset X$, there is the space $\mathcal{T}_k(U)$ of all translates or k-splines with knots inside U and its closure $\mathcal{W}_k(U)$, a Hilbert space called the native space of k on U:

Definition 1.

i) $\mathcal{T}_k(U) := \{s = \sum_{y \in U} a_y k(\cdot, y) : a_y = 0$ except for finitely many $y \in U\}$.

ii) $\mathcal{W}_k(U) := \overline{\mathcal{T}_k(U)}$, where the closure is with respect to the native norm $\|s\|_{k,U}^2 := \sum_{y,z \in U} a_y a_z k(y, z)$.

iii) *The inner product of* $s = \sum a_y k(\cdot, y)$ *and* $t = \sum b_z k(\cdot, z)$ *is*

$$[s, t]_U := \sum_{y,z \in U} a_y b_z k(y, z).$$

Here $\mathcal{T}_k, \mathcal{W}_k, \|s\|_k$, and $[s, t]$ all denote the case when $U = X$. When $U = X$, evaluation of Definition 1.iii at $t = k(\cdot, x)$, i.e. $b_z = \delta_x(z)$, yields

$$[s, k(\cdot, x)] = \sum_{y \in X} a_y k(y, x) = \sum_{y \in X} a_y k(x, y) = s(x). \tag{2}$$

Moreover, application of the Cauchy-Schwarz inequality on $\mathcal{T}_k$ implies

$$|s(x)| \leq k(x, x)^{1/2} \|s\|_k.$$

Thus, the inclusion $s \in \mathcal{T}_k \mapsto s \in \mathcal{C}(X)$ is continuous from the $\|\cdot\|_k$ topology to the topology of uniform convergence on compacta. In particular, this map extends to an inclusion of $\mathcal{W}_k$ into $\mathcal{C}(X)$, and (2) says k is a reproducing kernel for this space of continuous functions. Our key result is

Theorem 2. *Let* X *be a smooth Riemannian manifold with* d *a metric on* X. *Suppose* d *and the arclength metric are (locally) boundedly equivalent. Let* k *be a strictly positive definite kernel on* X, *and suppose* $k \in \mathcal{C}^{2r}$ *near the diagonal in* $X \times X$. *Then for any compact* $V \subset X$, *there is a constant* $C_{V,k}$, *depending only on* k *and* V *such that for any* $f \in \mathcal{W}_k$, *the error in uniformly approximating* f *by its interpolant* $s_Y(f)$ *on* V *satisfies*

$$\|f - s_Y(f)\|_{\infty,V} \leq C_{V,k} \rho^r \|f\|_k, \tag{3}$$

where $\rho = \mathrm{den}\,(Y, V) := \sup_{x \in V} \min_{y \in Y} d(x, y)$.

This result captures many of the local error estimates which have been proved for specific kernels on $\mathbb{R}^d$, S^{d-1}, and other manifolds [1,2,3,4,5]. Its proof is quite elementary, using only Taylor's theorem and the $\|\cdot\|_k$ minimizing properties of the spline interpolants. The next three sections give the details.

§2. Spline Interpolation, Orthogonal Projection and the Error in Interpolation of a Translate

First we recall that the key to uniform norm error estimates in a reproducing kernel space is the fact that interpolants are orthogonal projections. The relevant facts are summarized in

Proposition 3. *Let $s_Y(f)$ be the k-spline interpolant on Y to $f \in \mathcal{W}_k$. Then*

i) $s_Y(f)$ *is the orthogonal projection of f onto $\mathcal{W}_k(Y)$. That is* $[f - s_Y(f), s] = 0$ *for all $s \in \mathcal{W}_k(Y)$. In fact,*

$$[h, s] = 0 \text{ for any } h \in \mathcal{W}_k \text{ with } h(Y) = 0 \text{ and any } s \in \mathcal{W}_k(Y). \quad (4)$$

ii) $s_Y(f)$ *has the least norm of any $g \in \mathcal{W}_k$ which interpolates f on Y.*

Proof: The fact that any spline $s = \sum_{y \in Y} a_y k(\cdot, y)$ is orthogonal to any $h \in \mathcal{W}_k$ with $h(Y) = 0$ is immediate from the reproducing kernel property (2) of k

$$[h, s] = [h, \sum_{y \in Y} a_y k(\cdot, y)] = \sum_{y \in Y} a_y h(y) = 0.$$

Thus (4) holds. In particular when $h = f - s_Y(f)$ we have $[f - s_Y(f), s] = 0$ for all such s, which proves (i).

For (ii) note that for any $g \in \mathcal{W}_k$ which interpolates f, $h = g - s_Y(f)$ satisfies $h(Y) = 0$. So by (i) and Pythagoras

$$\begin{aligned} [g, g] &= [g - s_Y(f) + s_Y(f), g - s_Y(f) + s_Y(f)] \\ &= [g - s_Y(f), g - s_Y(f)] + [s_Y(f), s_Y(f)]. \end{aligned}$$

Hence $\|g\|_k \geq \|s_Y(f)\|_k$ and (ii) holds. $\square$

We can use (4) to produce an error estimate for k-spline interpolation in $\mathcal{W}_k$.

Proposition 4. *Let $s_Y(f)$ be the k-spline interpolant to $f \in \mathcal{W}_k$ at the points of Y. Then, for $x \in X$,*

$$|f(x) - s_Y(f)(x)| \leq P(x, Y)\|f\|_k, \quad (5)$$

where

$$P(x, Y) = \min_{\{a_y : y \in Y\} \subset \mathbf{R}} \left\| k(\cdot, x) - \sum_{y \in Y} a_y k(\cdot, y) \right\|_k$$

$$= \min_{a_y \in \mathbf{R}} \left\{ k(x, x) - 2 \sum_{y \in Y} a_y k(x, y) + \sum_{y, z \in Y} a_y a_z k(y, z) \right\}^{1/2} \quad (6)$$

$$= \| k(\cdot, x) - s_Y(k(\cdot, x)) \|_k.$$

Proof: The reproducing kernel property (2) and the orthogonality property (4) for any $s \in \mathcal{W}_k(Y)$ show

$$|f(x) - s_Y(f)(x)| = [f - s_Y(f), k(\cdot, x)] = [f - s_Y(f), k(\cdot, x) - s]$$

since $h = f - s_Y(f)$ vanishes on Y. Using Cauchy-Schwarz and Proposition 3, we get

$$|f(x) - s_Y(f)(x)| \leq \| f - s_Y(f) \|_k \| k(\cdot, x) - s \|_k \leq \| f \|_k \| k(\cdot, x) - s \|_k.$$

Now when $s = \sum_{y \in Y} a_y k(\cdot, y)$ the second factor on the right evaluates to the term inside the braces in (6). Also the minimum value of this factor over all $s \in \mathcal{W}_k(Y)$ is attained at the interpolant $s = s_Y(k(\cdot, x))$ by Proposition 3.ii. $\square$

Remark 5. The equality of the extreme terms in (6) becomes $P(x, Y) = |k(x, x) - s_Y(k(\cdot, x))(x)|^{1/2}$ using the reproducing kernel property (2). Thus, we have the not too surprising result that the interpolation error at x for a general $f \in \mathcal{W}_k$ is estimable in terms of the interpolation error at x for the reproducing kernel, $k(\cdot, x)$ itself.

§3. Localization of Error Estimates

The characterization of the error in (5) in terms of $P(x, Y)$ in (6) shows that it depends only on the behavior of k on $\{x, Y\}$. Moreover (6) also shows $P(x, Y) \leq P(x, Z)$ when $Z \subset Y$. This leads to an immediate localization principle for our main error estimate in (3).

Lemma 6. *Suppose for each* $z \in X$ *there exists a closed ball* $\overline{B(z, \rho_z)}$ *of radius* $\rho_z > 0$ *and density bound* $\rho_{0,z}$ *such that for any compact* $V_z \subset \overline{B(z, \rho_z)}$

$$\text{(3) holds for any finite } Y_z \subset \overline{B(z, \rho_z)} \text{ with den}(Y_z, V_z) \leq \rho_{0,z}. \quad (7)$$

Then (3) holds for arbitrary compact V and finite Y, provided $\rho = \operatorname{den}(Y, V)$ is sufficiently small.

Proof: For an arbitrary compact V, $V \subset \cup_1^s B(z_i, \rho_{z_i}/2)$ for some finite number of balls. Let $V_i = \overline{B(z_i, \rho_{z_i}/2)} \cap V$. Let $Y_i = Y \cap \overline{B(z_i, \rho_{z_i})}$. Suppose Y has density ρ, with

$$\rho \leq \rho_0 := \min_i \{\rho_{z_i}/2, \rho_{0,z_i} : i = 1, \ldots, s\}.$$

Then, for any $x \in V_i \subset V$, there is $y \in Y$ with $d(x, y) \leq \rho$. Such y must be in Y_i, since each $\overline{y} \notin Y_i$ satisfies

$$d(x, \overline{y}) \geq d(z_i, \overline{y}) - d(x, z_i) \geq \rho_{z_i}/2 > \rho.$$

So

$$\rho_i := \operatorname{den}(Y_i, V_i) \leq \rho \leq \rho_{0, z_i}.$$

Now the facts that (6) and the hypothesis (7) hold for $\overline{B(z_i, \rho_{z_i}/2)} \supset V_i$ yield

$$P(x, Y) \leq P(x, Y_i) \leq C_{V_i, k} \rho_i^r \leq \max_i \{C_{V_i, k}\} \rho^r, \quad x \in V_i.$$

Since the right side is independent of i, we get that

$$\sup_{x \in V} P(x, Y) \leq C_{V, k} \rho^r, \quad C_{V, k} = \max_i C_{V_i, k}. \quad \square$$

The required localized version of (3) will follow from the Euclidean space version of Theorem 2. The following lemma summarises some useful standard results concerning the scaling of Lagrange polynomials in Euclidean space. The proof of these may be found in e.g. [5].

Lemma 7. *Denote the polynomials of total degree less than $2r$ by Π_{2r}. Let Z be a Π_{2r} unisolvent set in $\mathbb{R}^d$ with $\operatorname{card}(Z) = \dim(\Pi_{2r})$. Let $\{p_z : z \in Z\}$ be the Lagrange polynomials based on Z, i.e., $p_z(y) = \delta_{yz}$ for $y \in Z$. Then for any $\lambda > 0$,*

(i) *$\lambda Z := \{\lambda z : z \in Z\}$ is unisolvent and the Lagrange polynomials based on λZ are $p_z^\lambda = p_z(\cdot/\lambda)$, $z \in Z$. In particular $p_z^\lambda(0) = p_z(0)$.*

(ii) *There exist positive constants d_0 and C_0 such that any set $S := \{\sigma_z : z \in Z\}$ with $\|z - \sigma_z\| \leq d_0$ is Π_{2r} unisolvent. Furthermore, if $\{p_z : z \in Z\}$ is a Lagrange basis for Π_{2r} based on S, then $|p_z(0)| \leq C_0$.*

Now the critical estimates for $P(x, Y)$ on Euclidean spaces follow directly from this lemma and Taylor series error estimates.

Proposition 8. *Suppose K is a strictly positive definite symmetric function on some compact neighborhood $\overline{U} \times \overline{U}$ of $(0,0)$ in $\mathbb{R}^d \times \mathbb{R}^d$. Further, suppose K is in $\mathcal{C}^{2r}(\overline{U} \times \overline{U})$. Then there exist constants $C > 0$ and ρ_0 such that for any finite set $Y \subset U$ with*

$$\mathrm{den}\,(Y,\overline{U}) := \sup_{v \in \overline{U}} \min_{y \in Y} \|v - y\| \leq \rho < \rho_0, \tag{8}$$

the following estimate holds:

$$P(0,Y)^2 = \min_{a_y \in \mathbf{R}, y \in Y} \left\{ K(0,0) - 2 \sum_{y \in Y} a_y K(0,y) + \sum_{y,z \in Y} a_y a_z K(y,z) \right\}$$
$$\leq C\rho^{2r}. \tag{9}$$

Proof: If we scale the set Z of Lemma 7 by $\lambda = \rho/d_0$, then we have a set λZ and around each $\eta \in \lambda Z$ there is a ball of radius ρ from which we may choose one of the interpolation points of $y_z \in Y$. Lemma 7 guarantees that the set $Y_Z := \{y_z : z \in Z\}$ is unisolvent and furthermore, if p_y are Lagrange polynomials based on Y_Z, then for some constant C_0, $|p_y(0)| \leq C_0$ for all $y \in Y_Z$. The number ρ_0 must be chosen small enough so that

$$\rho/d_0 Z \subset U \text{ for all } \rho \leq \rho_0. \tag{10}$$

The Lebesgue constant $\Lambda = \sum_{y \in Y_Z} |p_y(0)|$ for the associated Lagrange interpolation operator evaluated at 0 is always bounded by the fixed constant $\mathrm{card}\,(Z)C_0$, independent of the scaling.

So, for any $\rho < \rho_0$, the estimate (9) is obtained from Taylor's theorem as follows. The set Y_Z is contained in an open $C_1\rho$ ball about 0 for some $C_1 > 0$. The left hand side (LHS) of (9) is only increased by restricting the sums to be taken only over Y_Z instead of all of Y. But then if $p_y, y \in Y_Z$ are the Lagrange polynomials based on Y_Z, setting $a_y = p_y(0)$ and using the reproducing properties of $K(\cdot, w)$ yields

$$LHS \leq [K(\cdot,0) - \textstyle\sum_{y \in Y_Z} p_y(0)K(\cdot,y),\ K(\cdot,0) - \sum_{y \in Y_Z} p_y(0)K(\cdot,y)]$$

$$= \int_{|\cdot| \leq C_1\rho} (K(\cdot,0) - \sum_{y \in Y_Z} p_y(0)K(\cdot,y)) d(\delta_0 - \sum_{y \in Y_Z} p_y(0)\delta_y).$$

But any polynomial $p \in \Pi_{2r}$ is annihilated by the discrete measure in the integral above, since $p(0) = \sum_{y \in Y_Z} p_y(0)p(y)$ by the Lagrange properties of p_y. In particular if $T_{2r,w} \in \Pi_{2r}$ is the Taylor polynomial for $K(\cdot,w)$ centered at w, then subtracting $T_{2r,0} - \sum_{y \in Y_Z} p_y(0)T_{2r,y}$ from the last integrand changes nothing. Therefore,

$$LHS \leq \int_{|\cdot| \leq C_1 \rho} (K(\cdot, 0) - T_{2r,0}(\cdot)$$
$$- \sum_{y \in Y_Z} p_y(0)(K(\cdot, y) - T_{2r,y}(\cdot)))d(\delta_0 - \sum_{y \in Y_Z} p_y(0)\delta_y).$$

Since $K \in \mathcal{C}^{2r}(\overline{U} \times \overline{U})$ and $\max_{y,y' \in Y_Z \cup \{0\}}\{\|y\|, \|y - y'\|\} \leq 2C_1\rho$ each of the Taylor series remainder terms under the integral is bounded by some constant multiple of $(2C_1\rho)^{2r}$ on the support of the discrete measure, where the constant depends only on the $2r$'th derivatives of $K(\cdot, y)$ and $K(\cdot, 0)$. Hence,

$$\text{LHS} \leq (1 + \Lambda)C_r(2C_1\rho)^{2r}(1 + \Lambda),$$

where C_r is a constant depending only on the $2r$'th derivatives of K arising from the use of Taylor's formula. $\square$

Although this last proposition explicitly only bounds $P(0, Y)$, the proof actually shows that the same bound holds for $P(x, Y)$ for all x such that

$$\rho/d_0 Z + x \subset U \text{ for all } \rho \leq \rho_0, \tag{11}$$

a generalization of (10). The key to this is that translation of Z preserves unisolvency and the Lebesgue constants for the associated Lagrange polynomials. Since U is open and Z is finite, there will be a $\sigma_0 > 0$ such that (11) holds for $\|x\| \leq \sigma_0$. Thus, we have the full Euclidean localized version of our error estimates.

Corollary 9. *For $K \in \mathcal{C}^{2r}(\overline{U} \times \overline{U})$ as in Proposition 8, there exists a nontrivial closed ball $\overline{B(0, \sigma_0)}$ such that for Y, ρ_0 and C as in that proposition,*

$$P(x, Y)^2 \leq C\rho^{2r} \text{ for all } x, \|x\| \leq \sigma_0.$$

Also the full version of Theorem 2 holds for $X = U$.

Corollary 10. *For a strictly positive definite $K \in \mathcal{C}^{2r}(U \times U)$, $U \subset \mathbb{R}^d$ open, and any finite Y, compact V included in U, there exist constants $C_{V,K}$, such that for all $f \in \mathcal{W}_K(U)$,*

$$\|f - s_Y(f)\|_{\infty,V} \leq C_{V,K}\rho^r\|f\|_{K,U} \tag{12}$$

where $\rho = \text{den}\,(Y, V) := \sup_{x \in V} \min_{y \in Y} \|x - y\|$.

Proof: Corollary 9 shows the hypotheses for the localization Lemma 6 are true for the closed balls $\overline{B(z, \sigma_z)} \subset U$. $\square$

§4. Proof of Full Error Bounds on X

All the ingredients are at hand for the proof of Theorem 2, which we repeat for convenience.

Theorem 2. *Let X be a smooth Riemannian manifold with d a metric on X. Suppose d and the arclength metric are (locally) boundedly equivalent. Let k be a strictly positive definite kernel on X, and suppose $k \in C^{2r}$ near the diagonal in $X \times X$. Then for any compact $V \subset X$, there is a constant $C_{V,k}$, depending only on k and V such that for any $f \in \mathcal{W}_k$, the error in uniformly approximating f by its interpolant $s_Y(f)$ on V satisfies*

$$\|f - s_Y(f)\|_{\infty,V} \le C_{V,k} \rho^r \|f\|_k$$

where $\rho = \operatorname{den}(Y, V) := \sup_{x \in V} \min_{y \in Y} d(x,y)$.

Proof: We only need to prove the hypotheses of the localization Lemma 6 hold. But for any $z \in X$ we can use an appropriate smooth chart $\phi : U \longrightarrow X$ from an open neighborhood of $0 \in \mathbb{R}^d$ with $\phi(0) = z$ to do this. First, for any such chart, the function

$$K : U \times U \longrightarrow \mathbb{R} \quad \text{such that } K(x,y) = k(\phi(x), \phi(y))$$

is strictly positive definite, since $K_Y = k_{\phi(Y)}$ is strictly positive definite. Since k is C^{2r} near the diagonal in $X \times X$, we may shrink U until K is C^{2r}. Thus by Corollary 10 applied to K and U, (12) holds. Now, it is clear that the map

$$\phi_* : \mathcal{T}_K(U) \longrightarrow \mathcal{T}_k(\phi(U)) : \sum a_y K(\cdot, y) \mapsto \sum a_y k(\phi(\cdot), \phi(y))$$

extends to a linear isometry of the associated native spaces. This guarantees that $P(x, Y) = P(\phi(x), \phi(Y))$, i.e., the error functions are the same.

To finish the proof we merely need to bring in the local bounded equivalence of the metric d and arclength metric on X. This means that on any compact subset $B \subset \phi(U)$, there are constants C_3, C_4 with

$$C_3 d(x,y) \le \|\phi^{-1}(x) - \phi^{-1}(y)\| \le C_4 d(x,y), \text{ all } x, y \in B.$$

From these inequalities it follows that for any finite $Y \subset B$

$$\frac{1}{C_4} \operatorname{den}(\phi^{-1}(Y), \phi^{-1}(B)) \le \operatorname{den}(Y, B) \le \frac{1}{C_3} \operatorname{den}(\phi^{-1}(Y), \phi^{-1}(B)).$$

Then (12) on U, restricted to $\phi^{-1}(B)$ with $B = \overline{B(z, \rho_z)} \subset U$ and $Y_z \subset B$ with $\operatorname{den}(Y, B) < \rho_0 / C_4$ yields (3) on any compact $V_z \subset B$ with

$$C_{V_z, k} = C_{\phi^{-1}(V_z), K}(C_3)^r.$$

Thus the localization hypotheses of Proposition 6 are satisfied. This completes the proof. $\square$

Acknowledgments. EPSRC Grant GR/L36222 supported J. Levesley and NSF Grant DMS-9972004 partially supported D. L. Ragozin.

References

1. Golitschek, M. von, and W. A. Light, Interpolation by polynomials and radial basis functions on spheres, Constr. Approx. **17** (2001), 1–18.

2. Levesley, J., W. Light, D. L. Ragozin, and X. Sun, A simple approach to the variational theory for interpolation on spheres, in *New Developments in Approximation Theory*, M. W. Müller, M. D. Buhmann, D. H. Mache and M. Felten (eds), ISNM **132**, Birkhäuser (1999), 117–143.

3. Levesley, J., and D. L. Ragozin , Radial basis interpolation on compact homogeneous manifolds: Convergence rates, Technical Report 2001/04, University of Leicester.

4. Light, W. A., and H. Wayne, Error estimates for approximation by radial basis functions, in *Wavelet Analysis and Approximation Theory*, S.P. Singh and A. Carbone (eds), Kluwer Academic, Dordrecht, 1995, 215–246.

5. Madych, W. R., and S. A. Nelson, Multivariate interpolation and conditionally positive definite functions II, Math. Comp. **54** (1990), 211–230.

Jeremy Levesley
Department of Mathematics
University of Leicester
Leicester LE1 7RH
UK
jl1@mcs.le.ac.uk

David L. Ragozin
University of Washington
Department of Mathematics
Seattle, WA 98195-4350
USA
rag@math.washington.edu

Asymptotic Expressions for Multivariate Positive Linear Operators

Antonio-Jesús López-Moreno, J. Martínez-Moreno,
and F.-J. Muñoz-Delgado

Abstract. In this paper we study the asymptotic expansion of the partial derivatives of sequences of linear shape preserving operators. We fix sufficient conditions on the sequence of operators under which the asymptotic formula for the partial derivatives can be obtained from the expansion of the sequence by differentiating. We explicitly compute the asymptotic expression of a multivariate version of the Baskakov-Schurer operators, and we use our results to derive the asymptotic expansion of its partial derivatives. We carry out the analogous study for the multivariate Bernstein operators on the simplex. In both cases we compute Voronovskaja type formulas for all partial derivatives of the sequences.

§1. Introduction and Notation

Given $m \in \mathbb{N} = \{1, 2, \ldots\}$, $H \subset \mathbb{R}^m$ and the linear subspaces $W, \overline{W} \subseteq \mathbb{R}^H$, let us consider a sequence of linear positive operators $\{L_n : W \to \overline{W}\}_{n \in \mathbb{N}}$. It is of special interest to study the convergence properties of the sequence, that is, whether for a function $f \in W$ we have that $\lim_{n \to \infty} L_n f = f$ and the way in which such convergence takes place. The behavior of the sequence can be analyzed by means of inequalities bounding the approximation error $|L_n f(x) - f(x)|$, with $x \in H$, or by asymptotic expansions of the sequence. For $r \in \mathbb{N}$ the expression

$$L_n f(x) = f(x) + \sum_{i=1}^{r} n^{-i} K_i(f, r, x) + o(n^{-r})$$

is called an **asymptotic expansion** for $\{L_n\}_{n \in \mathbb{N}}$ of order r. Indeed, the fact of $\lim_{n \to \infty} L_n f(x) = f(x)$ is equivalent to the asymptotic expression of

Approximation Theory X: Abstract and Classical Analysis   287
Charles K. Chui, Larry L. Schumaker, and Joachim Stöckler (eds.), pp. 287–307.
Copyright © 2002 by Vanderbilt University Press, Nashville, TN.
ISBN 0-8265-1415-4.

order 0 given by $L_n f(x) = f(x) + o(1)$. An asymptotic formula of order 1 is called a **Voronovskaja formula** and it is usually written in the form

$$\lim_{n \to \infty} n \left(L_n f(x) - f(x) \right) = K(f, x).$$

In the univariate case, $m = 1$, the results of Sikkema [14] give us general approximation theorems to obtain asymptotic expansions of a sequence of linear positive operators. Through these results, Abel [1,2,3,5,6] obtains explicitly the complete asymptotic expression for several operators and, in certain cases, also for the derivatives.

In the multivariate case, Walz [15] and Abel [7] show complete asymptotic formulas for the multivariate Bernstein operators on the simplex, and in [4], Abel calculates expressions for the bivariate Bleimann, Butzer and Hahn operators.

Though the asymptotic expansions for the derivatives of certain operators have been obtained by Abel [3,5,6]; the computations were carried out in a particular way for every case and only for univariate operators.

In [12], López-Moreno et al. establish a general result in the univariate case which, under certain reasonable conditions on the sequence $\{L_n\}_{n \in \mathbb{N}}$, allows to obtain the asymptotic formula for the derivatives even if the operators of the sequence do not preserve classical convexities.

The aim of this paper is to prove general results in the multivariate case in order to compute the asymptotic expansion of the partial derivatives of sequences of linear operators. In Section 2 we establish such results and in Section 3 we use them to explicitly compute the complete asymptotic formulae for the partial derivatives of certain multivariate versions of the Bernstein and Baskakov operators.

The rest of this section is devoted to fix the notation that will be used throughout the paper. In the multivariate case it is of special importance to work with a suitable notation in order to present the results in a compact way.

For each real number, $a \in \mathbb{R}$, $[a]$ is the integer part of a and $|a|$ is the absolute value of a. Given $\alpha \in \mathbb{R}^m$, the i-th, $i = 1, \ldots, m$, component of the vector α is denoted by α_i, so that $\alpha = (\alpha_1, \ldots, \alpha_m)$. Let us denote $\mathbf{0} = (0, \ldots, 0) \in \mathbb{R}^m$ and $e_i = (0, \ldots, \overset{i)}{1}, \ldots, 0)$, $i = 1, \ldots, m$. Given $\alpha, \beta \in \mathbb{R}^m$ and $a \in \mathbb{R}$, as usual, we will use, whenever it makes sense, the following notation:

$$|\alpha| = \alpha_1 + \ldots + \alpha_m, \qquad \alpha! = \alpha_1! \cdots \alpha_m!,$$
$$\alpha^\beta = \alpha_1^{\beta_1} \ldots \alpha_m^{\beta_m}, \qquad \alpha^a = (\alpha_1^a, \ldots, \alpha_m^a),$$
$$a^\alpha = a^{|\alpha|}, \qquad a\alpha = (a\alpha_1, \ldots, a\alpha_m).$$

Notice that we use the bold face symbol $|\cdot|$ for the absolute value, while $|\cdot|$ denotes the sum of the components for elements of $\mathbb{R}^m$.

Given $\alpha, \beta \in \mathbb{R}^m$, we will write $\alpha \leq \beta$ whenever $\alpha_i \leq \beta_i$ for all $i = 1, \ldots, m$. We will also make use of the following notations for sums with several indices:

$$\sum_{i=\alpha}^{\beta} A_i = \sum_{i \in \mathbb{Z}^m, \alpha \leq i \leq \beta} A_i \quad \text{and} \quad \sum_{i=\alpha}^{\infty} A_i = \sum_{i \in \mathbb{Z}^m, \alpha \leq i} A_i.$$

Given $x, y \in \mathbb{R}^m$, let $[x, y]$ denote the set

$$[x, y] = \{z \in \mathbb{R}^m : \forall i \in \{1, \ldots, m\}, \ \min\{x_i, y_i\} \leq z_i \leq \max\{x_i, y_i\}\}.$$

Given $i, j \in \mathbb{N}_0 = \{0, 1, 2, \ldots, \}$, the **Stirling numbers** S_i^j and σ_i^j of the first and second kind are respectively defined by

$$x^{\underline{i}} = \sum_{j=0}^{i} S_i^j x^j \quad \text{and} \quad x^i = \sum_{j=0}^{i} \sigma_i^j x^{\underline{j}},$$

where $x^{\underline{i}} = x(x-1)\ldots(x-i+1)$ is the falling factorial and $x^{\underline{0}} = 1$.

If we take $\alpha, \beta \in \mathbb{N}_0^m$ and $a \in \mathbb{N}_0$, we will extend the definition of binomial and Stirling numbers as follows:

$$\sigma_\alpha^\beta = \sigma_{\alpha_1}^{\beta_1} \ldots \sigma_{\alpha_m}^{\beta_m}, \quad \binom{\alpha}{\beta} = \frac{\alpha!}{\beta!(\alpha - \beta)!} \quad \text{and} \quad \binom{a}{\beta} = \frac{a!}{\beta!(a - |\beta|)!}.$$

In this case, we also have $\binom{\alpha}{\beta} = \binom{\alpha_1}{\beta_1} \cdots \binom{\alpha_m}{\beta_m}$.

Several well-known results can be reproduced with the vectorial notation that we have just considered. We thus have, for any $\alpha, \beta \in \mathbb{R}_0^m$ and $r \in \mathbb{N}_0^m$, that Newton's binomial formula remains valid with the following formulation

$$(\alpha + \beta)^r = \sum_{i=0}^{r} \binom{r}{i} \alpha^i \beta^{r-i}. \tag{1}$$

We will denote by $t : \mathbb{R}^m \ni z \mapsto t(z) = z \in \mathbb{R}^m$ the identity map in $\mathbb{R}^m$, and by $t_i : \mathbb{R}^m \ni z = (z_1, \ldots, z_m) \mapsto t_i(z) = z_i \in \mathbb{R}$ the i-th projection. The restrictions of t and t_i to any subset of $\mathbb{R}^m$ will also be denoted by t and t_i, respectively.

Given $x \in \mathbb{R}^m$ and $\delta \in \mathbb{R}^+$, $B_\delta(x)$ is the Euclidean ball of radius δ centered at x, and for a subset $H \subseteq \mathbb{R}^m$, $\mathbf{Int}(H)$ is the set of all interior points of H and $\mathbf{CL}(H)$ is the closure of H. Given $\alpha \in \mathbb{N}_0^m$, we denote by D^α the partial derivative operator, α_1 times with respect to the first variable, $\ldots$, α_m times with respect to the m-th variable. Take $H \subseteq \mathbb{R}^m$ such that $H \subseteq \mathbf{CL}(\mathbf{Int}(H))$ and $x \in H$. A function $f \in \mathbb{R}^H$ is said to be **differentiable with continuity of order** α at x if there exists $\delta \in \mathbb{R}^+$

and $\overline{f} \in \mathbb{R}^{B_\delta(x)}$ with $\overline{f}|_{B_\delta(x) \cap H} = f|_{B_\delta(x) \cap H}$ such that $D^\alpha \overline{f}$ is defined and continuous in $B_\delta(x)$. In that case we define $D^\alpha f(x) = D^\alpha \overline{f}(x)$. Because of the condition $H \subseteq \mathbf{CL}(\mathbf{Int}(H))$ it is straightforward that the definition of $D^\alpha f(x)$ is independent of the chosen function $\overline{f}$. We also denote by $C^0(H)$ or $C(H)$ the set of all continuous functions in H, and by $C^r(H)$, $r \in \mathbb{N}$, the set of functions differentiable with continuity of order $\alpha \in \mathbb{N}_0^m$ at x, for every $x \in H$ and each α with $|\alpha| \leq r$; let $C^\infty(H) = \cap_{r \in \mathbb{N}} C^r(H)$. Moreover, given $\alpha \in \mathbb{N}_0^m$ and $r \in \mathbb{N}$, $\mathbb{P}_\alpha(H) = \mathbf{Span}\{t^\beta|_H : \beta \in \mathbb{N}_0^m, \beta \leq \alpha\}$ and $\mathbb{P}_r[n]$ is the set of all polynomials in n of degree at most r. Finally, $B(H, r) = \{f \in \mathbb{R}^H : \exists p \in \mathbb{P}_r(H)/|f| \leq |p|\}$ and $B(H, r, x) = \{f \in B(H, r) : f \text{ differentiable with continuity of order } r \text{ at } x\}$.

Let us take $H \subseteq \mathbb{R}^m$ and $f \in \mathbb{R}^H$. Given $x \in \mathbb{R}^m$, $\alpha \in \mathbb{N}_0$ and $h \in \mathbb{R}^+$, we will consider the forward difference $\Delta_h^\alpha f(x)$ defined by

$$\Delta_h^\alpha f(x) = \begin{cases} \sum_{i=0}^\alpha \binom{\alpha}{i} (-1)^{\alpha-i} f(x+hi), & \{x+hi : i \in \mathbb{N}_0^m, i \leq \alpha\} \subseteq H \\ 0, & \text{otherwise.} \end{cases}$$

If $[x, x+h\alpha] \subseteq H$, the mean value theorem implies that for f differentiable of sufficiently high order,

$$\Delta_h^\alpha f(x) = h^\alpha D^\alpha f(\xi_{h,x}), \tag{2}$$

for a point $\xi_{h,x} \in [x, x+h\alpha]$. From the already known relationship between univariate forward differences and second kind Stirling numbers [8, Section 24.1.4.II.C] we can establish the identity

$$\Delta_h^\alpha t^j(0) = \alpha! \sigma_j^\alpha h^j, \qquad h \in \mathbb{R}^+, \alpha, j \in \mathbb{N}_0^m. \tag{3}$$

We will say that a function $f \in \mathbb{R}^H$ is convex of order α, $\alpha \in \mathbb{N}_0^m$, whenever for all $h \in \mathbb{R}^+$, $\Delta_h^\alpha f \geq 0$. If f is differentiable with continuity of order α at any point of H, then from (2) it is immediate that f is convex of order α if and only if $D^\alpha f \geq 0$. A linear operator $L : W \to \overline{W}$ is convex of order α whenever

$$f \text{ convex of order } \alpha \Rightarrow Lf \text{ convex of order } \alpha.$$

In particular, an operator convex of order $\mathbf{0}$ is said to be positive.

In the univariate case, Leibniz' formula describes differentiation of the product of two functions. With the vectorial notation that we have just introduced, we can formulate a multivariate version of this formula. It is also possible to write the multivariate Taylor formula in a rather compact form.

Lemma 1.

i) *(Multivariate Leibniz' formula). If f, g are two functions defined in a domain of $\mathbb{R}^m$ and both of them differentiable with continuity of order $k \in \mathbb{N}_0^m$ at $x \in \mathbb{R}^m$, then*

$$D^k(f \cdot g)(x) = \sum_{\alpha=0}^{k} \binom{k}{\alpha} D^\alpha f(x) D^{k-\alpha} g(x). \tag{4}$$

ii) *(Taylor's formula). Let $H \subseteq \mathbb{R}^m$ be nonempty and such that $H \subseteq \mathbf{CL}(\mathbf{Int}(H))$. Given $x \in H$, $r \in \mathbb{N}$ and $f \in B(H, r, x)$, we have*

$$f(z) = \sum_{\substack{\alpha \in \mathbb{N}_0^m \\ |\alpha| \leq r}} \frac{D^\alpha f(x)}{\alpha!}(z - x)^\alpha + \sum_{\substack{\alpha \in \mathbb{N}_0^m \\ |\alpha| = r}} h_\alpha(z - x)(z - x)^\alpha \tag{5}$$

for all $z \in H$, where for each $\alpha \in \mathbb{N}_0^m$ with $|\alpha| = r$, h_α is a bounded function and $\lim_{z \to 0} h_\alpha(z) = 0$.

§2. Main Results

In [14] Sikkema proves a result by means of which the asymptotic expansion of a sequence of univariate linear positive operators $\{L_n\}_{n \in \mathbb{N}}$, defined over a domain in $\mathbb{R}$, can be obtained. For an even number r and a function f that is r–times differentiable, he proves that under suitable conditions,

$$L_n f(x) = \sum_{i=0}^{r} \frac{f^{(i)}(x)}{i!} L_n\left((t - x)^i\right)(x) + o(n^{-\frac{r}{2}}).$$

Abel [4] proves a two dimensional version of the Sikkema theorem. This result allows us to obtain the asymptotic expansion of bivariate positive linear operators also in terms of their central moments. We will extend Abel's result to an arbitrary dimension. Let us fix $m \in \mathbb{N}$ and a nonempty set $H \subseteq \mathbb{R}^m$ such that $H \subseteq \mathbf{CL}(\mathbf{Int}(H))$.

Theorem 2. *Given $r \in \mathbb{N}$ even, let $\{L_n : W \to C^\infty(H)\}_{n \in \mathbb{N}}$ be a sequence of linear positive operators defined on a linear subspace $W \subseteq \mathbb{R}^H$ such that $\mathbb{P}_{r+2}(H) \subseteq W$. Suppose that for some $x \in H$,*

$$L_n\left(|(t - x)^2|^s\right)(x) = O(\phi(n)^{-s}), \quad s = \frac{r}{2}, \frac{r}{2} + 1,$$

where ϕ is a nondecreasing strictly positive function with $\lim_{n \to \infty} \phi(n) = +\infty$. If $f \in B(H, r, x) \cap W$, then

$$L_n f(x) = \sum_{\alpha \in \mathbb{N}_0^m, |\alpha| \leq r} \frac{D^\alpha f(x)}{\alpha!} L_n\left((t - x)^\alpha\right)(x) + o(\phi(n)^{-\frac{r}{2}}).$$

Proof: Consider Taylor's formula (5) for the function f:

$$f(z) = \sum_{\alpha \in \mathbb{N}_0^m, |\alpha| \leq r} \frac{D^\alpha f(x)}{\alpha!}(z-x)^\alpha + \sum_{\alpha \in \mathbb{N}_0^m, |\alpha| = r} h_\alpha(z-x)(z-x)^\alpha, \quad z \in H,$$

where the functions h_α are bounded and $\lim_{z \to 0} h_\alpha(z) = 0$. Then, for any $\varepsilon \in \mathbb{R}^+$ there exists $A \in \mathbb{R}^+$ such that $|h_\alpha(z-x)| \leq \varepsilon + A|(z-x)^2|$ for every $\alpha \in \mathbb{N}_0^m$ with $|\alpha| = r$. It is easy to check that, for $\alpha \in \mathbb{N}_0^m$ with $|\alpha| = r$, $|(t-x)^\alpha| \leq |(t-x)^2|^{\frac{r}{2}}$. Now, if we let $A_1 = \sum_{\alpha \in \mathbb{N}_0^m, |\alpha| = r} 1$, we have

$$\left| f(z) - \sum_{\alpha \in \mathbb{N}_0^m, |\alpha| \leq r} \frac{D^\alpha f(x)}{\alpha!}(z-x)^\alpha \right| \leq \sum_{\alpha \in \mathbb{N}_0^m, |\alpha| = r} |h_\alpha(z-x)||(z-x)^\alpha|$$

$$\leq \sum_{\alpha \in \mathbb{N}_0^m, |\alpha| = r} \left(\varepsilon + A|(z-x)^2|\right)|(z-x)^2|^{\frac{r}{2}}$$

$$\leq \varepsilon A_1|(z-x)^2|^{\frac{r}{2}} + A A_1|(z-x)^2|^{\frac{r}{2}+1}, \quad z \in H.$$

Since the operators are positive, we can apply L_n to both sides of the inequality. By evaluating at x and taking into account that ε is arbitrary, we conclude the proof. $\square$

To establish our next result we will consider the integral operators

$$S^{e_i} f(x) = \int_0^{t_i(x)} f(x + (z - t_i(x))e_i)dz, \quad f \in C(H), x \in H,$$

for $i = 1, \ldots, m$, and their compositions $S^\alpha = (S^{e_1})^{\alpha_1} \circ \cdots \circ (S^{e_m})^{\alpha_m}$, where $\alpha \in \mathbb{N}_0^m$ and $(S^{e_i})^{\alpha_i}$ stands for $S^{e_i} \circ \overset{\alpha_i)}{\cdots} \circ S^{e_i}$. If we have $[0, z] \subseteq H$ for each $z \in H$, then it is easy to see that the operators S^α are well-defined for any $\alpha \in \mathbb{N}_0^m$.

We now establish the main theorem of the paper. Our next result enables us to study the asymptotic behavior of the partial derivatives of a sequence of linear positive operators provided this sequence satisfies certain natural conditions.

Theorem 3. *Let us suppose that $[0, z] \subseteq H$ holds for all $z \in H$. We consider a sequence of linear positive operators, $\{L_n : W \to C^\infty(H)\}_{n \in \mathbb{N}}$, defined on the linear subspace $W \subseteq \mathbb{R}^H$ such that $\{t^\alpha : \alpha \in \mathbb{N}_0^m\} \subseteq W$. For any $\alpha, \beta \in \mathbb{N}_0^m$, $n \in \mathbb{N}$ and $z \in H$, we will denote*

$$V_{n,\alpha,\beta}(z) = \left[D^\beta L_n\left((t-z)^\alpha\right)\right](z).$$

Assume that for all $\alpha \in \mathbb{N}_0^m$ and $z \in H$ the following conditions hold:

i) $D^{e_i}V_{n,\alpha,\beta}(z) = V_{n,\alpha,\beta+e_i}(z) - \alpha_i V_{n,\alpha-e_i,\beta}(z)$, for each $\beta \in \mathbb{N}_0^m$ and $i \in \{1,\ldots,m\}$ with $\alpha_i > 0$,

ii) L_n is convex of order α,

iii) There exists $\phi \in \mathbb{P}_1[n] - \mathbb{P}_0[n]$, independent of α, such that

$$\phi(n)^{|\alpha|} V_{n,\alpha,\mathbf{0}}(z) \in \mathbb{P}_{\left[\frac{|\alpha|}{2}\right]}[n].$$

Then, given $x \in H$, $r \in \mathbb{N}$ even, $k \in \mathbb{N}_0^m$ and $f \in W \cap C^{|k|}(H)$ such that $S^k D^k f \in W$ and $D^k f \in B(H, r, x)$, we have

$$D^k L_n f(x) = P_{f,k,r,n}(x) + o(\phi(n)^{-\frac{r}{2}}), \tag{6}$$

where $P_{f,k,r,n}(x)$ is a polynomial in $\phi(n)^{-1}$ of degree $\frac{r}{2}$. If, in addition, for a $s \in \mathbb{N}_0^m$, $f \in C^{|k+s|}(H)$, $S^{k+s}D^{k+s}f \in W$ and $D^{k+s}f \in B(H,r,x)$, then

$$D^s P_{f,k,r,n}(x) = P_{f,k+s,r,n}(x).$$

Proof: We will divide the proof into three steps:

1) From the shape preserving properties of L_n (hypotheses ii)) we deduce that, for any $\alpha \in \mathbb{N}_0^m$, $L_n(\mathbb{P}_\alpha(H)) \subseteq \mathbb{P}_\alpha(H)$. Then $V_{n,\alpha,\beta}$ is a polynomial, and in this case $V_{n,\alpha,\beta} \in C^\infty(H)$. Moreover, from i) and iii), we can prove by induction that

$$\phi(n)^\alpha V_{n,\alpha,\beta}(z) \in \mathbb{P}_{\left[\frac{|\alpha+\beta|}{2}\right]}[n], \quad z \in H.$$

2) Let us define the linear operators $\tilde{L}_n : \tilde{W} \to C^\infty(H)$ by $\tilde{L}_n = D^k L_n S^k$, where $\tilde{W} = \{g \in C(H) : S^k g \in W\}$. If $g \in \tilde{W}$ satisfies $g \geq 0$, then $S^k g$ is convex of order k, and therefore $\tilde{L}_n g = D^k L_n S^k g \geq 0$. We have thus shown that the operators $\tilde{L}_n$ are positive. Besides, we check at once that $\{t^\alpha : \alpha \in \mathbb{N}_0^m\} \subseteq \tilde{W}$. Furthermore,

a) If $g \in C^{|k|}(H) \cap W$ and $D^k g \in \tilde{W}$, then

$$\tilde{L}_n(D^k g) - D^k L_n g = D^k L_n \left(S^k D^k g - g\right).$$

We know that

$$D^k \left(S^k D^k g - g\right) = D^k g - D^k g = 0$$

and, since L_n is convex of order k, we have $D^k L_n(S^k D^k g - g) = 0$ and therefore

$$\tilde{L}_n(D^k g) = D^k L_n g.$$

b) By means of a), for any $\alpha \in \mathbb{N}_0^m$,

$$\tilde{L}_n \left((t-x)^\alpha\right)(x) = \frac{\alpha!}{(\alpha+k)!} \tilde{L}_n \left(D^k(t-x)^{\alpha+k}\right)(x)$$

$$= \frac{\alpha!}{(\alpha+k)!} V_{n,\alpha+k,k}(x).$$

c) From b) and **1)**, $\phi(n)^{\alpha+k} \tilde{L}_n \left((t-x)^\alpha\right)(x) \in \mathbb{P}_{\left[\frac{|\alpha+2k|}{2}\right]}[n]$, and then

$$\tilde{L}_n \left((t-x)^\alpha\right)(x) = O(\phi(n)^{-\left(|\alpha+k|-\left[\frac{|\alpha+2k|}{2}\right]\right)}) = O(\phi(n)^{-\left[\frac{|\alpha|+1}{2}\right]}),$$

where we have used that $\mathbb{P}_{\left[\frac{|\alpha+2k|}{2}\right]}[n] = \mathbb{P}_{\left[\frac{|\alpha+2k|}{2}\right]}[\phi(n)]$. Furthermore, we can also conclude that

$$\tilde{L}_n \left((t-x)^\alpha\right)(x), V_{n,\alpha+k,k}(x) \in \mathbb{P}_{|\alpha+k|}[\phi(n)^{-1}].$$

With these considerations in mind it is immediate that the operators $\tilde{L}_n$ satisfy the conditions of Theorem 2 that applied for the function $D^k f \in \tilde{W}$ yields the identity

$$D^k L_n f(x) = \tilde{L}_n(D^k f)(x)$$

$$= \sum_{i\in\mathbb{N}_0^m, |i|\leq r} \frac{D^{i+k} f(x)}{i!} \tilde{L}_n \left((t-x)^i\right)(x) + o(\phi(n)^{-\frac{r}{2}}). \tag{7}$$

If we consider the projection $\Pi : \bigcup_{s\in\mathbb{N}} \mathbb{P}_s[\phi(n)^{-1}] \to \mathbb{P}_{\frac{r}{2}}[\phi(n)^{-1}]$ given by

$$\Pi\left(\sum_{i=0}^s a_i \phi(n)^{-i}\right) = \sum_{i=0}^{\frac{r}{2}} a_i \phi(n)^{-i},$$

and we call

$$P_{f,k,r,n}(x) = \sum_{i\in\mathbb{N}_0^m, |i|\leq r} \frac{D^{i+k} f(x)}{(i+k)!} \Pi\left(V_{n,i+k,k}(x)\right), \tag{8}$$

from (7) and b), we have $D^k L_n f(x) = P_{f,k,r,n}(x) + o(\phi(n)^{-\frac{r}{2}})$ which proves (6). Finally, from the properties of $V_{n,i+k,k}(x)$ and (8) we conclude that $P_{f,k,r,n}(x)$ is a polynomial of degree $\frac{r}{2}$ in $\phi(n)^{-1}$.

3) Let us suppose now that $f \in C^{|k+s|}(H)$, $S^{k+s} D^{k+s} f \in W$ and $D^{k+s} f \in B(H,r,x)$. Following the same arguments for **2)** for the partial derivative of order $k+s$ we get that $D^{k+s} L_n f(x) = P_{f,k+s,r,n}(x) + o(\phi(n)^{-\frac{r}{2}})$, with $P_{f,k+s,r,n}(x)$ defined as in **2)**. Also, for any $j \in \mathbb{N}_0^m$, $0 \leq j \leq s$, we can consider $P_{f,k+j,r,n}(x)$ given by (8). We will see that there exists a recurrence relation between such expressions.

Since $D^{k+s}f \in B(H,r,x)$, from (8) it is immediate that $P_{f,k+j,r,n}$, $j \in \mathbb{N}_0^m$, $0 \leq j \leq s$, is differentiable of sufficiently high order at x. Let us calculate $D^{e_j}P_{f,k,r,n}(x)$, $j = 1, \ldots, m$. By differentiating in (8) we obtain

$$D^{e_j}P_{f,k,r,n}(x) = \sum_{i \in \mathbb{N}_0^m, |i| \leq r} \frac{D^{i+k+e_j}f(x)}{(i+k)!} \Pi\left(V_{n,i+k,k}(x)\right)$$

$$+ \sum_{i \in \mathbb{N}_0^m, |i| \leq r} \frac{D^{i+k}f(x)}{(i+k)!} \Pi\left(D^{e_j}V_{n,i+k,k}(x)\right)$$

$$= \sum_{i \in \mathbb{N}_0^m, |i| \leq r+1, 1 \leq i_j} \frac{D^{i+k}f(x)}{(i+k)!}(i_j + k_j)\Pi\left(V_{n,i+k-e_j,k}(x)\right)$$

$$+ \sum_{i \in \mathbb{N}_0^m, |i| \leq r+1, 1 \leq i_j} \frac{D^{i+k}f(x)}{(i+k)!} \Pi\left(D^{e_j}V_{n,i+k,k}(x)\right)$$

$$- \sum_{i \in \mathbb{N}_0^m, |i| = r+1, 1 \leq i_j} \frac{D^{i+k}f(x)}{(i+k)!} \Pi\left(D^{e_j}V_{n,i+k,k}(x)\right).$$

By means of the already mentioned properties of $V_{n,i+k,k}(x)$, it is possible to check that $V_{n,i+k,k}(x) = O(\phi(n)^{-\left[\frac{|i|+1}{2}\right]})$. Thus, in the last sum, since $|i| = r+1$, we have $V_{n,i+k,k}(x) = O(\phi(n)^{-(\frac{r}{2}+1)})$, $\Pi\left(V_{n,i+k,k}(x)\right) = 0$ and, in the same way, $\Pi\left(D^{e_j}V_{n,i+k,k}(x)\right) = 0$. Then, we can write

$$D^{e_j}P_{f,k,r,n}(x) =$$

$$= \sum_{|i| \leq r+1, 1 \leq i_j} \frac{D^{i+k}f(x)}{(i+k)!} \Pi\left((i_j + k_j)V_{n,i+k-e_j,k}(x) + D^{e_j}V_{n,i+k,k}(x)\right)$$

$$= \text{(by means of i))} = \sum_{i \in \mathbb{N}_0^m, |i| \leq r+1, 1 \leq i_j} \frac{D^{i+k}f(x)}{(i+k)!} \Pi\left(V_{n,i+k,k+e_j}(x)\right)$$

$$= \sum_{i \in \mathbb{N}_0^m, |i| \leq r} \frac{D^{i+k+e_j}f(x)}{(i+k+e_j)!} \Pi\left(V_{n,i+k+e_j,k+e_j}(x)\right) = P_{f,k+e_j,r,n}(x).$$

Following the same arguments, we can prove by induction that $D^s P_{f,k,r,n}(x) = P_{f,k+s,r,n}(x)$. $\square$

The hypotheses i), ii) and iii) of Theorem 3 are satisfied for a wide class of linear operators. In fact, point i) is satisfied by any interpolatory type operator and the properties ii) and iii) hold for important examples of classical linear positive operators. Now we present several applications of this result.

§3. Applications

3.1 Multivariate Baskakov-Schurer operators

The Baskakov operators [9] defined on a certain subspace of $\mathbb{R}^{[0,\infty)}$ were introduced in 1957. Later, Schurer [13] defines a sequence of linear positive operators depending on a parameter $p \in \mathbb{N}_0$ that in the case $p = 0$ yields the original Baskakov operators. In this section we will study a multivariate version of the Baskakov-Schurer operators defined on the set $H = (\mathbb{R}_0^+)^m$ as follows: given $n \in \mathbb{N}$, $p \in \mathbb{N}_0$, $x \in H$ and $f \in \mathbb{R}^H$,

$$
A_{n,p}f(x) = (1 + |x|)^{-(n+p)} \sum_{k=0}^{\infty} \binom{n+p+|k|-1}{k} \left(\frac{x}{1+|x|}\right)^k f\left(\frac{k}{n}\right). \tag{9}
$$

In the case $m = 1$, we obtain the univariate Baskakov-Schurer operators, and for $m > 1$ we have multivariate operators different from the tensor product construction.

The power series that defines $A_{n,p}$ is not always convergent, so we consider the subspace of $\mathbb{R}^H$, $\mathcal{D} = \{f \in \mathbb{R}^H : \forall x \in H, \forall n \in \mathbb{N}, |A_{n,p}f(x)| < +\infty\}$. Then $A_{n,p} : \mathcal{D} \to C^{\infty}(H)$ is a linear positive operator.

In the following results we will express the partial derivatives of $A_{n,p}$ in terms of forward differences. First of all, we will study partial derivatives of first order and later we extend it to any order.

Lemma 4. *Given* $n \in \mathbb{N}$, $p \in \mathbb{N}_0$, $f \in \mathcal{D}$ *and* $i \in \{1, \ldots, m\}$,

$$
D^{e_i} A_{n,p} f = \frac{1}{1 + |t|} A_{n,p} \left((n(|t| + 1) + p) \Delta_{\frac{1}{n}}^{e_i} f \right). \tag{10}
$$

Proof: Since

$$
D^{e_i} \left(\frac{t}{1+|t|}\right)^k = -\frac{|k|}{1+|t|} \left(\frac{t}{1+|t|}\right)^k + \frac{k_i}{1+|t|} \left(\frac{t}{1+|t|}\right)^{k-e_i}, \tag{11}
$$

we have

$D^{e_i} A_{n,p} f$

$$
= -(n+p)(1+|t|)^{-(n+p+1)} \sum_{k=0}^{\infty} \binom{n+p+|k|-1}{k} \left(\frac{t}{1+|t|}\right)^k f\left(\frac{k}{n}\right)
$$

$$
- (1+|t|)^{-(n+p+1)} \sum_{k=0}^{\infty} \binom{n+p+|k|-1}{k} |k| \left(\frac{t}{1+|t|}\right)^k f\left(\frac{k}{n}\right)
$$

$$
+ (1+|t|)^{-(n+p+1)} \sum_{k=0}^{\infty} \binom{n+p+|k|-1}{k} k_i \left(\frac{t}{1+|t|}\right)^{k-e_i} f\left(\frac{k}{n}\right).
$$

Make the change of index $k_i = k_i - 1$ in the last sum, and let us use the identity

$$\binom{n+p+|k+e_i|-1}{k+e_i}(k_i+1) = (n+p+|k|)\binom{n+p+|k|-1}{k}$$

to obtain

$$D^{e_i}A_{n,p}f = (1+|t|)^{-(n+p+1)}\sum_{k=0}^{\infty}\binom{n+p+|k|-1}{k}(n+p+|k|)\times$$

$$\times\left(\frac{t}{1+|t|}\right)^k\left[f\left(\frac{k+e_i}{n}\right)-f\left(\frac{k}{n}\right)\right].$$

Finally, as $\Delta_{\frac{1}{n}}^{e_i}f\left(\frac{k}{n}\right) = f\left(\frac{k+e_i}{n}\right) - f\left(\frac{k}{n}\right)$ and $(n|t|)(\frac{k}{n}) = |k|$, we complete the proof. $\square$

Theorem 5. *Given* $i \in \mathbb{N}_0^m$, $n \in \mathbb{N}$, $p \in \mathbb{N}_0$ *and* $f \in \mathcal{D}$,

$$D^i A_{n,p}f = \frac{1}{(1+|t|)^i}A_{n,p}\left((n(|t|+1)+p+|i|-1)^{\underline{|i|}}\Delta_{\frac{1}{n}}^i f\right). \qquad (12)$$

Proof: The proof is by induction on $|i|$. The case $|i| = 1$ is just Lemma 4. We suppose that the result is valid for $|i|$, and we will prove it for $i + e_j$, $j = 1, \ldots . m$, (i.e., for $|i| + 1$). We have

$$D^{i+e_j}A_{n,p}f = D^{e_j}\left(\frac{1}{(1+|t|)^i}A_{n,p}\left((n(|t|+1)+p+|i|-1)^{\underline{|i|}}\Delta_{\frac{1}{n}}^i f\right)\right)$$

$$= -\frac{|i|}{(1+|t|)^{|i|+1}}A_{n,p}\left((n(|t|+1)+p+|i|-1)^{\underline{|i|}}\Delta_{\frac{1}{n}}^i f\right)$$

$$+ \frac{1}{(1+|t|)^i}D^{e_j}A_{n,p}\left((n(|t|+1)+p+|i|-1)^{\underline{|i|}}\Delta_{\frac{1}{n}}^i f\right)$$

$$= -\frac{|i|}{(1+|t|)^{|i|+1}}A_{n,p}\left((n(|t|+1)+p+|i|-1)^{\underline{|i|}}\Delta_{\frac{1}{n}}^i f\right)$$

$$+ \frac{1}{(1+|t|)^{|i|+1}}A_{n,p}\Big((n(|t|+1)+p)$$

$$\Delta_{\frac{1}{n}}^{e_j}\left[(n(|t|+1)+p+|i|-1)^{\underline{|i|}}\Delta_{\frac{1}{n}}^i f\right]\Big),$$

$$(13)$$

where we have used the induction hypothesis and Lemma 4. On the other hand, for two arbitrary functions $g_1, g_2 \in \mathcal{D}$, it is clear that

$$\Delta_{\frac{1}{n}}^{e_j}(g_1 g_2)(z) = g_1(z)\Delta_{\frac{1}{n}}^{e_j}g_2(z) + g_2(z+\frac{e_j}{n})\Delta_{\frac{1}{n}}^{e_j}g_1(z), \quad z \in H,$$

and then

$$\Delta_{\frac{1}{n}}^{e_j}\left[(n(|t|+1)+p+|i|-1)^{\underline{|i|}}\Delta_{\frac{1}{n}}^{i}f\right] = (n(|t|+1)+p+|i|)^{\underline{|i|}}\Delta_{\frac{1}{n}}^{i+e_j}f$$
$$+\left[(n(|t|+1)+p+|i|)^{\underline{|i|}} - (n(|t|+1)+p+|i|-1)^{\underline{|i|}}\right]\Delta_{\frac{1}{n}}^{i}f$$
$$= (n(|t|+1)+p+|i|)^{\underline{|i|}}\Delta_{\frac{1}{n}}^{i+e_j}f + (n(|t|+1)+p+|i|-1)^{\underline{|i|-1}}|i|\Delta_{\frac{1}{n}}^{i}f.$$

In order to complete the proof it is sufficient to replace the last expression in (13). $\square$

An immediate consequence of the preceding results is the fact that the $A_{n,p}$ operators satisfy the following shape preserving properties: given $n \in \mathbb{N}$, $p \in \mathbb{N}_0$ and $k \in \mathbb{N}^m$:

i) $A_{n,p}(\mathbb{P}_k(H)) \subseteq \mathbb{P}_k(H)$. That is, $A_{n,p}$ preserves algebraic polynomials and their degrees,

ii) $A_{n,p}$ is convex of order k.

Now let us fix $p \in \mathbb{N}_0$ and study the moments of the $A_{n,p}$ operators. As in Theorem 3, we will use the notation

$$V_{n,\alpha,\beta}(x) = D^\beta A_{n,p}\left((t-x)^\alpha\right)(x), \quad \alpha, \beta \in \mathbb{N}_0^m, x \in H.$$

To abbreviate, we will also write $V_{n,\alpha} = V_{n,\alpha,0}$. Our first objective is to prove hypothesis iii) of Theorem 3 for $A_{n,p}$. In order to get it, we are going to study the behaviour of $V_{n,\alpha}$ as a polynomial in n^{-1}. As usual (for instance see [10] for the classical Bernstein operators on $[0,1]$), we will make use of the recursive formula that we prove in our next result.

Lemma 6. Given $i \in \{1,\ldots,m\}$ and $\alpha \in \mathbb{N}_0^m$,

$$n(1+|t|-t_i)V_{n,\alpha+e_i} - nt_i\sum_{j=1,j\neq i}^{m} V_{n,\alpha+e_j}t_i(1+|t|)D^{e_i}V_{n,\alpha}$$
$$+ pt_iV_{n,\alpha} + \alpha_it_i(1+|t|)V_{n,\alpha-e_i},$$

where we have assumed that $V_{n,\alpha} = 0$ if $\alpha_i < 0$ for some $i = 1,\ldots,m$.

Proof: For $x \in H$ with $t_i(x) = 0$, it is easy to prove that $V_{n,\alpha+e_i}(x) = 0$ and the result follows immediately. Therefore, we can suppose that $x \in H$ and $t_i \neq 0$. We have

$$n^{|\alpha|}V_{n,\alpha} = (1+|t|)^{-(n+p)}\sum_{k=0}^{\infty}\binom{n+p+|k|-1}{k}(k-nt)^\alpha\left(\frac{t}{1+|t|}\right)^k.$$

If $\alpha_i \neq 0$,

$$n^{|\alpha|} D^{e_i} V_{n,\alpha} =$$

$$= -(n+p)(1+|t|)^{-(n+p+1)} \sum_{k=0}^{\infty} (k-nt)^{\alpha} \binom{n+p+|k|-1}{k} \left(\frac{t}{1+|t|}\right)^k$$

$$- (1+|t|)^{-(n+p)} \sum_{k=0}^{\infty} n\alpha_i (k-nt)^{\alpha-e_i} \binom{n+p+|k|-1}{k} \left(\frac{t}{1+|t|}\right)^k$$

$$+ (1+|t|)^{-(n+p)} \sum_{k=0}^{\infty} (k-nt)^{\alpha} \binom{n+p+|k|-1}{k}$$

$$\left[-\frac{|k|-k_i}{1+|t|} \left(\frac{t}{1+|t|}\right)^k + k_i \frac{1+|t|-t_i}{(1+|t|)^2} \left(\frac{t}{1+|t|}\right)^{k-e_i} \right],$$

$$(14)$$

where we have used, in the third sum, a slight modification of identity (11). In the last sum it is also clear that

$$|k| - k_i = \sum_{j=1, j\neq i}^{m} (k_j - nt_j + nt_j)$$

and that $k_i = k_i - nt_i + nt_i$. Then we can write

$$n^{|\alpha|} D^{e_i} V_{n,\alpha} =$$

$$= -(n+p)(1+|t|)^{-(n+p+1)} \sum_{k=0}^{\infty} (k-nt)^{\alpha} \binom{n+p+|k|-1}{k} \left(\frac{t}{1+|t|}\right)^k$$

$$- n\alpha_i (1+|t|)^{-(n+p)} \sum_{k=0}^{\infty} (k-nt)^{\alpha-e_i} \binom{n+p+|k|-1}{k} \left(\frac{t}{1+|t|}\right)^k$$

$$- \sum_{j=1, j\neq i}^{m} (1+|t|)^{-(n+p+1)} \sum_{k=0}^{\infty} (k-nt)^{\alpha+e_j} \binom{n+p+|k|-1}{k} \left(\frac{t}{1+|t|}\right)^k$$

$$- \sum_{j=1, j\neq i}^{m} (1+|t|)^{-(n+p+1)} nt_j \sum_{k=0}^{\infty} (k-nt)^{\alpha} \binom{n+p+|k|-1}{k} \left(\frac{t}{1+|t|}\right)^k$$

$$+ (1+|t|)^{-(n+p+1)} n(1+|t|-t_i) \sum_{k=0}^{\infty} (k-nt)^{\alpha} \binom{n+p+|k|-1}{k} \left(\frac{t}{1+|t|}\right)^k$$

$$+ (1+|t|)^{-(n+p+1)} \frac{1+|t|-t_i}{t_i} \times$$

$$\times \sum_{k=0}^{\infty} (k-nt)^{\alpha+e_i} \binom{n+p+|k|-1}{k} \left(\frac{t}{1+|t|}\right)^k.$$

In this last expression the fifth sum cancels with the fourth and the factor n of the first one. After that it is easy to obtain the result for the points $x \in H$ such that $t_i(x) = x_i \neq 0$. If $\alpha_i = 0$, the proof is the same, but in (14) the second sum does not appear. $\square$

This recurrence relation allows us to study the degree of $V_{n,\alpha}(x)$ as a polynomial in n^{-1}. In the univariate case such consequence is immediate (see for example [14, pag. 335]). For the multivariate case, it is necessary to be more careful so that we present the proof in a separate theorem.

Theorem 7. For any $\alpha \in \mathbb{N}_0^m$ and $x \in H$, $n^{|\alpha|} V_{n,\alpha}(x) \in \mathbb{P}_{\left\lceil \frac{|\alpha|}{2} \right\rceil}[n]$.

Proof: The proof is by induction on $|\alpha|$. The result is evident for $|\alpha| = 0$. Let us suppose that the result is valid for every $\beta \in \mathbb{N}_0^m$ such that $|\beta| \leq |\alpha|$ and prove it for $\alpha + e_i$, $i = 1, \ldots, m$ (i. e. for $|\alpha| + 1$). Consider the matrix

$$A(x) = \begin{pmatrix} 1 + |x| - x_1 & -x_1 & \cdots & -x_1 \\ -x_2 & 1 + |x| - x_2 & \cdots & -x_2 \\ \vdots & \vdots & \ddots & \vdots \\ -x_m & -x_m & \cdots & 1 + |x| - x_m \end{pmatrix}.$$

It is easily checked that $\det(A(x)) = (1 + |x|)^{m-1} \neq 0$ and also that Lemma 6 can be written in matrix form as

$$n^{|\alpha|+1} A(x) \begin{pmatrix} V_{n,\alpha+e_1}(x) \\ V_{n,\alpha+e_2}(x) \\ \vdots \\ V_{n,\alpha-e_m}(x) \end{pmatrix} = n^{|\alpha|} \cdot$$

$$\cdot \begin{pmatrix} x_1(1 + |x|)D^{e_1}V_{n,\alpha}(x) + px_1 V_{n,\alpha}(x) + \alpha_1 x_1(1 + |x|)V_{n,\alpha-e_1}(x) \\ x_2(1 + |x|)D^{e_2}V_{n,\alpha}(x) + px_2 V_{n,\alpha}(x) + \alpha_2 x_2(1 + |x|)V_{n,\alpha-e_2}(x) \\ \vdots \\ x_m(1 + |x|)D^{e_m}V_{n,\alpha}(x) + px_m V_{n,\alpha}(x) + \alpha_m x_m(1 + |x|)V_{n,\alpha-e_m}(x) \end{pmatrix}.$$

The induction hypotheses implies that all entries of the column of the right-hand side of the above identity, when they are multiplied by $n^{|\alpha|}$, have degree in n at most $\left\lceil \frac{|\alpha|+1}{2} \right\rceil$. Since $A(x)$ is a nonsingular matrix, by using the inverse matrix of $A(x)$, we can compute $n^{|\alpha|+1} V_{n,\alpha+e_i}(x)$, $i = 1, \ldots, m$, as a linear combination of the expressions of the right-hand side of the equality and therefore its degree in n is also less than or equal to $\left\lceil \frac{|\alpha|+1}{2} \right\rceil$. $\square$

Moreover, as a consequence of the above theorem, for any $\alpha \in \mathbb{N}_0^m$ and $x \in H$, $V_{n,\alpha}(x)$ is a polynomial in n^{-1} in which all monomials have degree greater than or equal to $|\alpha| - \left[\frac{|\alpha|}{2}\right] = \left[\frac{|\alpha|+1}{2}\right]$, and so

$$A_{n,p}\left((t-x)^\alpha\right)(x) = V_{n,\alpha}(x) = O(n^{-\left[\frac{|\alpha|+1}{2}\right]}). \tag{15}$$

From the preceding results, we know that the sequence $\{A_{n,p}\}_{n\in\mathbb{N}}$ satisfies conditions i), ii) and iii) of Theorem 3 and also the hypotheses of Theorem 2. We begin studying the asymptotic expansion of the sequence of operators by means of Theorem 2.

Theorem 8. *Let* $r \in \mathbb{N}$ *even,* $x \in H$ *and* $f \in B(H, r, x)$. *Then,*

$$A_{n,p}f(x) = P_{f,\mathbf{0},r,n}(x) + o(n^{-\frac{r}{2}}),$$

where $P_{f,\mathbf{0},r,n}(x)$ *is given by*

$$P_{f,\mathbf{0},r,n}(x) = \sum_{\delta=0}^{\frac{r}{2}} n^{-\delta} \sum_{\alpha\in\mathbb{N}_0^m, \delta\le|\alpha|\le 2\delta} \frac{D^\alpha f(x)}{\alpha!} \sum_{\beta=\mathbf{0}, |\beta|\le\delta}^{\alpha} x^{\alpha-\beta} H(\alpha, \delta, \beta),$$

with

$$H(\alpha, \delta, \beta) = \sum_{\nu=\beta, |\nu|\ge\delta}^{\alpha} \binom{\alpha}{\nu} (-1)^{\alpha-\nu} \sigma_\nu^{\nu-\beta} \times$$

$$\times \sum_{j=|\nu|-\delta}^{|\nu-\beta|} S_{|\nu-\beta|}^j \binom{j}{|\nu|-\delta} (p + |\nu-\beta| - 1)^{j+\delta-|\nu|}.$$

Proof: Given $\nu \in \mathbb{N}_0^m$, we know that $A_{n,p}(t^\nu) \in \mathbb{P}_\nu(H)$ and in this case Taylor's formula (5) yields the identity

$$A_{n,p}(t^\nu) = \sum_{i=\mathbf{0}}^\nu \frac{D^i A_{n,p}(t^\nu)(\mathbf{0})}{i!} t^i.$$

From Theorem 5 and (3), since for every function $g \in \mathcal{D}$, $A_{n,p}g(\mathbf{0}) = g(\mathbf{0})$, we have

$$D^i A_{n,p}(t^\nu)(\mathbf{0}) = (n+p+|i|-1)^{\underline{|i|}} \Delta_{\frac{1}{n}}^i t^\nu(\mathbf{0}) = \frac{(n+p+|i|-1)^{\underline{|i|}} i! \sigma_\nu^i}{n^\nu},$$

and, by the definition of the Stirling numbers,

$$(n + p + |i| - 1)^{\underline{|i|}} = \sum_{j=0}^{|i|} S_{|i|}^j (n + p + |i| - 1)^j$$

$$= \sum_{j=0}^{|i|} S_{|i|}^j \sum_{\beta=0}^{j} \binom{j}{\beta} (p + |i| - 1)^{j-\beta} n^\beta.$$

If we put this all together, we obtain

$$A_{n,p}(t^\nu) = \sum_{i=0}^{\nu} \frac{\sigma_\nu^i}{n^\nu} t^i \sum_{j=0}^{|i|} S_{|i|}^j \sum_{\beta=0}^{j} \binom{j}{\beta} (p + |i| - 1)^{j-\beta} n^\beta$$

$$= \sum_{i=0}^{\nu} \frac{\sigma_\nu^i}{n^\nu} t^i \sum_{j=0}^{|i|} S_{|i|}^j \sum_{\delta=|\nu|-j}^{|\nu|} \binom{j}{|\nu| - \delta} (p + |i| - 1)^{j-|\nu|+\delta} n^{|\nu|-\delta}$$

$$= \sum_{\delta=0}^{|\nu|} n^{-\delta} \sum_{i=0, |\nu|-\delta \leq |i|}^{\nu} \sigma_\nu^i t^i \sum_{j=|\nu|-\delta}^{|i|} S_{|i|}^j \binom{j}{|\nu| - \delta} (p + |i| - 1)^{j-|\nu|+\delta},$$

$$\tag{16}$$

where we have replaced β by $\delta = |\nu| - \beta$ and changed the order of the sums. Given $\alpha \in \mathbb{N}_0^m$, by using Newton's formula (1), we have

$$A_{n,p}\left((t - x)^\alpha\right)(x) = \sum_{\nu=0}^{\alpha} \binom{\alpha}{\nu} (-x)^{\alpha-\nu} A_{n,p}(t^\nu)(x).$$

Using (16) and making the change of the index i by $\beta = \nu - i$, it will suffice to rearrange the order of the sums to obtain

$$A_{n,p}\left((t - x)^\alpha\right)(x) = \sum_{\delta=\left[\frac{|\alpha|+1}{2}\right]}^{|\alpha|} n^{-\delta} \sum_{\beta=0, |\beta| \leq \delta}^{\alpha} x^{\alpha-\beta} H(\alpha, \delta, \beta),$$

where the lower limit of the sum for δ is a consequence of (15). Finally, Theorem 7 guarantees that we can apply Theorem 2 to complete the proof. □

In the last theorem we have explicitly obtained the complete asymptotic expression for the sequence $\{A_{n,p}\}_{n\in\mathbb{N}}$. We have already mentioned that the shape preserving properties of the operators and the suitable behavior of their moments enable us to make use of Theorem 3 in order to compute the asymptotic expansion for the partial derivatives of the sequence.

Theorem 9. Let $r \in \mathbb{N}$ even, $k \in \mathbb{N}_0^m$, $x \in H$ and $f \in C^{|k|}(H) \cap B(H, r)$ such that $D^k f \in B(H, r, x)$. Then,

$$D^k A_{n,p} f(x) = P_{f,k,r,n}(x) + o(n^{-\frac{r}{2}}), \tag{17}$$

where $P_{f,k,r,n}(x) = D^k P_{f,0,r,n}(x)$ and $P_{f,0,r,n}(x)$ is defined as in Theorem 8. Furthermore, the expansion $P_{f,k,r,n}(x)$ can be expressed as

$$P_{f,k,r,n}(x) = k! \sum_{\delta=0}^{\frac{r}{2}} n^{-\delta} \sum_{\substack{\gamma=k \\ \delta \leq |\gamma| \leq |k|+2\delta}}^{\infty} \frac{D^\gamma f(x)}{\gamma!} \times$$

$$\times \sum_{\substack{\alpha=\gamma-k, \\ \delta \leq |\alpha| \leq 2\delta}}^{\gamma} \binom{\gamma}{\alpha} \sum_{\substack{\beta=0 \\ |\beta| \leq \delta}}^{\gamma-k} \binom{\alpha-\beta}{k+\alpha-\gamma} x^{\gamma-\beta-k} H(\alpha, \delta, \beta),$$

where $H(\alpha, \delta, \beta)$ is given in Theorem 8.

Proof: Let us take $W = \cup_{r \in \mathbb{N}} B(H, r)$. It is easy to check that both W and H satisfy the conditions of Theorem 3. It is also immediate that $\{t^\alpha : \alpha \in \mathbb{N}_0^m\} \subseteq W$ and that $f \in W \cap C^{|k|}(H)$ implies that $S^k D^k f \in W$.

We will apply Theorem 3 but first we will check that the conditions of the theorem are satisfied for the $A_{n,p}$ operators. Hypothesis i) and iii) are evident and, from Theorem 7, ii) holds if we take $\phi(n) = n$. Therefore, we can use Theorem 3 to obtain (17) and the properties with respect to the differentiation of the expansion $P_{f,k,r,n}(x)$. The explicit formulae for $P_{f,k,r,n}(x)$ are obtained by differentiating the expression given for $P_{f,0,r,n}(x)$ in Theorem 8. We only need to use Leibniz' formula (4) to compute

$$D^k[D^\alpha f \cdot x^{\alpha-\beta}](x) = \sum_{\substack{\gamma=0 \\ \gamma \geq k-\alpha+\beta}}^{k} \binom{k}{\gamma} D^{\alpha+\gamma} f(x) x^{\alpha-\beta+\gamma-k} \frac{(\alpha-\beta)!}{(\alpha-\beta+\gamma-k)!}$$

$$= \sum_{\substack{\gamma=\alpha \\ \gamma \geq k+\beta}}^{k+\alpha} \binom{k}{\gamma-\alpha} D^\gamma f(x) x^{\gamma-\beta-k} \frac{(\alpha-\beta)!}{(\gamma-\beta-k)!},$$

where we have made the change of index $\gamma = \alpha + \gamma$ to obtain the last expression. Later, it suffices to use this identity in the formula for $P_{f,0,r,n}(x)$, to rearrange the order of the sums and to simplify to finally obtain the explicit expression of the theorem. $\square$

As a corollary, we can obtain the Voronovskaja formula for all partial derivatives of the sequence $\{A_{n,p}\}_{n \in \mathbb{N}}$.

Corollary 10. Let $k \in \mathbb{N}_0^m$, $x \in H$ and $f \in C^{|k|}(H) \cap B(H, 2)$ such that $D^k f \in B(H, 2, x)$. Then,

$$\lim_{n \to \infty} 2n \left(D^k A_{n,p} f(x) - D^k f(x) \right) = \left((2p-1)|k| + |k|^2 \right) D^k f(x)$$

$$+ \sum_{i=1}^m \left[(x_i + x_i^2) D^{k+2e_i} f(x) + (k_i + 2x_i(p + |k|)) D^{k+e_i} f(x) \right]$$

$$+ \sum_{\substack{i,j=1 \\ i \neq j}}^m x_i x_j D^{k+e_i+e_j} f(x).$$

Proof: Theorem 8 gives us the following expression for $P_{f,\mathbf{0},2,n}(x)$:

$$P_{f,\mathbf{0},2,n}(x) = f(x) + \frac{1}{2n} \sum_{i=1}^m \left[(x_i + x_i^2) D^{2e_i} f(x) + 2p x_i D^{e_i} f(x) \right]$$

$$+ \frac{1}{2n} \sum_{\substack{i,j=1 \\ i \neq j}}^m x_i x_j D^{e_i+e_j} f(x).$$

From Theorem 9 we know that $P_{f,k,2,n}(x) = D^k P_{f,\mathbf{0},2,n}(x)$, so we only need to differentiate in the above identity to reach the result. $\square$

3.2. The Bernstein operators on the simplex

For $m \in \mathbb{N}$, let us consider the simplex $H = \{x \in \mathbb{R}^m : |x| \leq 1, 0 \leq x\}$. The multivariate Bernstein operators on the simplex H are defined by

$$B_n f(x) = \sum_{k \in \mathbb{N}_0^m, |k| \leq n} \binom{n}{k} x^k (1 - |x|)^{n-|k|} f\left(\frac{k}{n}\right).$$

Recently, several papers [7,11,15] have been devoted to the study of the asymptotic behavior of the sequence $\{B_n\}_{n \in \mathbb{N}}$. We are going to compute the asymptotic expansion for the partial derivatives. In order to apply our results, recall that it is well-known that the Bernstein operators are convex of any order $\alpha \in \mathbb{N}_0^m$. Moreover, Abel [7, Proposition 4 and Lemma 6], for $\alpha \in \mathbb{N}_0^m$ and $x \in H$, proves that

$$n^{|\alpha|} V_{n,\alpha}(x) \in \mathbb{P}_{\left[\frac{|\alpha|}{2}\right]}[n], \tag{18}$$

where $V_{n,\alpha}(x) = B_n\left((t-x)^\alpha\right)(x)$. In fact, we can affirm that the operators B_n meet the conditions of Theorem 3 so we can use such result to derive the asymptotic expansion of their partial derivatives. Our starting point will be the explicit asymptotic formula given by Abel [7, Theorem 1] for the sequence $\{B_n\}_{n\in\mathbb{N}}$ in the following way: given $r \in \mathbb{N}$ even, $x \in H$ and $f \in B(H, r, x)$,

$$B_n f(x) = P_{f,0,r,n}(x) + o(n^{-\frac{r}{2}}),$$

where

$$P_{f,0,r,n}(x) = f(x) + \sum_{\delta=1}^{\frac{r}{2}} n^{-\delta} \sum_{\substack{\alpha \in \mathbb{N}_0^m, \\ \delta < |\alpha| \le 2\delta}} \frac{D^\alpha f(x)}{\alpha!} \sum_{\beta=0}^{\alpha} a(\delta, \alpha, \beta) x^{\alpha-\beta} \quad (19)$$

and $a(\delta, \alpha, \beta) = \displaystyle\sum_{\substack{\nu=\beta \\ \delta \le |\nu|}}^{\alpha} (-1)^{\alpha-\nu} S_{|\nu-\beta|}^{|\nu|-\delta} \binom{\alpha}{\nu} \sigma_\nu^{\nu-\beta}$. Notice that if we denote

by $B(H)$ the set of all bounded functions on H, in this case, for any $r \in \mathbb{N}$, we have $B(H, r) = B(H)$. Then we can prove the following:

Theorem 11. *Let $r \in \mathbb{N}$ even, $k \in \mathbb{N}_0^m$, $x \in H$ and a bounded function $f \in C^{|k|}(H)$ such that $D^k f \in B(H, r, x)$. Then,*

$$D^k B_n f(x) = P_{f,k,r,n}(x) + o(n^{-\frac{r}{2}}),$$

where $P_{f,k,r,n}(x) = D^k P_{f,0,r,n}(x)$ and $P_{f,0,r,n}(x)$ is given by (19). Furthermore, the expansion $P_{f,k,r,n}(x)$ can be expressed as

$$P_{f,k,r,n}(x) = k! \sum_{\delta=0}^{\frac{r}{2}} n^{-\delta} \sum_{\substack{\gamma=k \\ \delta < |\gamma| \le |k|+2\delta}}^{\infty} \frac{D^\gamma f(x)}{\gamma!} \times$$

$$\times \sum_{\substack{\alpha=\gamma-k, \\ \delta < |\alpha| \le 2\delta}}^{\gamma} \binom{\gamma}{\alpha} \sum_{\beta=0}^{\gamma-k} \binom{\alpha-\beta}{k+\alpha-\gamma} x^{\gamma-\beta-k} a(\delta, \alpha, \beta),$$

with $a(\delta, \alpha, \beta)$ as defined above.

Proof: Take $W = B(H)$. It is obvious that both W and H satisfy the conditions of Theorem 3 and from (18) and already known properties we can assert that the Bernstein operators satisfy the conditions i), ii) and iii) of Theorem 3. Therefore, we can apply such a theorem in much the same way we have done before for the Baskakov operators in the proof of Theorem 9. $\square$

Abel [7] also obtains the Voronovskaja formula: given $x \in H$ and a function $f \in B(H, 2, x)$,

$$\lim_{n\to\infty} 2n \left(B_n f(x) - f(x)\right) = \sum_{i=1}^{m} x_i D^{2e_i} f(x) - \sum_{i,j=1}^{m} x_i x_j D^{e_i+e_j} f(x). \quad (20)$$

We use this identity and the preceding theorem in order to compute the Voronovskaja formula for all partial derivatives of the sequence $\{B_n\}_{n\in\mathbb{N}}$.

Corollary 12. Let $k \in \mathbb{N}_0^m$, $x \in H$ and a bounded function $f \in C^{|k|}(H)$ such that $D^k f \in B(H, 2, x)$. Then,

$$\lim_{n \to \infty} 2n \left(D^k B_n f(x) - D^k f(x) \right) = \left(|k| - |k|^2 \right) D^k f(x)$$

$$+ \sum_{i=1}^{m} \left[(x_i - x_i^2) D^{k+2e_i} f(x) + (k_i - 2|k|x_i) D^{k+e_i} f(x) \right]$$

$$- \sum_{\substack{i,j=1 \\ i \neq j}}^{m} x_i x_j D^{k+e_i+e_j} f(x).$$

Proof: We consider the value of $P_{f,0,2,n}(x)$ given by (20) and then we use the relation $P_{f,k,2,n}(x) = D^k P_{f,0,2,n}(x)$ of Theorem 11. $\square$

References

1. Abel, U., On the asymptotic approximation with operators of Bleimann, Butzer and Hahn, Indag. Mathem., New Ser. **7** (1996), 1–9.

2. Abel, U., The complete asymptotic expansion for the Meyer-König and Zeller operators, J. Math. Anal. Appl. **208** (1997), 109–119.

3. Abel, U., Asymptotic approximation with Kantorovich polynomial, Approx. Theory Appl. **3** 14 (1998), 106–116.

4. Abel, U., On the asymptotic approximation with bivariate operators of Bleimann, Butzer and Hahn, J. Approx. Theory **97** (1999), 181–198.

5. Abel, U., Asymptotic approximation by Bernstein-Durrmeyer operators and their derivatives, Approx. Theory Appl. **16** 2 (2000), 1–12.

6. Abel, U. and M. Ivan, Asymptotic expansion of the Jakimovski-Leviatan operators and their derivatives, Proc. of the F. Alexits Conference (Budapest, 1999), to appear.

7. Abel, U. and M. Ivan, Asymptotic expansion of the multivariate Bernstein polynomials on a simplex, Approx. Theory Appl. **16:3** (2000), 85–93.

8. Abramowitz, M. and A. Stegun, *Handbook of Mathematical Functions with Formulas, Graphs and Mathematical Tables*, Dover Publications, Inc. New York, 1972.

9. Baskakov, V. A., An example of a sequence of linear positive operators in the space of continuous functions, Dokl. Akad. Nauk SSSR **113** (1957), 249–251.

10. DeVore, R. A. and G. G. Lorentz, *Constructive Approximation,* A Series of Comprehensive Studies in Mathematics, 303, Springer-Verlag, Berlin-Heidelberg, 1993.

11. Lai, M. J., Asymptotic formulae of multivariate Bernstein approximation, J. Approx. Theory **70** (1992), 229–242.

12. López-Moreno, A. J., J. Martínez-Moreno and F. J. Muñoz-Delgado, Asymptotic expression of derivatives of Bernstein operators, Rendiconti del Circolo Matematico di Palermo, (to appear).

13. Schurer, F., *On Linear Positive Operators in Approximation Theory,* Dissertation, University of Delft, 1965.

14. Sikkema, P. C., On some linear positive operators, Indag. Math. **32** (1970), 327–337.

15. Walz, G., Asymptotic expansions for multivariate polynomial approximation, J. Comput. Appl. Math. **122** (2000), 317–328.

A.-J. López-Moreno
Departamento de Matemáticas
Universidad de Jaén, Spain
ajlopez@ujaen.es

J. Martínez-Moreno
Departamento de Matemáticas
Universidad de Jaén, Spain
jmmoreno@ujaen.es

F. J. Muñoz-Delgado
Departamento de Matemáticas
Universidad de Jaén, Spain
fdelgado@ujaen.es

Approximation by Penalized Least Squares

José L. Martínez-Morales

Abstract. Let M be a compact Riemannian manifold and let $\{\mathbf{x}_i\}_{i=1}^{p}$ be vectors in a Hermitian vector bundle over M. The approximation problem here considered is to find a section f on M which approximates the data $\{\mathbf{x}_i\}_{i=1}^{p}$ in the **penalized least squares** sense, that is, by minimizing a functional defined as the sum of the mean squared distance to the data plus a term that measures the "energy" of the section f. It is demonstrated that the approximation problem has a unique solution and a formula for the solution is obtained. As an illustration, concrete examples are developed in detail.

§1. Introduction

Let M be a compact Riemannian manifold, and suppose that $\{x_i\}_{i=1}^{p}$ is a set of scattered points lying on M. In this paper we are interested in the following problem:

Problem 1. *Given vectors $\{\mathbf{x}_i\}_{i=1}^{p}$, find a (smooth) vector-valued function f defined on M which approximates the data in the sense that*

$$f(x_i) \approx \mathbf{x}_i, \qquad i = 1, ..., p.$$

Data fitting problems where the underlying domain is a manifold M arise in many areas, including e.g. geophysics and meteorology where a sphere is taken as a model of the earth. The question of whether interpolation or approximation should be carried out depends on the setting, although in practice measured data are almost always noisy, in which case approximation is probably more appropriate.

One approach to Problem 1 is to create a least squares fit: find a function f which minimizes $L(f) = \sum_{i=1}^{p} |f(x_i) - \mathbf{x}_i|^2$.

Approximation Theory X: Abstract and Classical Analysis
Charles K. Chui, Larry L. Schumaker, and Joachim Stöckler (eds.), pp. 309–324.

When the data are especially noisy, it may be useful to replace this discrete least squares problem by a **penalized least squares problem**. The idea is to minimize a combination

$$I(f) = L(f) + \lambda \mathcal{E}(f),$$

where $\mathcal{E}(f)$ is a measure of energy defined by

$$\int_M |Af(x)|^2 dx,$$

where A is an operator acting on f. The parameter λ controls the trade-off between these two quantities, and is typically chosen to be a small positive number, see [4].

One way to define the energy functional is to take A as an appropriate differential operator, such as

$$A = (-\Delta + P)^{m/2}, \tag{1}$$

where m is an even integer, Δ is the **Laplace-Beltrami operator** on M, and P is the projection on the constants. The definition of A for the case where m is odd is more complicated and can be found in [12]. The functionals (1) are similar to the functionals which are minimized in defining **spherical thin plate splines**, see [14,15,16,17].

The penalized least squares approach to Problem 1 is a generalization of the smoothing spline, ridge regression, and generalized cross-validation problems ([5]), which have appeared in the statistical literature. References to the work done in this area are [3,4,10,13,18].

This paper analyses the problem from the approximation theory point of view. A general penalized least squares problem can be formulated as follows:

Let X be a linear space. On X there are given two semi-definite inner products $\langle \cdot, \cdot \rangle$ and $(\cdot, \cdot)$ with associated semi-norms $|\cdot|$ and $||\cdot||$, respectively. For given $F \in X$ and $\lambda > 0$ find $f \in X$ so that

$$|F - f|^2 + \lambda ||f||^2 = \inf_{\phi \in X} \{|F - \phi|^2 + \lambda ||\phi||^2\}. \tag{2}$$

We consider a special case of (2).

To mention results about the general problem (2), it is easy to prove the uniqueness of the minimal f in (2) if

$$[\cdot, \cdot]_\lambda := \langle \cdot, \cdot \rangle + \lambda(\cdot, \cdot)$$

is an inner product on X, and the existence of f if X is a Hilbert space for $[\cdot, \cdot]_\lambda$. Since the existence and uniqueness of such a minimal f are our main results, it would suffice to prove that our $(X, [\cdot, \cdot]_\lambda)$ is a Hilbert space. Instead, we give a constructive proof of the existence and uniqueness of the minimizing function. As a byproduct of this proof, we obtain an explicit formula for the minimizing function. This explicit formula can be used in applications of the type (1), and an estimate of the approximation rate of the corresponding numerical scheme can be obtained. Such estimates of the approximation rate will be the subject of a future paper.

In the present paper an application of penalized least squares fitting is developed by letting the manifold M be the n-dimensional sphere. In particular we consider the case of the two-sphere, and a figure is included to visually show the quality of the fitting. For a number of other numerical examples involving penalized least squares fitting on the sphere, see [1], where figures and tables related to storage, exactness, and computational time are included.

The paper is organized as follows: in Section 2 some basic definitions and preliminary results are given. In Section 3, the existence and uniqueness of a global minimum of the functional I are demonstrated, and an explicit formula for such a minimum is obtained. Finally, in Section 4 we apply our method to the n-dimensional sphere case and explicit formulas are obtained.

§2. Basic Definitions and Preliminary Results

This section is intended to give a theoretical formulation of the penalized least squares scheme. We begin by defining the space in which the minimization process will be carried out, and also introduce the notation that will be used throughout the paper. The description of our setting in this section might be too short. For more on the various definitions used here, see [2,7].

Let M be a compact n-dimensional C^∞ Riemannian manifold such that if $\partial M \neq \emptyset$, then the boundary ∂M is a closed submanifold and near its boundary M is a direct product $\partial M \times [0, \delta), \delta > 0$. To avoid technical difficulties we use also a manifold $\tilde{M} = ((-\infty, 0] \times \partial M) \cup M$.

For a finite dimensional hermitian vector bundle $\mathcal{V}_M$ over M, let $\Gamma(\mathcal{V}_M)$ be the space of all C^∞ sections of $\mathcal{V}_M$. Denote by $\langle , \rangle$ the Hermitian inner product and set $|\cdot| = \sqrt{\langle \cdot, \cdot \rangle}$. For $f \in \Gamma(\mathcal{V}_M)$, we define norms $||f||_k$ by

$$||f||_k^2 = \sum_{l \leq k} \int_M |(D^l f)(x)|^2 dx,$$

where dx is the volume element defined by the Riemannian metric, and Df denotes covariant differentiation with respect to the Riemannian connection. Let $\Gamma^k(\mathcal{V}_M)$ be the completion of $\Gamma(\mathcal{V}_M)$ with respect to the norm $||f||_k$. To simplify the notation we write $|| \cdot ||$ instead of $|| \cdot ||_0$.

Denote by $\pi : \mathcal{V}_M \to M$ the projection of the bundle. Suppose that $\{x_i\}_{i=1}^p$ is a set of scattered points lying on M, and consider vectors $\{\mathbf{x}_i\}_{i=1}^p \subset \mathcal{V}_M$ such that $\pi \mathbf{x}_i = x_i$. For $k \geq [\frac{n}{2}]+1$ consider an isomorphism $A : \Gamma^k(\mathcal{V}_M) \to \Gamma^0(\mathcal{V}_M)$, and define the functional

$$I(f) = \sum_{i=1}^p |\mathbf{x}_i - f(x_i)|^2 + \lambda\|Af\|^2. \tag{3}$$

By the Sobolev lemma (see [11], p. 116), every section in $\Gamma^k(\mathcal{V}_M)$ is continuous. Therefore, the domain of the functional I is set to be $\Gamma^k(\mathcal{V}_M)$.

The map $\mathcal{K}$

In the following section the existence and uniqueness of a global minimum of the functional I are demonstrated. Such a minimum is given in terms of the inverse of a matrix which contains all of the information supplied by the vectors $\{\mathbf{x}_i\}_{i=1}^p$. We call this matrix the "system matrix".

Next we develop some technical machinery. We start by defining the system matrix and other fiber maps. Let E be the fiber of $\mathcal{V}_M$ and let $Hom(E)$ denote the space of linear homomorphisms of E into itself. Let $\mathcal{V}_{M \times M}$ be the C^∞ vector bundle over $M \times M$ with fiber $Hom(E)$. By Theorem 4 in [9], the operator $A^{-1} : \Gamma^0(\mathcal{V}_M) \to \Gamma^k(\mathcal{V}_M)$ is associated with a kernel $K(\cdot,\cdot) \in \Gamma^0(\mathcal{V}_{M \times M})$.

For $x, y \in M$ and $H \in Hom(E_x, E_y)$, denote by $H^* \in Hom(E_y, E_x)$ the adjoint of H defined by

$$\langle e_x, H^* e_y \rangle = \langle H e_x, e_y \rangle, \qquad e_x \in E_x, \, e_y \in E_y.$$

Definition 1. *For $i, j = 1, ..., p$, let*

$$\mathcal{K}_{ij} \in Hom(E_{x_j}, E_{x_i}),$$

$$P_i \in Hom(E_{x_1} \times \cdots \times E_{x_p}, E_{x_i}),$$

$$\mathcal{K} \in Hom(E_{x_1} \times \cdots \times E_{x_p}),$$

be the maps defined by

$$\mathcal{K}_{ij} = \int_M K(x_i, x) K(x_j, x)^* dx, \tag{4}$$

$$P_i(e_{x_1}, ..., e_{x_p}) = e_{x_i}, \tag{5}$$

$$P_i \mathcal{K} = \sum_{j=1}^p \mathcal{K}_{ij} P_j. \tag{6}$$

Definition 2. *Let* $\langle , \rangle$ *be the inner product in* $E_{x_1} \times \cdots \times E_{x_p}$ *defined by*

$$\langle (e_{x_1}, ..., e_{x_p}), (e'_{x_1}, ..., e'_{x_p}) \rangle = \sum_{i=1}^{p} \langle e_{x_i}, e'_{x_i} \rangle, \qquad e_{x_i}, e'_{x_i} \in E_{x_i}. \quad (7)$$

The map $\mathcal{K}$ has the following properties: (i) it is self-adjoint with respect to the inner product (7), (ii) it is positive semidefinite and, if we denote by $\mathcal{I}$ the identity map in $E_{x_1} \times \cdots \times E_{x_p}$, then (iii) the map $\lambda \mathcal{I} + \mathcal{K}$ is invertible. These assertions are consequences of the following theorem.

Theorem 1. *For* $(e_{x_1}, ..., e_{x_p}), (e'_{x_1}, ..., e'_{x_p}) \in E_{x_1} \times \cdots \times E_{x_p}$, *we have*

(i) $\langle (e_{x_1}, ..., e_{x_p}), \mathcal{K}(e'_{x_1}, ..., e'_{x_p}) \rangle = \langle \mathcal{K}(e_{x_1}, ..., e_{x_p}), (e'_{x_1}, ..., e'_{x_p}) \rangle.$

(ii) $\langle (e_{x_1}, ..., e_{x_p}), \mathcal{K}(e_{x_1}, ..., e_{x_p}) \rangle = \left\| \sum_{i=1}^{p} K(x_i, \cdot)^* e_{x_i} \right\|^2.$

(iii) $\langle (e_{x_1}, ..., e_{x_p}), (\lambda \mathcal{I} + \mathcal{K})(e_{x_1}, ..., e_{x_p}) \rangle \geq \lambda \langle (e_{x_1}, ..., e_{x_p}), (e_{x_1}, ..., e_{x_p}) \rangle.$

In order to prove this theorem, a technical lemma is needed. Let

$$(f, g) = \int_M \langle f(x), g(x) \rangle dx, \qquad (8)$$

be the inner product of two sections $f, g \in \Gamma^0(\mathcal{V}_M)$.

Lemma 1. *Let* $1 \leq i \leq p$ *and* $(e_{x_1}, ..., e_{x_p}) \in E_{x_1} \times \cdots \times E_{x_p}$. *Then, for* $e'_{x_i} \in E_{x_i}$,

$$\langle e'_{x_i}, P_i \mathcal{K}(e_{x_1}, ..., e_{x_p}) \rangle = \sum_{j=1}^{p} \left(K(x_i, \cdot)^* e'_{x_i}, K(x_j, \cdot)^* e_{x_j} \right).$$

Proof: We have

$$P_i \mathcal{K}(e_{x_1}, ..., e_{x_p}) = \sum_{j=1}^{p} \mathcal{K}_{ij} P_j(e_{x_1}, ..., e_{x_p}) \qquad \text{(by (6))}$$

$$= \sum_{j=1}^{p} \mathcal{K}_{ij} e_{x_j} \qquad \text{(by (5))}$$

$$= \sum_{j=1}^{p} \int_M K(x_i, x) K(x_j, x)^* dx \, e_{x_j} \qquad \text{(by (4))}.$$

Taking the inner product with e'_{x_i} gives

$$\langle e'_{x_i}, P_i \mathcal{K}(e_{x_1}, ..., e_{x_p}) \rangle = \sum_{j=1}^{p} \int_M \langle e'_{x_i}, K(x_i, x) K(x_j, x)^* e_{x_j} \rangle dx$$

$$= \sum_{j=1}^{p} \int_M \big\langle K(x_i,x)^* e'_{x_i}, K(x_j,x)^* e_{x_j} \big\rangle dx$$

$$\text{(by definition of } K(x_i,x)^*)$$

$$= \sum_{j=1}^{p} \big(K(x_i,\cdot)^* e'_{x_i}, K(x_j,\cdot)^* e_{x_j} \big) \qquad \text{(by (8)).} \quad \square$$

In the sequel, purely formal calculations such as the preceding one will sometimes be omitted.

Proof of Theorem 1. (i) We have

$$\langle (e_{x_1},...,e_{x_p}), \mathcal{K}(e'_{x_1},...,e'_{x_p}) \rangle = \sum_{i=1}^{p} \langle e_{x_i}, P_i \mathcal{K}(e'_{x_1},...,e'_{x_p}) \rangle$$

$$\text{(by Definition 2)}$$

$$= \sum_{i=1}^{p} \sum_{j=1}^{p} \big(K(x_i,\cdot)^* e_{x_i}, K(x_j,\cdot)^* e'_{x_j} \big) \quad (9)$$

$$\text{(by Lemma 1)}$$

$$= \sum_{j=1}^{p} \overline{\langle e'_{x_j}, P_j \mathcal{K}(e_{x_1},...,e_{x_p}) \rangle}$$

(again by Lemma 1, where the upper line denotes complex conjugation)

$$= \sum_{j=1}^{p} \langle P_j \mathcal{K}(e_{x_1},...,e_{x_p}), e'_{x_j} \rangle$$

$$= \langle \mathcal{K}(e_{x_1},...,e_{x_p}), (e'_{x_1},...,e'_{x_p}) \rangle.$$

For part (ii) we have

$$\langle (e_{x_1},...,e_{x_p}), \mathcal{K}(e_{x_1},...,e_{x_p}) \rangle = \sum_{i=1}^{p} \sum_{j=1}^{p} \big(K(x_i,\cdot)^* e_{x_i}, K(x_j,\cdot)^* e_{x_j} \big)$$

$$\text{(by (9))}$$

$$= \bigg(\sum_{i=1}^{p} K(x_i,\cdot)^* e_{x_i}, \sum_{j=1}^{p} K(x_j,\cdot)^* e_{x_j} \bigg)$$

$$= \bigg\| \sum_{i=1}^{p} K(x_i,\cdot)^* e_{x_i} \bigg\|^2.$$

Finally part (iii) follows from (ii). $\square$

§3. The Minimum of I

In this section we show that there exists a unique global minimum of the functional I in $\Gamma^k(\mathcal{V}_M)$. We start by defining our candidate for this.

Definition 3. *Let* f_{min} *be the section defined by*

$$f_{min} = A^{-1} \sum_{i=1}^{p} K(x_i, \cdot)^* P_i (\lambda \mathcal{I} + \mathcal{K})^{-1}(\mathbf{x}_1, ..., \mathbf{x}_p). \tag{10}$$

Notice that for any $e_{x_i} \in E_{x_i}$, $K(x_i, \cdot)^* e_{x_i} \in \Gamma^0(\mathcal{V}_M)$ since $K(\cdot, \cdot) \in \Gamma^0(\mathcal{V}_{M \times M})$. Therefore $A^{-1} K(x_i, \cdot)^* e_{x_i} \in \Gamma^k(\mathcal{V}_M)$. In particular $f_{min} \in \Gamma^k(\mathcal{V}_M)$.

As a preliminary result, we compute the value of the functional I at f_{min}.

Theorem 2.

$$I(f_{min}) = \sum_{i=1}^{p} \langle \mathbf{x}_i - f_{min}(x_i), \mathbf{x}_i \rangle. \tag{11}$$

An obvious consequence of this theorem is that if f_{min} interpolates the data $\{\mathbf{x}_i\}_{i=1}^{p}$, then $I(f_{min})$ is zero. To prove Theorem 2 we need two technical lemmas.

Lemma 2. *For* $\mathbf{x} \in \mathcal{V}_M$ *and* $f \in \Gamma^0(\mathcal{V}_M)$, *we have*

$$\langle f(\pi\mathbf{x}), \mathbf{x} \rangle = \big(f, A^* K(\pi\mathbf{x}, \cdot)^* \mathbf{x}\big),$$

where A^* *denotes the adjoint of* A.

Proof: We have

$$\langle f(\pi\mathbf{x}), \mathbf{x} \rangle = \langle (A^{-1}Af)(\pi\mathbf{x}), \mathbf{x} \rangle = \left\langle \int_M K(\pi\mathbf{x}, x) Af(x) dx, \mathbf{x} \right\rangle$$

$$= \int_M \langle K(\pi\mathbf{x}, x) Af(x), \mathbf{x} \rangle dx = \int_M \langle Af(x), K(\pi\mathbf{x}, x)^* \mathbf{x} \rangle dx$$

$$= \big(Af, K(\pi\mathbf{x}, \cdot)^* \mathbf{x}\big) = \big(f, A^* K(\pi\mathbf{x}, \cdot)^* \mathbf{x}\big). \quad \square$$

Lemma 3. *The section* f_{min} *satisfies*

$$\lambda A f_{min} + \sum_{i=1}^{p} K(x_i, \cdot)^* f_{min}(x_i) = \sum_{i=1}^{p} K(x_i, \cdot)^* \mathbf{x}_i. \tag{12}$$

Proof: We first compute

$$f_{min}(x_i) = A^{-1} \sum_{j=1}^{p} K(x_j, x_i)^* P_j (\lambda \mathcal{I} + \mathcal{K})^{-1}(\mathbf{x}_1, ..., \mathbf{x}_p) \qquad \text{(by (10))}$$

$$= \int_M K(x_i, x) \sum_{j=1}^{p} K(x_j, x)^* P_j (\lambda \mathcal{I} + \mathcal{K})^{-1}(\mathbf{x}_1, ..., \mathbf{x}_p) dx$$

$$= \sum_{j=1}^{p} \int_M K(x_i, x) K(x_j, x)^* dx P_j (\lambda \mathcal{I} + \mathcal{K})^{-1}(\mathbf{x}_1, ..., \mathbf{x}_p)$$

$$= \sum_{j=1}^{p} \mathcal{K}_{ij} P_j (\lambda \mathcal{I} + \mathcal{K})^{-1}(\mathbf{x}_1, ..., \mathbf{x}_p) \qquad \text{(by (4))}$$

$$= P_i \mathcal{K} (\lambda \mathcal{I} + \mathcal{K})^{-1}(\mathbf{x}_1, ..., \mathbf{x}_p) \qquad \text{(by (6))}.$$

Then

$$\sum_{i=1}^{p} K(x_i, \cdot)^* f_{min}(x_i) = \sum_{i=1}^{p} K(x_i, \cdot)^* P_i \mathcal{K} (\lambda \mathcal{I} + \mathcal{K})^{-1}(\mathbf{x}_1, ..., \mathbf{x}_p). \quad (13)$$

On the other hand,

$$\lambda A f_{min} = \lambda \sum_{i=1}^{p} K(x_i, \cdot)^* P_i (\lambda \mathcal{I} + \mathcal{K})^{-1}(\mathbf{x}_1, ..., \mathbf{x}_p),$$

by (10). Adding this to (13)) gives

$$\lambda A f_{min} + \sum_{i=1}^{p} K(x_i, \cdot)^* f_{min}(x_i)$$

$$= \sum_{i=1}^{p} K(x_i, \cdot)^* P_i (\lambda \mathcal{I} + \mathcal{K})(\lambda \mathcal{I} + \mathcal{K})^{-1}(\mathbf{x}_1, ..., \mathbf{x}_p)$$

$$= \sum_{i=1}^{p} K(x_i, \cdot)^* P_i (\mathbf{x}_1, ..., \mathbf{x}_p)$$

$$= \sum_{i=1}^{p} K(x_i, \cdot)^* \mathbf{x}_i. \quad \square$$

Proof of Theorem 2. We start by computing

$$|\mathbf{x}_i - f_{min}(x_i)|^2$$
$$= |\mathbf{x}_i|^2 - \langle \mathbf{x}_i, f_{min}(x_i) \rangle - \langle f_{min}(x_i), \mathbf{x}_i \rangle + \langle f_{min}(x_i), f_{min}(x_i) \rangle$$
$$= |\mathbf{x}_i|^2 - \langle \mathbf{x}_i, f_{min}(x_i) \rangle - \langle f_{min}(x_i), \mathbf{x}_i \rangle + \left(A^* K(x_i, \cdot)^* f_{min}(x_i), f_{min} \right),$$

by Lemma 2, where in this case $\mathbf{x} = f_{min}(x_i)$ and $f = f_{min}$. Substituting this and $||Af_{min}||^2 = (A^*Af_{min}, f_{min})$ into (3), gives

$$I(f_{min}) = \sum_{i=1}^{p} \left(|\mathbf{x}_i|^2 - \langle \mathbf{x}_i, f_{min}(x_i) \rangle - \langle f_{min}(x_i), \mathbf{x}_i \rangle \right.$$
$$\left. + \left(A^*K(x_i, \cdot)^* f_{min}(x_i), f_{min} \right) \right) + \lambda(A^*Af_{min}, f_{min})$$

$$= \sum_{i=1}^{p} \left(|\mathbf{x}_i|^2 - \langle \mathbf{x}_i, f_{min}(x_i) \rangle - \langle f_{min}(x_i), \mathbf{x}_i \rangle \right)$$
$$+ \left(A^* \left(\sum_{i=1}^{p} K(x_i, \cdot)^* f_{min}(x_i) + \lambda A f_{min} \right), f_{min} \right)$$

(by factoring f_{min} and A^*)

$$= \sum_{i=1}^{p} \left(|\mathbf{x}_i|^2 - \langle \mathbf{x}_i, f_{min}(x_i) \rangle - \langle f_{min}(x_i), \mathbf{x}_i \rangle \right)$$
$$+ \left(A^* \sum_{i=1}^{p} K(x_i, \cdot)^* \mathbf{x}_i, f_{min} \right) \qquad \text{(by (12))}$$

$$= \sum_{i=1}^{p} \left(|\mathbf{x}_i|^2 - \langle \mathbf{x}_i, f_{min}(x_i) \rangle - \langle f_{min}(x_i), \mathbf{x}_i \rangle \right) + \sum_{i=1}^{p} \langle \mathbf{x}_i, f_{min}(x_i) \rangle$$

(by Lemma 2, where in this case $\mathbf{x} = \mathbf{x}_i$ and $f = f_{min}$)

$$= \sum_{i=1}^{p} \langle \mathbf{x}_i - f_{min}(x_i), \mathbf{x}_i \rangle. \quad \square$$

Theorem 3. *For all $f \in \Gamma^k(\mathcal{V}_M)$,*

$$I(f) - I(f_{min}) = \sum_{i=1}^{p} |(f - f_{min})(x_i)|^2 + \lambda ||A(f - f_{min})||^2. \qquad (14)$$

Corollary 1. *The section given by (10) is the unique minimum of the functional I.*

Proof: This is a consequence of Theorem 3. If $I(f_{min}) = I(f)$ then $A(f - f_{min}) = 0$, by (14). Hence $f = f_{min}$, since A is injective. $\quad \square$

Proof of Theorem 3. We begin by computing the right hand side of (14). By Lemma 2,

$$|(f - f_{min})(x_i)|^2 = |f(x_i)|^2 - \langle f(x_i), f_{min}(x_i) \rangle - \langle f_{min}(x_i), f(x_i) \rangle$$
$$+ \langle f_{min}(x_i), f_{min}(x_i) \rangle =$$

$$
= |f(x_i)|^2 - \big(f, A^* K(x_i, \cdot)^* f_{min}(x_i)\big)
$$
$$
- \big(A^* K(x_i, \cdot)^* f_{min}(x_i), f\big)
$$
$$
+ \big(f_{min}, A^* K(x_i, \cdot)^* f_{min}(x_i)\big), \tag{15}
$$

On the other hand,

$$
||A(f - f_{min})||^2 = ||Af||^2 - (Af, Af_{min}) - (Af_{min}, Af) + (Af_{min}, Af_{min})
$$
$$
= ||Af||^2 - (f, A^* Af_{min}) - (A^* Af_{min}, f)
$$
$$
+ (f_{min}, A^* Af_{min}). \tag{16}
$$

By (15) and (16),

$$
\sum_{i=1}^{p} |(f - f_{min})(x_i)|^2 + \lambda ||A(f - f_{min})||^2
$$

$$
= \sum_{i=1}^{p} \Big(|f(x_i)|^2 - \big(f, A^* K(x_i, \cdot)^* f_{min}(x_i)\big) - \big(A^* K(x_i, \cdot)^* f_{min}(x_i), f\big)
$$
$$
+ \big(f_{min}, A^* K(x_i, \cdot)^* f_{min}(x_i)\big) \Big) + \lambda \big(||Af||^2 - (f, A^* Af_{min})
$$
$$
- (A^* Af_{min}, f) + (f_{min}, A^* Af_{min}) \big)
$$

$$
= \sum_{i=1}^{p} |f(x_i)|^2 - \left(f, A^* \left(\sum_{i=1}^{p} K(x_i, \cdot)^* f_{min}(x_i) + \lambda Af_{min} \right) \right)
$$
$$
- \left(A^* \left(\sum_{i=1}^{p} K(x_i, \cdot)^* f_{min}(x_i) + \lambda Af_{min} \right), f \right)
$$
$$
+ \left(f_{min}, A^* \left(\sum_{i=1}^{p} K(x_i, \cdot)^* f_{min}(x_i) + \lambda Af_{min} \right) \right) + \lambda ||Af||^2
$$

(by factoring f_{min}, f and A^* in the inner products)

$$
= \sum_{i=1}^{p} |f(x_i)|^2 - \left(f, A^* \left(\sum_{i=1}^{p} K(x_i, \cdot)^* \mathbf{x}_i \right) \right) - \left(A^* \left(\sum_{i=1}^{p} K(x_i, \cdot)^* \mathbf{x}_i \right), f \right)
$$
$$
+ \left(f_{min}, A^* \left(\sum_{i=1}^{p} K(x_i, \cdot)^* \mathbf{x}_i \right) \right) + \lambda ||Af||^2 \tag{by (12)}
$$

$$
= \sum_{i=1}^{p} \Big(|f(x_i)|^2 - \big(f, A^* K(x_i, \cdot)^* \mathbf{x}_i\big) - \big(A^* K(x_i, \cdot)^* \mathbf{x}_i, f\big)
$$
$$
+ \big(f_{min}, A^* K(x_i, \cdot)^* \mathbf{x}_i\big) \Big) + \lambda ||Af||^2
$$

$$
= \sum_{i=1}^{p} \Big(|f(x_i)|^2 - \langle f(x_i), \mathbf{x}_i \rangle - \langle \mathbf{x}_i, f(x_i) \rangle + \langle f_{min}(x_i), \mathbf{x}_i \rangle \Big) + \lambda ||Af||^2
$$

(by Lemma 2)

$$= \sum_{i=1}^{p} \left(|f(x_i)|^2 - \langle f(x_i), \mathbf{x}_i \rangle - \langle \mathbf{x}_i, f(x_i) \rangle + \langle f_{min}(x_i), \mathbf{x}_i \rangle \right) + \lambda ||Af||^2$$

$$+ \sum_{i=1}^{p} |\mathbf{x}_i|^2 - \sum_{i=1}^{p} |\mathbf{x}_i|^2$$

$$= \sum_{i=1}^{p} |f(x_i) - \mathbf{x}_i|^2 + \lambda ||Af||^2 - \sum_{i=1}^{p} \langle \mathbf{x}_i - f_{min}(x_i), \mathbf{x}_i \rangle$$

$$= I(f) - I(f_{min}),$$

by (3) and (11). □

§4. Penalized Least Squares Fitting on the Sphere

In this section we apply our method to the case when M is the n-dimensional unit sphere. Let $\mathcal{V}_M$ be the normal bundle over the n-dimensional unit sphere. We define the energy functional by taking A as

$$A = -\Delta + P,$$

where Δ is the Laplace-Beltrami operator on M, and P is the projection on the constants. The eigenvalues of the elliptic operator A are the same as those of $-\Delta$, except that one of them has been changed from zero to 1. Therefore $A : \Gamma^2(\mathcal{V}_M) \to \Gamma^0(\mathcal{V}_M)$ is an isomorphism.

The minimizing section (10) is given in terms of the kernel K of the operator A^{-1}. In the following theorem we compute a formula for K.

Theorem 4. *Let H_0, H_1... be an orthogonal sequence of n-dimensional spherical harmonics. Then*

$$K(x,y) = H_0(x)H_0(y)/||H_0|| + \sum_{j=1}^{\infty} [n_j(n_j + n - 1)]^{-1} H_j(x)H_j(y)/||H_j||,$$

$$(17)$$

where n_j denotes the order of H_j.

Proof: Consider a function $\gamma \in \Gamma^0(\mathcal{V}_M)$. We solve the equation

$$(-\Delta + P)f = \gamma, \qquad f \in \Gamma^2(\mathcal{V}_M), \qquad (18)$$

as

$$f(x) = \int_{S^n} K(x,y)\gamma(y)dx.$$

The solution f has the following expansion in terms of the H_j,

$$f = \sum_{j=0}^{\infty} f_j H_j. \tag{19}$$

By Theorem 3.2.11 in [6],

$$\Delta f = -\sum_{j=0}^{\infty} n_j(n_j + n - 1) f_j H_j. \tag{20}$$

On the other hand

$$P f = f_0 H_0. \tag{21}$$

Assume that γ has a finite expansion in terms of the spherical harmonics,

$$\gamma = \sum_{j=0}^{J} c_j H_j, \tag{22}$$

with

$$c_j = (\gamma, H_j)/\|H_j\|. \tag{23}$$

Substituting (21) through (22) into (18) gives

$$\sum_{j=0}^{\infty} n_j(n_j + n - 1) f_j H_j + f_0 H_0 = \sum_{j=0}^{J} c_j H_j,$$

therefore

$$f_0 = c_0 \quad \text{and} \quad n_j(n_j + n - 1) f_j = \begin{cases} c_j, & 0 < j \le J, \\ 0, & J < j. \end{cases}$$

Solving for f_j and substituting into (21) give

$$f = c_0 H_0 + \sum_{j=1}^{J} [n_j(n_j + n - 1)]^{-1} c_j H_j$$

$$= (\gamma, H_0)/\|H_0\| H_0 + \sum_{j=1}^{J} [n_j(n_j + n - 1)]^{-1} (\gamma, H_j)/\|H_j\| H_j \quad \text{(by (23))}$$

$$= \int_M \gamma(x) H_0(x) dx/\|H_0\| H_0$$

$$+ \sum_{j=1}^{J} [n_j(n_j + n - 1)]^{-1} \int_M \gamma(x) H_j(x) dx/\|H_j\| H_j \quad \text{(by (8))}$$

$$= \int_M \left(H_0(x)/\|H_0\| H_0 + \sum_{j=1}^{J} [n_j(n_j + n - 1)]^{-1} H_j(x)/\|H_j\| H_j \right) \gamma(x) dx.$$

Since J is arbitrary,

$$K(\cdot,x) = H_0(x)/\|H_0\|H_0 + \sum_{j=1}^{\infty}[n_j(n_j+n-1)]^{-1}H_j(x)/\|H_j\|H_j. \quad \square$$

The minimizing section can be expressed as a series of Legendre polynomials. The coefficients in this series depend on the dimension of the sphere and the degree of the polynomials.

Since the kernel K is square integrable with respect to the two independent variables, it is square integrable with respect to any one of the variables, keeping the other one fixed. In the proof of the following lemma, the expansion (17) of the kernel K in spherical harmonics is used, as well as the *addition theorem* for spherical harmonics. Let $x \cdot y$ denote the dot product in $\mathbb{R}^n$.

Lemma 4. *Let*

$$N(l,n) = \frac{(2l+n-1)(l+n-2)!}{l!(n-1)!}.$$

Let σ_n *denote the surface area of* M, *and let* P_l^n *denote the Legendre polynomial of dimension* n *and degree* l. *Then*

$$\int_M K(x,y)K(y,z)^*dy = \frac{1}{\sigma_n}\left(1 + \sum_{l=1}^{\infty}[l(l+n-1)]^{-2}N(l,n)P_l^n(x\cdot z)\right).$$

Proof: By Theorem 4 in [9], $K(\cdot,\cdot) \in \Gamma^0(\mathcal{V}_{M\times M})$; in particular,

$$\forall x \in M : K(x,\cdot) \in \Gamma^0(\mathcal{V}_M).$$

By (17) and Theorem 3.2.10 in [6],

$$\int_M K(x,z)K(z,y)^*dz = H_0(x)H_0(y) + \sum_{j=1}^{\infty}[n_j(n_j+n-1)]^{-2}H_j(x)H_j(y).$$

By Theorem 3.3.3 in [6],

$$H_0(x)H_0(y) + \sum_{j=1}^{\infty}[n_j(n_j+n-1)]^{-2}H_j(x)H_j(y)$$

$$= \frac{1}{\sigma_n}P_0^n(x\cdot y) + \sum_{l=1}^{\infty}[l(l+n-1)]^{-2}\frac{N(l,n)}{\sigma_n}P_l^n(x\cdot y)$$

$$= \frac{1}{\sigma_n} + \sum_{l=1}^{\infty}[l(l+n-1)]^{-2}\frac{N(l,n)}{\sigma_n}P_l^n(x\cdot y),$$

since $P_0^n(t) = 1$ ([6], p. 85). $\square$

Consider the case of the sphere of dimension two. Then, the series of Legendre polynomials can be expressed in terms of the dilogarithm function.

Corollary 2. *Let* $n=2$. *Then*

$$\int_M K(x,y)K(y,z)^* dy = \frac{1}{4\pi}\left(-1 + \frac{\pi^2}{7} - \mathrm{dilog}(x \cdot z)\right). \quad (24)$$

Proof: By Theorem 5 in [8],

$$\int_M K(x,y)K(y,z)^* dy = \frac{1}{\sigma_n}\left(1 - \frac{1}{7}(12 - \pi^2) - \mathrm{dilog}(x \cdot z)\right),$$

where $\sigma_n=4\pi$. $\square$

Since $\mathcal{V}_M$ is the normal bundle on M, the vectors $\{\mathbf{x}_i\}_{i=1}^p$ are, in fact, scalars.

Proposition 1. *Let*

$$\mathbf{x} = (\mathbf{x}_1, ..., \mathbf{x}_p). \quad (25)$$

Consider the function $f{:}M \times M \to \mathbb{R}$ *defined as*

$$f(x,y) = \frac{1}{4\pi}\left(-1 + \frac{\pi^2}{7} - \mathrm{dilog}(x \cdot y)\right). \quad (26)$$

Consider the matrix $\mathrm{M}=(f(x_i, x_j))$, *where* $\{x_i\}_{i=1}^p$ *is the set of scattered points lying on* M. *Consider the function* $F{:}M \to \mathbb{R}^p$ *defined as*

$$F(x) = (f(x,x_1), ..., f(x,x_p)). \quad (27)$$

Denote by $\mathrm{I}_{p\times p}$ *the* $p \times p$ *identity matrix. Then*

$$f_{min} = F \cdot (\lambda\mathrm{I}_{p\times p} + \mathrm{M})^{-1}\mathbf{x}.$$

Proof: Since the fiber of $\mathcal{V}_M$ is one-dimensional, from Definition 1 and (24), it follows that the map $\mathcal{K}$ is the matrix M. By (10) and (25),

$$f_{min} = A^{-1}\sum_{i=1}^p K(x_i, \cdot)^* P_i(\lambda\mathrm{I}_{p\times p} + \mathrm{M})^{-1}\mathbf{x}$$

$$= \int_M K(\cdot, x)\sum_{i=1}^p K(x_i, x)^* P_i(\lambda\mathrm{I}_{p\times p} + \mathrm{M})^{-1}\mathbf{x}\, dx$$

$$= \sum_{i=1}^p f(\cdot, x_i)P_i(\lambda\mathrm{I}_{p\times p} + \mathrm{M})^{-1}\mathbf{x} \qquad \text{(by (24) and (26))}$$

$$= F \cdot (\lambda\mathrm{I}_{p\times p} + \mathrm{M})^{-1}\mathbf{x}$$

by (6) and (27). $\square$

Figure 1 illustrates an application to data fitting on the sphere. In plot (1), a spherical harmonic of fifth order is plotted on the normal bundle of the sphere. In plot (2), a total of 150 points are sampled from the original surface and a little noise is added. Plots (3) and (4) show the approximation obtained. As a side comment, it is remarkable how the quality of the approximation is maintained in regions where there are few sampled points.

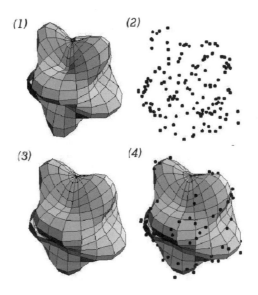

Fig. 1. An example of sphere fitting.

References

1. Alfeld, P., M. Neamtu, and L. L. Schumaker, Fitting scattered data on sphere-like surfaces using spherical splines, J. Comput. Appl. Math. **73** (1996), 5–43.

2. Clemens, H., *Introduction to Hodge Theory*, Appunti dei Corsi Tenuti da Docenti della Scuola, Scuola Normale Superiore, Pisa, 1998.

3. Eubank, R. L. and R. F. Gunst, Diagnostics for penalized least-squares estimators, Statist. Probab. Lett. **4** (1986), 265–272.

4. von Golitschek, M., and L. L. Schumaker, Data fitting by penalized least squares, in *Algorithms for Approximation II*, J. C. Mason and M. G. Cox (eds.), Chapman & Hall, London, 1990, 210–227.

5. Good, P. I., *Resampling Methods. A Practical Guide to Data Analysis*, Birkhäuser Boston, 1999.

6. Groemer, H., *Geometric Applications of Fourier Series and Spherical Harmonics*, Cambridge University Press, New York, 1996.

7. Lee, J. M., *Riemannian Manifolds. An Introduction to Curvature*, Springer Verlag, New York, 1997.

8. Martínez-Morales, J. L., Generalized Legendre series and the fundamental solution of the Laplacian on the n-sphere, preprint.

9. Martínez-Morales, J. L., The Kernel Theorem of Hilbert-Schmidt operators, preprint. Available at www.matcuer.unam.mx/~martinez/ Artículos_publicados_o_sometidos/Kernel.ps.

10. Nowak, R. D., Penalized least squares estimation of Volterra filters and higher order statistics, IEEE Trans. Signal Process **46** (1998), 419–428.

11. Omori, H., *Infinite-Dimensional Lie Groups*, American Mathematical Society, Providence, 1997.

12. Seeley, R. T., Complex powers of an elliptic operator, A.M.S. Proc. Symp. Pure Math. **10** (1967), 288–307. Corrections in: The resolvent of an elliptic boundary problem. Am. J. Math. **91**, 917–919 (1969).

13. Sun, W., A penalized least squares method for image reconstruction and its stability analysis, Information 3 **1** (2000), 43–49.

14. Wahba, G., Spline interpolation and smoothing on the sphere, SIAM J. Sci. Statist. Comput. **2** (1981), 5–16.

15. Wahba, G., Errata: Spline interpolation and smoothing on the sphere, SIAM J. Sci. Statist. Comput. **3** (1982), 385–386.

16. Wahba, G., Vector splines on the sphere, with application to the estimation of vorticity and divergence from discrete, noisy data, in *Multivariate Approximation Theory II*, W. Schempp and K. Zeller, (eds.), Birkhäuser, Basel, 1982, 407–429.

17. Wahba, G., Surface fitting with scattered noisy data on Euclidean *d*-space and on the sphere, Rocky Mountain J. Math. **14** (1984), 281–299.

18. Weyrich, N., Bivariate spline approximation by penalized least squares, in *Mathematical Methods in Computer Aided Geometric Design, II*, T. Lyche and L. Schumaker (eds.), Academic Press, Boston, 1992, 607–614.

José L. Martínez-Morales
Instituto de Matemáticas
Universidad Nacional Autónoma de México
Ap. Post. 6-60
Cuernavaca, Mor. 62131
MEXICO
martinez@matcuer.unam.mx

Conditions of Regularity for Bernstein-Rogosinski-type Means of Double Fourier Series

Yuri Nosenko

Abstract. Conditions of regularity for Bernstein-Rogosinski-type (or Rogosinski-type) means of double Fourier series of continuous functions are obtained. These means were first introduced by R.M. Trigub. Such means depend on different numerical parameters, an appropriate borelian measure and its support, and a shape of a domain which generates partial sums of Fourier series. We investigate the case when the partial sums are generated by different polygons (a rectangle, a rhombus) with the origin inside every polygon and the measure, which is uniformly distributed on the areas of correspondent polygons.

§1. Introduction

Let $T^2 = (-\pi, \pi]^2$. Suppose $f \in C(T^2)$ is 2π-periodic with respect to each variable,

$$f(x_1, x_2) \sim \sum_{(k_1, k_2)} c_{k_1, k_2} e^{i(k_1 x_1 + k_2 x_2)} \tag{1}$$

be the Fourier series of f with c_{k_1, k_2} the Fourier coefficients of f w.r.t. the trigonometric system $\{ e^{i(k_1 x_1 + k_2 x_2)}, (k_1, k_2) \in \mathbb{Z}^2 \}$.

Let W_0 be some domain in $\mathbb{R}^2$ containing the origin, and let pW_0 be a homothetic transform of W_0 with the coefficient of homothety $p > 0$, namely

$$pW_0 = \{ (x_1, x_2) : (x_1/p, x_2/p) \in W_0 \} .$$

Then, for $n \in \mathbb{N}$,

$$S_n(f; W_0, x_1, x_2) = \sum_{(k_1, k_2) \in nW_0} c_{k_1, k_2} e^{i(k_1 x_1 + k_2 x_2)}$$

are the partial sums of (1) corresponding to W_0 (or generated by W_0).

Approximation Theory X: Abstract and Classical Analysis   325
Charles K. Chui, Larry L. Schumaker, and Joachim Stöckler (eds.), pp. 325–332.
Copyright Ⓒ 2002 by Vanderbilt University Press, Nashville, TN.
ISBN 0-8265-1415-4.

The means

$$R_n(f; x_1, x_2) = \int_{\mathbb{R}^2} S_n \left(f; W_0; x_1 - \frac{\gamma u}{n}, x_2 - \frac{\gamma v}{n} \right) d\mu(u, v) \qquad (2)$$

were first introduced in this most general form by R. M. Trigub [5], and were investigated in different directions there. They are known as Bernstein-Rogosinski-type (or Rogosinski-type) means. Here $n \in \mathbb{N}, \gamma \in \mathbb{R}$, and μ is a finite and normalized Borel measure on $\mathbb{R}^2$. Generally speaking, the means (2) depend on the numerical parameter γ, the choice of the measure μ and the shape of the domain W_0.

In the discrete case, when the measure α_k ($\sum_k \alpha_k = 1$ because of normalization of the measure) is concentrated at the points (u_k, v_k), the means (2) can be reduced to

$$R_n(f; x_1, x_2) = \sum_k \alpha_k S_n \left(f; W_0; x_1 - \frac{\gamma u_k}{n}, x_2 - \frac{\gamma v_k}{n} \right). \qquad (3)$$

Note that the partial sums of the series (1) corresponding to W_0 can be obtained from (2) or (3) by putting $\gamma = 0$ or concentrating the measure at the origin. In our results $\gamma \neq 0$.

There are two main problems in the theory of linear means for Fourier series. The first problem is that of regularity, which describes convergence of such means to a generating function. In our case, it is the convergence of $R_n(f; x_1, x_2)$ to $f(x_1, x_2)$ itself in the uniform metric. The second problem deals with approximation properties of such means. From now on, we will restrict ourselves to the problem of regularity.

Very special classical cases of the means (2) were first introduced and investigated by W. Rogosinski and S. N. Bernstein. They considered the case of uniformly distributed measure at two points (see [5] for references) in the one-dimensional setting (a measure $1/2$ is concentrated at each point). Regularity and approximation properties of these classical means and some interesting generalizations were investigated by different authors. The means (2) and (3) were studied in these directions as well. Regularity of $R_n(f; x_1, x_2)$ was investigated in different cases of homogeneous and nonhomogeneous, discrete and continuous distributions of the measure (see [2,4,5] for results and references).

As in [5], the means (2) can be represented in the equivalent form

$$R_n(f; x_1, x_2) = \sum_{(k_1, k_2) \in \mathbb{Z}^2} \phi \left(\frac{k_1}{n}, \frac{k_2}{n} \right) c_{k_1, k_2} e^{i(k_1 x_1 + k_2 x_2)}, \qquad (4)$$

where

$$\phi(x_1, x_2) = \chi_{W_0}(x_1, x_2) \int_{\mathbb{R}^2} e^{-i\gamma(x_1 u + x_2 v)} d\mu(u, v). \qquad (5)$$

Here χ_{W_0} is the characteristic function (indicator) of W_0.

A necessary condition of regularity of (2) in $C(T^2)$ for a Jordan domain W_0 is

$$I = \int_{\mathbf{R}^2} e^{-i\gamma(x_1 u + x_2 v)} d\mu(u,v) = 0 \qquad (6)$$

for every point $(x_1, x_2) \in \partial W_0$, the boundary of W_0, see [5]. Conditions of regularity of (2) in $C(T^2)$ can be divided into three main groups:

1) regularity of (3) for a discrete distribution of the measure μ,

2) regularity of (2) for uniform (continuous) distribution of the measure μ along perimeters of different curves in $\mathbf{R}^2$,

3) regularity of (2) for uniform (continuous) distribution of the measure μ on areas of different figures in $\mathbf{R}^2$.

The problems of the first group are practically solved in [1,3], as conditions of regularity (both positive and negative results) for (3) are obtained for arbitrary right-angle polygons W_0 centered at the origin, and positive results are obtained for almost arbitrary polygons W_0.

On the other hand, there are only isolated results for the groups 2) and 3) (say, if W_0 is the unit circle or the unit square as in [5]; see also [3] for other results and references).

§2. Main Results

We include here some of the positive results on regularity of Bernstein-Rogosinski-type means (2) for the case where μ is the uniform distribution on some polygon W^* and W_0, which generate partial sums of (1), is also a polygon (a rectangle, a rhombus). All these results are unified under the principal idea of *perpendicularity* of W_0 and W^*.

This idea first appeared in the discrete case in [1] where W_0 is a parallelogram. Here is the result.

Theorem A. *Let W_0 be a parallelogram that is symmetric with respect to the origin. Then there exists a measure μ whose support contains at least four points, such that the means (3) are regular in $C(T^2)$. Moreover, if the support of μ contains exactly four points, then the means (3) are regular in $C(T^2)$ if and only if these points are vertices of a new parallelogram W^*, whose sides are perpendicular to the sides of W_0, the measure μ is uniformly distributed at these points, and the ratio of the product of lengths of two nonparallel sides of W_0 and W^* is equal to the ratio of two odd numbers.*

This idea of perpendicularity for W_0 and W^* is very fruitful for the analysis of regularity of the means (2). We demonstrate this idea in our new results, Theorems 1–3 which we now state.

Theorem 1. Let W_0 be the rectangle with sides of length $2a$ and $2b$ parallel to the coordinate axes and which is centered at the origin. Then the means (2) are regular in $C(T^2)$ if and only if the measure μ is uniformly distributed on the area of a new rectangle W^* whose sides are parallel to the coordinate axes, have length $2c$ and $2d$, and the ratio of the product of two mutually perpendicular sides of W_0 and W^* is equal to the ratio of two natural numbers, and $\gamma = \frac{\pi k}{ac}$ for some $k \in \mathbb{N}$.

Theorem 2. Let W_0 be a rhombus with diagonals of lengths $2a$ and $2b$ and with vertices on the coordinate axes. Then the means (2) are regular in $C(T^2)$ if and only if the measure μ is uniformly distributed on the area of the rhombus W^*, which is the rotation by $\pi/2$ of pW_0, $p > 0$, around the origin, and $\gamma = \frac{2\pi k}{pab}$ for some $k \in \mathbb{N}$.

R. M. Trigub has proved in [5] that for partial sums of (1) corresponding to the unit square W_0 and the support of the measure μ as the perimeter of W_0 (∂W_0), the means (3) are irregular in $C(T^2)$. One can extend this result very easily from the unit square to rectangles which are symmetric with respect to the origin and whose sides are parallel to the coordinate axes (*i.e.*, the means (3) remain irregular in this case).

By taking into account these considerations and our Theorem A, we come to the following conclusion. Let W_0 be a rectangle, symmetric with respect to the origin and sides parallel to the coordinate axes. Suppose that the support of the measure μ is another rectangle W^*. Then the Bernstein-Rogosinski-type means (2) or (3)

1) can be regular if the measure is homogeneously concentrated at the vertices of W^* (Theorem A),

2) are irregular if the support of the homogeneous measure is the perimeter of W^* [5],

3) can be regular once again if the homogeneous measure is concentrated over an area of W^* (Theorem 1).

The following problems arise in this regard. How can one reduce the support of the measure from the area of W_0 in the direction of its perimeter and the means (2) remain regular? Or how can one enlarge the support of a measure from the perimeter of W^* in the direction of the area of W_0 and the means (2) become regular? The next theorem clarifies this problem.

Theorem 3. Let W_0, W^*, W^{**} be rectangles which are symmetric with respect to the origin, whose sides are parallel to the coordinate axes and have the lengths $2a$ and $2b$ (for W_0), $2c$ and $2d$ (for W^*), and $2(c - \epsilon)$ and $2(d - \epsilon)$ (for W^{**}). If the measure μ is uniformly distributed over the area of $W^* \setminus W^{**}$ (a union of corresponding strips), the ratio of the products of mutually perpendicular sides of W_0 and W^* is rational,

$$\gamma = \frac{\pi k}{bd} \qquad \left(\gamma = \frac{\pi k}{ac}\right)$$

for some $k \in \mathbb{N}$, and ϵ is a rational multiple of c or d, then the means (2) are regular in $C(T^2)$.

Remark. Theorem 3 remains valid for different values of ϵ along different coordinate axes (say, ϵ_1 and ϵ_2). Of course, there is a connection between these different values of ϵ.

§3. Auxiliary Results

Let us take a matrix $\Lambda = \| \lambda_{k_1,k_2}^{(m,n)} \|$ with elements depending on $m \in \mathbb{N}, n \in \mathbb{N}$. We form the linear means for (1) using this matrix, namely,

$$\tau_{m,n}(f;\Lambda) = \tau_{m,n}(f;\Lambda,x_1,x_2)$$

$$\sim \sum_{k_1=-\infty}^{\infty} \sum_{k_2=-\infty}^{\infty} \lambda_{k_1,k_2}^{(m,n)} c_{k_1,k_2} e^{i(k_1 x_1 + k_2 x_2)}. \tag{7}$$

It is well known, that these means are regular if $\lim_{m\to\infty,n\to\infty} \lambda_{k_1,k_2}^{(m,n)} = 1$ and corresponding Lebesgue constants (norms of operators (7)) are uniformly bounded in m and n.

To check regularity of the means (7) (the means (2) or (3) are the special cases of (7), see (4)) we will use two results due to R. M. Trigub [5] and V. P. Zastavnyi [6] given here as Lemmas 1 and 2.

Lemma 1. *If the elements of the matrix Λ are the values of the finite function $\phi(u,v)$ for $u = \frac{k_1}{m}, v = \frac{k_2}{n}$ with the support $[-1,1]^2$ and*

1) *ϕ, as a function of u, is in Lip ($\frac{1}{2} + \epsilon$) for some $\epsilon > 0$ (uniformly in v),*

2) *ϕ, as a function of v, is in Lip ($\frac{1}{2} + \epsilon$) for some $\epsilon > 0$ (uniformly in v),*

3) *the number of points of inflection of ϕ, as a function of u, is bounded,*

4) *the number of points of inflection of ϕ, as a function of uv, is bounded,*

then the Lebesgue constants, corresponding to Λ, are bounded in m and n.

Lemma 2. *Let μ be a finite complex Borel measure in $\mathbb{R}^2$ with complex support, and suppose the function g is defined by*

$$g(x,y) = \int_{\mathbb{R}^2} e^{-i(ux+vy)} d\mu(u,v).$$

Then for any $a > 0$, the number of zeroes of the functions $Re\ g(x,y)$ and $Im\ g(x,y)$, restricted to $[-a,a]^2$ and regarded as functions of x or y, is uniformly bounded (with respect to the other variable).

§4. Proofs of Main Results

Proofs of the stated theorems will be realized in two stages, namely

1) using the conditions of a theorem we find W_0 and γ satisfying the condition (6),

2) we check the boundedness of the corresponding Lebesgue constants for the shapes of W_0 and values of γ obtained in the first step.

Proof of Theorem 1. Let f, W_0, μ, W^* be as in Theorem 1. Evaluation of the integral in (6) (we will omit some technical details) in our case gives us

$$I = \left(cd\gamma^2 x_1 x_2\right)^{-1} \sin(\gamma x_1 c) \sin(\gamma x_2 d).$$

Condition (6) applied to the boundary ∂W_0 of W_0, namely to the part $x_2 = |b|$, gives $\sin(\gamma bd) = 0$. Arguing in the same way for the other part of ∂W_0 gives $\sin(\gamma ac) = 0$. Using these relations, we obtain

$$\frac{ac}{bd} = \frac{n}{k}, \qquad \gamma = \frac{\pi n}{ac},$$

for some $n, k \in \mathbb{N}$.

Now we will prove regularity of the means (2) for our case. In accordance with (4), (5), (7), the elements of the matrix Λ which defines our linear means are values of the function (here χ_{W_0} is a characteristic function of W_0)

$$\phi(x_1, x_2) = \chi_{W_0} \left(cd\gamma^2 x_1 x_2\right)^{-1} \sin(\gamma x_1 c) \sin(\gamma x_2 d), \quad (x_1 x_2) \neq 0,$$

$$\phi(0, x_2) = \chi_{W_0} (\gamma dx_2)^{-1} \sin(\gamma dx_2), \quad x_2 \neq 0,$$

$$\phi(x_1, 0) = \chi_{W_0} (\gamma cx_1)^{-1} \sin(\gamma cx_1), \quad x_1 \neq 0,$$

$$\phi(0, 0) = 1.$$

By taking the value of ϕ at $(0,0)$ into account, we notice that it remains to prove the boundedness of the Lebesgue constants corresponding to $\phi(x_1, x_2)$ This will complete the proof of the theorem. Let us return to Lemma 1. As $|\phi'_{x_1}(x_1, x_2)| \leq C_1$ and $|\phi'_{x_2}(x_1, x_2)| \leq C_2$ for $(x_1, x_2) \in W_0 \setminus \partial W_0$, it follows that $\phi \in Lip\, 1$ in x_1 for every x_2 and in x_2 for every x_1. Here C_1 and C_2 are absolute constants. This means that the conditions 1) and 2) of Lemma 1 hold with $\epsilon = \frac{1}{2}$. The conditions 3) and 4) of the same lemma are practically evident. In order to prove them, one has to use Lemma 2 for the function $g(x_1, x_2) = f''_{x_1 x_1}(x_1, x_2)$ and $g(x_1, x_2) = f''_{x_2 x_2}(x_1, x_2)$. Therefore, we can apply Lemma 1 and obtain the boundedness of the Lebesgue constants. $\square$

The boundedness of the corresponding Lebesgue constants can also be shown by another approach based on conditions of integrability of trigonometric series as in [1].

Proof of Theorem 2. Let f, W_0, μ, W^* be as in Theorem 2. In this case W^* has diagonals pb and pa corresponding to diagonals of W_0. Evaluation of the integral in (6) for these data gives us (we will omit some technical details) for $x_1 b \neq \pm x_2 a$

$$I = C(1 - e^{i\gamma p(x_1 b + x_2 a)})(1 - e^{i\gamma(-x_1 b + x_2 a)}) e^{-i\gamma p a x_2} (x_1^2 b^2 - x_2^2 a^2)^{-1}$$

(constants C here and below are independent of x_1 and x_2).

As $I = 0$ at the boundary ∂W_0 of W_0 in accordance with (6), we obtain that for (x_1, x_2) on the part of the boundary $x_1 b + x_2 a = ab$ the relation (6) is possible if $e^{i\gamma p a b} = 1$. This relation gives us $\gamma = \frac{2\pi k}{pab}$ for some $k \in \mathbb{N}$. A similar situation occurs for the other parts of ∂W_0.

The function values

$$\phi(x_1, x_2) = C\chi_{W_0}(1 - e^{i\gamma p(x_1 b + x_2 a)})(1 - e^{i\gamma(-x_1 b + x_2 a)})$$

$$\cdot e^{-i\gamma p a x_2}(x_1^2 b^2 - x_2^2 a^2)^{-1}$$

are the elements of the matrix which defines the corresponding linear means (7) or (4). Here, the function values of ϕ at points of indeterminacy are to be defined by continuity. The same technique (Lemmas 1 and 2) can be used in order to complete the proof of the theorem. $\square$

Proof of Theorem 3. Let f, W_0, μ, W^* be as before. Evaluation of the integral in (6) for these data gives

$$I = C x_1^{-1} x_2^{-1} (\sin(\gamma x_2(d - \epsilon)) \sin(\gamma x_1(c - \epsilon)) - \sin(\gamma x_1 c) \sin(\gamma x_2 d)),$$

where $x_1 x_2 \neq 0$ and C does not depend on x_1 and x_2. I is defined by continuity for $x_1 x_2 = 0$. In order to find γ, we will now use (6). The equations of ∂W_0 are $|x_1| = a, |x_2| = b$. For these equations we have the systems

$$\sin(\gamma a(c - \epsilon)) = 0, \qquad \sin(\gamma a c) = 0$$

and

$$\sin(\gamma b(d - \epsilon)) = 0, \qquad \sin(\gamma b d) = 0,$$

respectively. These systems lead to

$$\gamma a c = \pi k, \quad \gamma a \epsilon = \pi l, \quad \gamma b d = \pi m, \quad \gamma b \epsilon = \pi n$$

for some $k, l, m, n \in \mathbb{N}$, and this gives

$$\frac{bd}{ac} = \frac{m}{n}, \quad \epsilon = \frac{lc}{k} \text{ or } \epsilon = \frac{nd}{m},$$

and finally

$$\gamma = \frac{\pi n}{bd} \text{ or } \gamma = \frac{\pi k}{ac}.$$

The function values

$$\phi(x_1, x_2) = C \chi x_1^{-1} x_2^{-1} (\sin \gamma x_1 (c - \epsilon)) \sin(\gamma x_2 (d - \epsilon))$$
$$- \sin(\gamma x_1 c) \sin(\gamma x_2 d), \qquad x_1 x_2 \neq 0,$$

and continuitity of ϕ for $x_1 x_2 = 0$ define the elements of the matrix that defines the corresponding linear means (7) or (4). The same technique as in the previous proofs, with an application of Lemmas 1 and 2, completes the proof of this theorem. $\square$

Acknowledgments. The author expresses his thanks to the referee for valuable remarks and to Joachim Stoeckler for valuable remarks, advice, discussion, and encouragement.

References

1. Nosenko, Yu. L., Summation of double Fourier series by methods of Bernstein-Rogosinski type, Theory of Mappings and Approximation of Functions, "Naukova Dumka", Kiev, 1983, 89–97 (in Russian).

2. Nosenko, Yu. L., Regularity and approximation properties of means of Bernstein-Rogosinski type of double Fourier series, Ukrain. M. Zh. **37** (1985), no. 5, 599–604 (in Russian); English transl. in Ukr. math. J. **37** (1985), no. 5, 485–490.

3. Nosenko, Yu. L., Regularity of Bernstein-Rogosinski-type means, Theory of Approx. of funct. Proc. of the Inst. of Applied Math. and Mech., Donetsk, (1998), no. 3, 182–186 (in Russian).

4. Nosenko, Yu. L., Regularity conditions for Bernstein-Rogosinski-type means of double Fourier series, Buletinul Stiintific Univ. Baia Mare (Romania), Ser. B **16** (2000), no. 1, 93–98.

5. Trigub, R. M., Absolute convergence of Fourier integrals, summability of Fourier series, and polynomial approximation of functions on the torus, Izv. AN SSSR, Ser mathem. **44** (1980), no. 6, 1378–1409 (in Russian); English transl. in Math. USSR Izvestija **17** (1981), no. 3, 567–593.

6. Zastavnyi, V. P., Concerning a set of zeroes for Fourier transform of a measure and summability of double Fourier series by Bernstein-Rogosinski type methods, Ukrain. M. Zh. **36** (1984), no. 5, 615–621 (in Russian).

Yuri L. Nosenko
Dept. of Mathematics
Donetsk National Technical University
58, Artema str.
Donestk, 83000, Ukraine
yuri@nosenko.dgtu.donetsk.ua

Some Remarks on Zero-Increasing Transformations

Allan Pinkus

Abstract. Let T be a linear operator which maps polynomials of degree n to polynomials of degree n, for every n. We discuss the problem of trying to characterize the set of all such operators which satisfy

$$Z_I(p) \le Z_J(Tp)$$

for every real-valued polynomial p, where Z_I counts the number of zeros on the interval I of $\mathbb{R}$.

§1. Introduction

Let Π denote the space of all real-valued polynomials, and π_n the subspace of all polynomials of degree at most n. In this paper we will consider linear operators $T : \Pi \to \Pi$ for which

$$T : \pi_n \to \pi_n$$

for each $n \in \mathbb{Z}_+$. For each polynomial p, we let $Z_I(p)$ denote the number of zeros of p, counting multiplicity, on the interval $I \subseteq \mathbb{R}$.

The problem we will discuss is that of characterizing those T for which

$$Z_I(p) \le Z_J(Tp)$$

for all $p \in \Pi$. We call such operators **zero-increasing** (ZI). Problems of this type have a long and distinguished history. The interested reader may wish to browse in the books of Marden [18], Obreschkoff [19], and Pólya–Szegő [23].

Approximation Theory X: Abstract and Classical Analysis
Charles K. Chui, Larry L. Schumaker, and Joachim Stöckler (eds.), pp. 333–352.

Let us start with a few simple examples of such operators in order to convince ourselves that they do exist and can be non-trivial. Perhaps the simplest (but trivial) example is

$$(Tp)(x) = (x - a)p'(x),$$

where $a \in I$. From Rolle's Theorem (the most basic and fundamental of the zero counting methods on $\mathbb{R}$) we have

$$Z_I(p) \leq Z_I(Tp)$$

for every polynomial p and any interval I. Similarly it is easy to generalize this example to

$$(Tp)(x) = q_k(x)p^{(k)}(x),$$

where q_k is some fixed polynomial of degree k with k zeros in I. A somewhat more complicated example is the following. Let H_k denote the kth degree Hermite polynomial with leading coefficient 2^k. Then

$$Z_{\mathbb{R}}\left(\sum_{k=0}^{n} a_k x^k\right) \leq Z_{\mathbb{R}}\left(\sum_{k=0}^{n} a_k H_k(x)\right)$$

for every choice of $\{a_k\}$ and n. We will "prove" this inequality later. We also have

$$Z_{[-1,1]}\left(\sum_{k=0}^{n} a_k x^k\right) \leq Z_{[-1,1]}\left(\sum_{k=0}^{n} a_k T_k(x)\right)$$

and

$$Z_{[-1,1]}\left(\sum_{k=0}^{n} a_k x^k\right) \leq Z_{[-1,1]}\left(\sum_{k=0}^{n} a_k U_k(x)\right),$$

where T_k and U_k are the Chebyshev polynomials of the first and second kind, respectively, *i.e.*,

$$T_k(\cos\theta) = \cos k\theta$$

and

$$U_k(\cos\theta) = \frac{\sin(k+1)\theta}{\sin\theta}.$$

This and similar results are sometimes stated only for the special case where

$$\sum_{k=0}^{n} a_k x^k \text{ has only real zeros}$$

implies

$$\sum_{k=0}^{n} a_k H_k(x) \text{ has only real zeros}.$$

Or

$$\sum_{k=0}^{n} a_k x^k \text{ has all its zeros in } [-1,1]$$

implies

$$\sum_{k=0}^{n} a_k T_k(x) \text{ (or } \sum_{k=0}^{n} a_k U_k(x)) \text{ has all its zeros in } [-1, 1].$$

We will later discuss this variation on the general problem. In this form, the result for Hermite polynomials and for the two Chebyshev polynomials appears in Iserles and Saff [13], although their proof actually gives the more general result. In fact these results, in the general case, were already in Burbea [4]. Examples of other ZI operators can be found in the above two references and in Iserles, Nørsett [11], and Iserles, Nørsett and Saff [12].

We can list more and more of these transformations. But we will not. Rather we will try to develop the general theory, if at all possible. We will mainly consider the cases where I is the whole real line or one of its rays.

§2. Some Classic Theorems

We start with a result called the Hermite–Poulain Theorem. It is based on questions posed by Hermite [8], and answered by Poulain [24].

Hermite–Poulain Theorem. *Assume*

$$g(x) = \sum_{k=0}^{m} b_k x^k$$

is a polynomial with all real zeros and $b_0 \neq 0$. Then for any polynomial p,

$$Z_{\mathbb{R}}(p) \leq Z_{\mathbb{R}} \left(\sum_{k=0}^{m} b_k p^{(k)} \right).$$

Proof. The main idea in the proof of this result consists in noting that

$$\sum_{k=0}^{m} b_k p^{(k)} = (g(D)p)(x),$$

where

$$g(D) = \sum_{k=0}^{m} b_k D^k = \sum_{k=0}^{m} b_k \frac{d^k}{dx^k}.$$

That is, $\sum_{k=0}^{m} b_k p^{(k)}$ is obtained from p by the application of a constant coefficient ordinary differential equation. In addition, this ordinary differential equation is what we now call **disconjugate**. In this case, this simply means that g has only real zeros, *i.e.*,

$$g(D) = b_m \prod_{j=1}^{m} (D - \alpha_j I)$$

and $\alpha_j \in \mathbb{R}$. We also impose the condition $b_0 \neq 0$ which implies $\alpha_j \neq 0$. Otherwise the resulting polynomial loses degree, and the theorem is not correct in the stated form.

As the $D - \alpha_j I$ *commute*, to prove the theorem it suffices to verify

$$Z_{\mathbb{R}}(p) \leq Z_{\mathbb{R}}((D - \alpha I)p)$$

for $\alpha \in \mathbb{R}\backslash\{0\}$.

This may be done in a variety of ways. The most elegant I have seen is that due to Poulain himself. He writes

$$(D - \alpha I)p = e^{\alpha x} D(e^{-\alpha x}p).$$

As $\alpha \neq 0$ and p is a polynomial, $e^{-\alpha x}p(x)$ has an additional "zero" at ∞ or $-\infty$ depending upon the sign of α. We now apply Rolle's Theorem. $\square$

Note, in fact, that if $\alpha > 0$ then $D(e^{-\alpha x}p)$ has a zero interlacing those of p and to the right thereof. Thus, refining the above a bit, we also have:

Hermite–Poulain Theorem (II). *Let*

$$g(x) = \sum_{k=0}^{m} b_k x^k$$

be a polynomial with all positive zeros. Then

$$Z_{[A,\infty)}(p) \leq Z_{[A,\infty)}\left(\sum_{k=0}^{m} b_k p^{(k)}\right),$$

for every polynomial p, and $A \in \mathbb{R}$.

Disconjugate ordinary differential equations have an inverse given by integrating against a Green's function with certain desirable qualities. We will consider the inverse of $g(D)$, operating as a map from polynomials to polynomials of the same degree. The operator $g(D)$ formally exists since $b_0 \neq 0$ and

$$g(D)x^n = b_0 x^n + \sum_{k=0}^{n-1} b_{n,k} x^k.$$

But this is not an especially useful form of g from which to derive its inverse. Let us start with the simplest case $g(D) = D - \alpha I$, *i.e.*, the operator

$$(D - \alpha I)p = q.$$

This is equivalent to

$$e^{\alpha x} D(e^{-\alpha x}p) = q,$$

which then easily translates into

$$e^{-\alpha x} p(x) = \int e^{-\alpha y} q(y) \, dy$$

for some appropriate antiderivative.

For the above to make sense, in that it takes polynomials to polynomials, it follows that

$$p(x) = \begin{cases} -\int_x^\infty e^{\alpha(x-y)} q(y) \, dy, & \text{for } \alpha > 0, \\ \int_{-\infty}^x e^{\alpha(x-y)} q(y) \, dy, & \text{for } \alpha < 0. \end{cases}$$

In other words, if

$$(Sq)(x) = \begin{cases} -\int_x^\infty e^{\alpha(x-y)} q(y) \, dy, & \alpha > 0 \\ \int_{-\infty}^x e^{\alpha(x-y)} q(y) \, dy, & \alpha < 0 \end{cases}$$

then $S = T^{-1}$ and

$$Z_{I\!R}(Sq) \le Z_{I\!R}(q)$$

for every polynomial q.

This is not a surprising result. S not only takes polynomials of degree n to polynomials of degree n, but it is an integral equation with a **totally positive difference kernel** (called a **Pólya frequency function**). It was proven by Schoenberg [25] (see also Karlin [14]) that an operator given by such an integral equation is **variation diminishing**, *i.e.*, satisfies

$$Z_{I\!R}(Sq) \le Z_{I\!R}(q)$$

for all continuous functions for which the above integral makes sense (and not only polynomials) if and only if the kernel is totally positive (or to be more precise **sign regular**). (We might possibly call this property **zero decreasing**. However, the term variation diminishing is well established. What is actually meant by the term is that the number of **variations in sign** of the Sq is less than or equal that of q.)

What we just did applies to the simple operator $D - \alpha I$. However, multiplying such operators is equivalent (for the inverse) of convolving the difference kernels. Thus, the appropriate inverse operator to

$$\prod_{j=1}^m (D - \alpha_j I)$$

is an integral equation whose kernel is obtained by convolving totally positive kernels, and thus is also totally positive. There is one further fact which we wish to highlight using this simple example. The kernel of the above S, *i.e.*, of $(D - \alpha I)^{-1}$, is for $\alpha > 0$

$$K_\alpha(x) = \begin{cases} 0, & x > 0 \\ -e^{\alpha x}, & x < 0, \end{cases}$$

and has the two-sided Laplace transform

$$\frac{1}{s - \alpha} = \frac{1}{g(s)} \text{ for } s < \alpha.$$

For $\alpha < 0$ the kernel is given by

$$K_\alpha(x) = \begin{cases} e^{\alpha x}, & x > 0 \\ 0, & x < 0, \end{cases}$$

and has the two-sided Laplace transform

$$\frac{1}{s - \alpha} = \frac{1}{g(s)} \text{ for } s > \alpha.$$

Note that in both cases the two-sided Laplace transform equals $1/g(s)$, *i.e.*, is intimately connected with the ordinary differential equation, and exists in some neighborhood of the origin.

In general, if

$$g(D) = b_m \prod_{j=1}^{m} (D - \alpha_j I),$$

then the two-sided Laplace transform of the difference kernel of the inverse operator is given by

$$\frac{1}{b_m \prod_{j=1}^{m} (s - \alpha_j)} = \frac{1}{g(s)},$$

and exists for all s in some neighborhood of the origin. It actually exists in the strip

$$\left\{ z : \max_{\alpha_j < 0} \alpha_j < \operatorname{Re} z < \min_{\alpha_j > 0} \alpha_j \right\}.$$

We will shortly return to the Hermite–Poulain Theorem. However, we first pick up another thread. There is an additional class of operators which lead to operators with the ZI property. This is a result due to another outstanding 19th century French analyst, namely Laguerre [16].

Laguerre's Theorem. *Assume g is a polynomial with only real zeros, all of which lie outside the interval $[0, n]$. Then for every polynomial*

$$p(x) = \sum_{k=0}^{n} a_k x^k$$

of degree at most n, we have

$$Z_{\mathbb{R}}\left(\sum_{k=0}^{n} a_k x^k\right) \leq Z_{\mathbb{R}}\left(\sum_{k=0}^{n} a_k g(k) x^k\right).$$

These are different types of operators from those in the Hermite–Poulain Theorem. They are "diagonal" in the sense that

$$T x^k = g(k) x^k,$$

for $k = 0, 1, \ldots, n$. Note that there is a restriction here in that these operators only have the ZI property for polynomials of degree at most n, although there is no restriction on the degree of g. Thus, these operators in themselves need not be ZI operators on all of Π. However, we will use them to construct other ZI operators.

Proof. The idea of this proof is again linked to ordinary differential equations. If

$$g(x) = b \prod_{j=1}^{m} (x - \alpha_j),$$

then

$$(Tp)(x) = \sum_{k=0}^{n} a_k g(k) x^k = b \prod_{j=1}^{m} (xD - \alpha_j I) p(x)$$

simply because

$$b \prod_{j=1}^{m} (xD - \alpha_j I) x^k = b \prod_{j=1}^{m} (k - \alpha_j) x^k = g(k) x^k,$$

for each k. Thus,

$$Tp = g(xD)p.$$

The ordinary differential equation $g(xD)$ is an Euler equation. Again, as the

$$xD - \alpha_j I$$

commute, it suffices to prove that

$$Z_{\mathbb{R}}(p) \leq Z_{\mathbb{R}}((xD - \alpha I)p)$$

for $\alpha \notin [0, n]$.

Why is this true? We consider separately zeros of p which are at zero, positive and negative. As p and $(xD - \alpha I)p$ have the same order zero at $x = 0$, there is no problem with zeros at zero. For $x > 0$

$$(xD - \alpha I)p = x^{\alpha+1} D(x^{-\alpha} p(x)).$$

Now $x^{-\alpha}p(x)$ has a zero at zero for $\alpha < 0$, while for $\alpha > n$, $x^{-\alpha}p(x)$ has a "zero" at ∞ (since p is of degree at most n). In both cases, applying Rolle's Theorem we get

$$Z_{(0,\infty)}(p) \le Z_{(0,\infty)}((xD - \alpha I)p).$$

For $x < 0$, a similar argument holds. $\square$

From the above we also have

Laguerre's Theorem (II). *Let g be a polynomial with all negative zeros. Then*

$$Z_{[0,A]}\left(\sum_{k=0}^{n} a_k x^k\right) \le Z_{[0,A]}\left(\sum_{k=0}^{n} a_k g(k) x^k\right),$$

for every polynomial p, and $A > 0$ (and similarly on $[-A, 0]$).

If there is a zero of g in $[0, n]$, then neither the above argument nor the results are valid. However, if we agree to count the number of zeros of the zero polynomial as infinite, then the above result will also hold for polynomials g with all real zeros outside $[r, n]$, r a positive integer, if $g(0) = \cdots = g(r-1) = 0$, see Craven and Csordas [6].

Laguerre's Theorem (III). *For any $r \in \mathbb{Z}_+$, let h be a polynomial of the form*

$$h(x) = x(x-1)(x-2)\cdots(x-r)q(x)g(x),$$

where g is as in Laguerre's Theorem, and q is any polynomial with all real zeros in $[0, r+1)$. Then

$$Z_{\mathbb{R}}\left(\sum_{k=0}^{n} a_k x^k\right) \le Z_{\mathbb{R}}\left(\sum_{k=0}^{n} a_k h(k) x^k\right).$$

Proof. It suffices to prove this result for h of the form

$$h(x) = x(x-1)(x-2)\cdots(x-r)q(x).$$

We then apply Laguerre's Theorem to obtain the full result. As previously, for $p(x) = \sum_{k=0}^{n} a_k x^k$ we have

$$\sum_{k=0}^{n} a_k h(k) x^k = h(xD)p.$$

Now it is readily checked that

$$xD(xD - I)(xD - 2I)\cdots(xD - rI)p = x^{r+1}p^{(r+1)}(x).$$

From Rolle's Theorem, we have

$$Z_{\mathbb{R}}(p) \leq r + 1 + Z_{\mathbb{R}}(p^{(r+1)}) = Z_{\mathbb{R}}(x^{r+1}p^{(r+1)}(x)).$$

Let us now consider

$$(xD - \alpha I)xD(xD - I)(xD - 2I)\cdots(xD - rI)p.$$

As the $(xD - \alpha I)$ commute, the order here is unimportant. From the previous analysis it therefore follows that this equals

$$(xD - \alpha I)x^{r+1}p^{(r+1)}(x),$$

and a simple calculation shows that this is equivalent to

$$x^{r+1}(xD - (\alpha - r - 1)I)p^{(r+1)}(x).$$

Thus, we have

$$(xD - \alpha_1 I)\cdots(xD - \alpha_k I)xD(xD - I)(xD - 2I)\cdots(xD - rI)p(x) =$$

$$x^{r+1}(xD - (\alpha_1 - r - 1)I)\cdots(xD - (\alpha_k - r - 1)I)p^{(r+1)}(x).$$

As each $\alpha_i \in [0, r+1)$, we have $\alpha_i - r - 1 < 0$ and thus from Laguerre's Theorem and Rolle's Theorem

$$Z_{\mathbb{R}}(p) \leq r + 1 + Z_{\mathbb{R}}(p^{(r+1)}(x))$$
$$\leq r + 1 + Z_{\mathbb{R}}((xD - (\alpha_1 - r - 1)I)\cdots(xD - (\alpha_k - r - 1)I)p^{(r+1)}(x))$$
$$= Z_{\mathbb{R}}(x^{r+1}(xD - (\alpha_1 - r - 1)I)\cdots(xD - (\alpha_k - r - 1)I)p^{(r+1)}(x))$$
$$= Z_{\mathbb{R}}((xD - \alpha_1 I)\cdots(xD - \alpha_k I)xD(xD - I)(xD - 2I)\cdots(xD - rI)p(x)),$$

which is exactly the result we wanted. $\square$

Can we easily invert $g(xD)$? Obviously yes. The inverse is simply given by the diagonal operator taking x^k to $x^k/g(k)$, $k = 0, 1, \ldots, n$. But this form of the inverse is not really useful. Another form of the inverse is more insightful. We first state it for $g(xD) = xD - \alpha I$. For $\alpha < 0$

$$\frac{1}{k - \alpha} = \int_0^1 t^{k-\alpha-1}dt = \int_0^1 t^{-\alpha-1}t^k dt,$$

and thus for $x > 0$

$$\frac{x^k}{k - \alpha} = \int_0^x \left(\frac{y}{x}\right)^{-\alpha-1}\frac{1}{x}y^k dy.$$

Similarly for $\alpha > n$ (and $0 \le k \le n$) it follows that

$$\frac{x^k}{k - \alpha} = -\int_x^\infty \left(\frac{y}{x}\right)^{-\alpha-1} \frac{1}{x} y^k \, dy \, .$$

Thus for each p of degree at most n, the operator $S_\alpha = (xD - \alpha I)^{-1}$ is given by

$$(S_\alpha p)(x) = \int_0^x \left(\frac{y}{x}\right)^{-\alpha-1} \frac{1}{x} p(y) \, dy$$

for $\alpha < 0$, and by

$$(S_\alpha p)(x) = -\int_x^\infty \left(\frac{y}{x}\right)^{-\alpha-1} \frac{1}{x} p(y) \, dy$$

for $\alpha > n$, and for such p we have

$$Z_{(0,\infty)}(S_\alpha p) \le Z_{(0,\infty)}(p) \, .$$

(The above integrals do not converge for all $p \in \pi_n$ if $0 \le \alpha \le n$.) A similar result holds on $(-\infty, 0)$. In some sense we have just complicated things. However, two things are worth noting. Firstly,

$$\left(\left(\prod_{j=1}^m S_{\alpha_j}\right) p\right)(x)$$

also has the form

$$\int_0^\infty L\left(\frac{y}{x}\right) \frac{1}{x} p(y) \, dy$$

for some appropriate L. Secondly, if we substitute $y = e^u$ and $x = e^v$, then the "kernel" of the integral operator S_α has the form

$$e^{\alpha(v-u)} e^{-u} \, ,$$

where it is not zero. This is "essentially" the same kernel we saw previously, in reference to the inverse of $(D - \alpha I)$.

§3. Consequences of the Hermite–Poulain Theorem

The question we now pose is the following. Are the two sets of operators we have so far discussed, *i.e.*, those obtained from the Hermite–Poulain Theorem and those from Laguerre's Theorem, in some sense the essential building blocks of all linear ZI operators taking polynomials of degree n to polynomials of degree n, all n?

Assuming we are interested in "all linear operators" then we must consider not only the specific operators given by the Hermite–Poulain Theorem and Laguerre's Theorem, but also those operators in their "closure". (And we must understand which operators are in their closure.) Note, for example, that neither the translation nor the dilation operator is contained in what we have so far discussed, and these operators obviously must be considered.

The Hermite–Poulain Theorem tells us that an operator

$$T = g(D)$$

determined by any polynomial g with all real zeros has the desired property. From continuity considerations the same will be true for all operators which are obtained as limits of $g(D)$'s of the above form. So what is the appropriate limit of the set of polynomials with all real zeros?

This and related questions were examined by Laguerre [15] who was the first to characterize those functions which can be uniformly approximated by polynomials whose zeros are all real, or whose zeros are all positive, and also by Pólya [20,21] who proved and generalized some of these results. These facts may be found in a variety of texts such as Hille [9], Hirschman and Widder [10], and Levin [17]. The set of functions

$$\mathcal{A}_1 = \left\{ C x^m e^{-c^2 x^2 + a x} \prod_{k=1}^{\infty} (1 + \alpha_k x) e^{-\alpha_k x} \right\},$$

where $C, c, a, \alpha_k \in I\!R$, $m \in Z\!\!Z_+$, and $\sum_{k=1}^{\infty} \alpha_k^2 < \infty$, are those functions which may be obtained as the uniform limit (on $[-A, A]$ for any $A > 0$) of polynomials having only real zeros. Defined on C, these are entire functions. (Note that this is a rather restricted set, and it should be contrasted with what the Weierstrass Theorem tells us about approximation by polynomials.) We sometimes call this set of functions the first Pólya–Laguerre (or Laguerre–Pólya) class.

In the Hermite–Poulain Theorem we also need the fact that $g(0) \neq 0$. This simply implies that $m = 0$ (and $C \neq 0$) in the above. Thus, we have that if

$$g(D) = C e^{-c^2 D^2 + a D} \prod_{k=1}^{\infty} (I + \alpha_k D) e^{-\alpha_k D},$$

then

$$Z_{I\!R}(p) \leq Z_{I\!R}(g(D)p)$$

for every polynomial p. We understand the $e^{-c^2 D^2}$ and $e^{a D}$ as given by their power series expansion and as such $g(D)$ is well-defined on the set of polynomials.

What are each of the operators $e^{-c^2 D^2}$, e^{aD}, and $(I + \alpha_k D)$? The third operator we have already discussed. It is also readily checked that

$$(e^{aD} p)(x) = p(x + a),$$

i.e., e^{aD} is just the shift operator, and obviously

$$Z_{I\!R}(p) = Z_{I\!R}(e^{aD} p).$$

The operator $e^{-c^2 D^2}$ is more interesting. It may be shown, see Carnicer, Peña, Pinkus [5], that

$$e^{-c^2 D^2} x^k = c^k H_k \left(\frac{x}{2c} \right),$$

where H_k is the kth degree Hermite polynomial (with leading coefficient 2^k). Thus,

$$Z_{I\!R} \left(\sum_{k=0}^{n} a_k x^k \right) \le Z_{I\!R} \left(\sum_{k=0}^{n} a_k H_k(x) \right),$$

as claimed in the introduction of this paper.

What about the converse? That is, if T is a linear operator taking π_n to π_n for all n and satisfying

$$Z_{I\!R}(p) \le Z_{I\!R}(Tp)$$

for every polynomial p, does this imply that

$$T = g(D)$$

for some $g \in \mathcal{A}_1$?

The answer is no!! For $g \in \mathcal{A}_1$ $(m = 0)$ we have $g(0) = C(\ne 0)$ which immediately implies that

$$g(D)x^n = Cx^n + \sum_{k=0}^{n-1} b_{n,k} x^k.$$

Thus, $g(D)$ is an operator on the set of all polynomials which acts, on the monomials, as a triangular matrix with "constant diagonal entries".

What if we therefore ask the same question, but restrict ourselves to operators of this particular form? That is, assume T is a linear operator of the form

$$Tx^n = Cx^n + \sum_{k=0}^{n-1} b_{n,k} x^k,$$

which satisfies

$$Z_{I\!R}(p) \le Z_{I\!R}(Tp)$$

for every polynomial p. Does this imply that $T = g(D)$ for some $g \in \mathcal{A}_1$? The answer now is yes. This result may be found in Carnicer, Peña, Pinkus [5].

What can we say about the inverse of such $T = g(D)$? As $g \in \mathcal{A}_1$ $(g(0) \ne 0)$ is the limit of the previously considered polynomials with all real roots, we can identify the inverse operator $S = T^{-1} = (g(D))^{-1}$. There are two possibilities. If $T = g(D) = Ce^{aD}$, then up to the constant C, T is just a shift, and thus S is the reverse shift and is given by $(1/C)e^{-aD}$. If, however, $g \in \mathcal{A}_1$, $g(0) \ne 0$, and $g(x) \ne Ce^{ax}$, then there exists a kernel K whose two-sided Laplace transform is $1/g$ in some neighborhood of the origin, *i.e.*,

$$\frac{1}{g(s)} = \int_{-\infty}^{\infty} e^{-sx} K(x) \, dx \,,$$

in some neighborhood of the origin, and

$$g(D)p = q$$

if and only if

$$p(x) = \int_{-\infty}^{\infty} K(x-y)q(y) \, dy \,.$$

This result is actually due to Pólya [21] who was motivated to consider this question for the same reasons which we have so far described. That is, he started with the Hermite–Poulain Theorem, considered the limiting operators obtained from functions in $\mathcal{A}_1$, and then considered the inverse of these operators. The fact that K, as a difference kernel, is totally positive and other related facts were proven by Schoenberg [25,26] (also in Schoenberg [27]). It is for exactly these reasons that Schoenberg called such kernels Pólya frequency functions. These results are also discussed in Hirschman and Widder [10], Karlin [14], and Widder [28].

What about the second version of the Hermite–Poulain Theorem? That is, we noted that if g has all positive zeros, then

$$Z_{[A,\infty)}(p) \le Z_{[A,\infty)}(g(D)p)$$

for every polynomial p and $A \in I\!R$. Again we should ask for the appropriate closure of this set of polynomials. It is

$$\mathcal{A}_2^+ = \left\{ Cx^m e^{-ax} \prod_{k=1}^{\infty} (1 - \alpha_k x) \right\},$$

where $C \in \mathbb{R}$, $a, \alpha_k \geq 0$, $m \in \mathbb{Z}_+$, and $\sum_{k=1}^{\infty} \alpha_k < \infty$. The difference between this and $\mathcal{A}_1$ is that there is no $e^{-c^2 x^2}$ and there is a more restrictive summability condition on the $\{\alpha_k\}$, in addition to the sign conditions. That is, this is the class obtained as the uniform limit, on some interval, of polynomials having only positive zeros. We let $\mathcal{A}_2^-$ be the uniform limit, on some interval, of polynomials having only negative zeros. Note that $g(x) \in \mathcal{A}_2^-$ if and only if $g(-x) \in \mathcal{A}_2^+$.

Assuming $g(0) \neq 0$ ($m = 0$ and $C \neq 0$ in the above) then if

$$g(D) = Ce^{-aD} \prod_{k=1}^{\infty} (I - \alpha_k D)$$

for $a, \alpha_k \geq 0$ and $\sum_{k=1}^{\infty} \alpha_k < \infty$, then

$$Z_{[A,\infty)}(p) \leq Z_{[A,\infty)}(g(D)p)$$

for every polynomial p and $A \in \mathbb{R}$. The converse also holds. If T has the form

$$Tx^n = Cx^n + \sum_{k=0}^{n-1} b_{n,k} x^k,$$

and

$$Z_{[A,\infty)}(p) \leq Z_{[A,\infty)}(Tp)$$

for every polynomial p and $A \in \mathbb{R}$, then $T = g(D)$ for some g as above. It actually suffices to consider only $A \geq 0$ or, $A = -\infty$ and $A = 0$. This result, however, is not true if we only demand the above ZI property to hold for one value of A, see Carnicer, Peña, Pinkus [5].

§4. The Pólya–Schur Property

Before relating to you the parallel generalization of Laguerre's Theorem, let us recall the following. Pólya, Schur [22] in 1914 characterized those linear operators Γ which are "diagonal" on the monomials, *i.e.*, operators of the form

$$\Gamma x^k = \gamma_k x^k, \qquad k = 0, 1, \dots,$$

and have the property that if p is a polynomial with all its zeros real, then Γp has all its zeros real. As we mentioned in the introduction, this problem is related to the one we have been considering. If T is a ZI operator, then for p with only real zeros it follows that Tp has only real zeros. That is, ZI operators have this Pólya–Schur property. However, it transpires that an operator Γ having this Pólya–Schur property does not necessarily have the

ZI property. If $\gamma_k = 1 + 4k^2$, then the associated Γ has the Pólya–Schur property, but not the ZI property. A different example appears in Bakan, Golub [3]. What Pólya and Schur proved is the following:

Pólya–Schur Theorem. *The above operator Γ has the Pólya–Schur property if and only if*

$$\sum_{k=0}^{n} \binom{n}{k} \gamma_k x^k \qquad (= \Gamma\left((1+x)^n\right))$$

has n real zeros all of one sign for all n or, equivalently, the function

$$\phi(x) = \sum_{k=0}^{\infty} \frac{\gamma_k}{k!} x^k$$

is either in $\mathcal{A}_2^+$ or in $\mathcal{A}_2^-$.

These are the same classes $\mathcal{A}_2^{\pm}$ which we considered previously, but the associated operators are totally different. Pólya and Schur also characterized those "diagonal" operators Γ with the property that if p is a polynomial with all its zeros real and of one sign, then Γp has all its zeros real. Such operators are characterized by the fact that the above ϕ is then in $\mathcal{A}_1$.

What if we now ask to characterize all operators of the form

$$\Gamma x^n = C x^n + \sum_{k=0}^{n-1} b_{n,k} x^k,$$

with this Pólya–Schur property. The ZI operators of this form have the Pólya–Schur property and we know how to characterize them. But are they all the operators with this property? Here the answer is in the affirmative. That is, if Γ has the above form of a "triangular operator with constant diagonal entries" and satisfies the Pólya–Schur property, then $\Gamma = g(D)$ for some $g \in \mathcal{A}_1$.

Thus with respect to the Pólya–Schur property, we know how to characterize all "diagonal operators" and also all "triangular operators with constant diagonal entries". This begs the question of how to characterize all such "triangular operators". Unfortunately these two types of operators do not commute. The problem remains open and difficult. Are all such operators simply (possibly infinite) products of operators of the above form?

§5. Consequences of Laguerre's Theorem

We now return to our original problem and to the operators given by Laguerre's Theorem, namely

$$Tx^k = g(k)x^k\,, \qquad k = 0, 1, \dots, n,$$

where g is a polynomial all of whose zeros are real and lie outside $[0, n]$. These operators have the ZI property on polynomials of degree at most n.

What is the appropriate closure of these polynomials as $n \to \infty$? It is neither $\mathcal{A}_1$ nor $\mathcal{A}_2^-$, but somewhere in between. It is given by

$$\mathcal{A}_3 = \left\{ Cx^m e^{-c^2 x^2 + ax} \prod_{k=1}^{\infty} (1 + \alpha_k x) e^{-\alpha_k x} \right\},$$

where $C, c, a \in \mathbb{R}$, $\alpha_k \geq 0$, $m \in \mathbb{Z}_+$, and $\sum_{k=1}^{\infty} \alpha_k^2 < \infty$. That is, it is exactly $\mathcal{A}_1$ except for the restriction $\alpha_k \geq 0$. It is definitely not $\mathcal{A}_2^-$. If $g \in \mathcal{A}_3$, and $g(0) \neq 0$, then

$$Z_{\mathbb{R}} \left(\sum_{k=0}^{n} a_k x^k \right) \leq Z_{\mathbb{R}} \left(\sum_{k=0}^{n} a_k g(k) x^k \right),$$

for every set of $\{a_k\}$ and now also for every n.

For example, if $g(x) = e^{ax}$, then $g(k) = b^k$ where $b = e^a$ is any positive value. This gives us the dilation map

$$Tx^k = (bx)^k\,,$$

which exactly preserves the number of real (and positive) zeros. If $g(x) = e^{-c^2 x^2}$, then $g(k) = q^{k^2}$ for some $q \in (0, 1)$. Thus,

$$Z_{\mathbb{R}} \left(\sum_{k=0}^{n} a_k x^k \right) \leq Z_{\mathbb{R}} \left(\sum_{k=0}^{n} a_k q^{k^2} x^k \right),$$

for any $q \in (0, 1)$. (The same holds on $(0, \infty)$.) This g is in $\mathcal{A}_3$, but is not in $\mathcal{A}_2^{\pm}$. We also have the classic example of

$$Z_{\mathbb{R}} \left(\sum_{k=0}^{n} a_k x^k \right) \leq Z_{\mathbb{R}} \left(\sum_{k=0}^{n} a_k \binom{n}{k} x^k \right),$$

which comes from the fact that the operator

$$Tx^k = \frac{x^k}{k!}$$

has the desired ZI property since

$$\frac{1}{\Gamma(x+1)} = e^{\gamma x} \prod_{k=1}^{\infty} \left(1 + \frac{x}{k}\right) e^{-x/k},$$

where $\Gamma(\cdot)$ is the Gamma function, and γ is the Euler constant. This function is in $\mathcal{A}_3$, but not in $\mathcal{A}_2^-$ (even discounting the $e^{\gamma x}$) because $\sum_{k=1}^{\infty} 1/k = \infty$. Both the above examples are due to Laguerre.

The question or conjecture which immediately surfaces is whether every diagonal operator

$$Tx^k = c_k x^k, \qquad k = 0, 1, \dots$$

(assume $c_k > 0$ for all k) with the ZI property necessarily satisfies

$$c_k = g(k), \qquad k = 0, 1, \dots$$

for some $g \in \mathcal{A}_3$. This is still an open conjecture. The result is not known in general. However, various cases are known based on work of Bakan, Craven, Csordas and Golub [2], see also [1,6,7].

They proved, for example, that if the above T is such that

$$Z_{[0,A]}(p) \leq Z_{[0,bA]}(Tp)$$

for some $b > 0$ and all $A > 0$, then

$$c_k = g(k), \qquad k = 0, 1, \dots$$

for some g with $e^{-ax}g(x) \in \mathcal{A}_2^-$ for some fixed $a \in \mathbb{R}$. They also proved that if

$$\overline{\lim_{k \to \infty}} \, c_k^{1/k} > 0,$$

then again

$$c_k = g(k), \qquad k = 0, 1, \dots$$

for some g for which $e^{-ax}g(x) \in \mathcal{A}_2^-$ for some fixed $a \in \mathbb{R}$. (Of course, neither q^{k^2} nor $1/k!$ satisfy these conditions.)

If this conjecture is valid, then one of its consequences is the following. Consider $g \in \mathcal{A}_3$ satisfying $g(0) \neq 0$. Then one possibility is that g is of the form $g(x) = Ce^{ax}$ in which case g generates the dilation operator, up to multiplication by the constant C, which exactly preserves the number of real (and positive) zeros. If g does not have this form then $S = T^{-1}$, which corresponds to the operator with diagonal entries $\{1/g(k)\}$, is also given on $(0, \infty)$ by

$$(Sp)(x) = \int_0^{\infty} L\left(\frac{y}{x}\right) \frac{1}{x} p(y) \, dy$$

for some appropriate totally positive L. The cases considered by Bakan, Craven, Csordas and Golub correspond to the case where L has finite support.

Even assuming that this conjecture is true, there still remains the intriguing problem (as with the Pólya–Schur property) of characterizing all T with the ZI property on $\mathbb{R}$. We seem far from a complete characterization. Even more difficult are these same problems on intervals other than $\mathbb{R}$, and/or with respect to polynomials of a fixed degree.

References

1. Bakan, A., T. Craven, and G. Csordas, Interpolation and the Laguerre-Pólya class, preprint.

2. Bakan, A., T. Craven, G. Csordas, and A. Golub, Weakly increasing zero-diminishing sequences, Serdica Math. J. **22** (1996), 1001–1024.

3. Bakan, A. G., and A. P. Golub, Some negative results on multiplier sequences of the first kind, Ukrainskii Mat. Z. **44** (1992), 305–309 = Ukrainian Math. J. **44** (1992), 264–268.

4. Burbea, J., Total positivity of certain reproducing kernels, Pacific J. Math. **67** (1976), 101–130.

5. Carnicer, J. M., J. M. Peña, and A. Pinkus, On some zero-increasing operators, Acta Math. Hungar. **94** (2002), 173–190.

6. Craven, T., and G. Csordas, Complex zero decreasing sequences, Methods Appl. Anal. **2** (1995), 420–441.

7. Craven, T., and G. Csordas, Problems and theorems in the theory of multiplier sequences, Serdica Math. J. **22** (1996), 515–524.

8. Hermite, Ch., Questions 777, 778, 779, Nouv. Ann. Math. **5** (1866), 432 and 479.

9. Hille, E., *Analytic Function Theory*, Volume II, Ginn, Boston, 1962.

10. Hirschman, I. I., and D. V. Widder, *The Convolution Transform*, Princeton University Press, Princeton, 1955.

11. Iserles, A., and S. P. Nørsett, Zeros of transformed polynomials, SIAM J. Math. Anal. **21** (1990), 483–509.

12. Iserles, A., S. P. Nørsett, and E. B. Saff, On transformations and zeros of polynomials, Rocky Mountain J. Math. **21** (1991), 331–357.

13. Iserles, A., and E. B. Saff, Zeros of expansions in orthogonal polynomials, Math. Proc. Camb. Phil. Soc. **105** (1989), 559–573.

14. Karlin, S., *Total Positivity*, Stanford University Press, Stanford, 1968.

15. Laguerre, E., Sur les fonctions du genre zéro et du genre un, C. R. Acad. Sci. **95** (1882), 828–831. Also to be found in *Oeuvres de Laguerre*, Vol I, Ch. Hermite, H. Poincaré, and E. Rouché (eds.), Gauthier-Villars et fils, Paris, 1898, 174–177.

16. Laguerre, E., Sur quelques points de la théorie des équations numériques, Acta Math. **4** (1884), 97–120. Also to be found in *Oeuvres de Laguerre*, Vol I, Ch. Hermite, H. Poincaré, and E. Rouché (eds.), Gauthier-Villars et fils, Paris, 1898, 184–206.

17. Levin, B. Ia., *Distribution of Zeros of Entire Functions*, AMS Transl. Math. Monographs, Vol. 5, Providence, 1964.

18. Marden, M., *Geometry of Polynomials*, AMS Math. Surveys, Volume 3, Providence, 1966.

19. Obreschkoff, N., *Verteilung und Berechnung der Nullstellen reeller Polynome*, VEB Deutscher Verlag der Wissenschaften, Berlin, 1963.

20. Pólya, G., Über Annäherung durch Polynome mit lauter reellen Wurzeln, Rend. Circ. Mat. Palermo **36** (1913), 279–295.

21. Pólya, G., Algebraische Untersuchungen über ganze Funktionen vom Geschlechte Null und Eins, J. Reine Angewandte Math. **145** (1915), 224–249.

22. Pólya, G., and J. Schur, Über zwei Arten von Faktorenfolgen in der Theorie der algebraischen Gleichungen, J. Reine Angew. Math. **144** (1914), 89–113.

23. Pólya, G., and G. Szegő, *Problems and Theorems in Analysis*, Volume II, Springer, Berlin, 1976.

24. Poulain, A., Théorèmes généraux sur les équations algébriques, Nouv. Ann. Math. **6** (1867), 21–33.

25. Schoenberg, I. J., On Pólya frequency functions. II: Variation diminishing integral operators of the convolution type, Acta Sci. Math. (Szeged) **12** (1950), 97–106.

26. Schoenberg, I. J., On Pólya frequency functions. I. The totally positive functions and their Laplace transform, J. Analyse Math. **1** (1951), 331–374.

27. Schoenberg, I. J., *I. J. Schoenberg: Selected Papers*, Ed. C. de Boor, Volume 2, Birkhäuser, Basel, 1988.

28. Widder, D. V., *An Introduction to Transform Theory*, Academic Press, NY, 1971.

Allan Pinkus
Dept. of Mathematics
Technion – Israel Institute of Technology
Haifa, 32000, ISRAEL
pinkus@tx.technion.ac.il
http://www.math.technion.ac.il/~pinkus/

The Aitken–Neville Scheme in Several Variables

Thomas Sauer and Yuesheng Xu

Abstract. We consider a geometric extension of the Aitken–Neville scheme for the evaluation of interpolation polynomials and characterize the point configurations which allow for the application of this scheme. Moreover, we derive a divided difference for these configurations which even coincides with the classical divided difference in the univariate case.

§1. Introduction

A major challenge in multivariate polynomial interpolation consists in the extension of constructions and properties of univariate polynomial interpolation to the multivariate case, see, for example, [2,5,12,16]. It is our goal in this paper to describe such an extension of the well–known Aitken–Neville scheme for the univariate polynomial interpolation which is based on geometric considerations, more precisely, on the crucial idea of repeated linear interpolation.

We begin with a brief review of the Aitken–Neville scheme for univariate polynomial interpolation. The Aitken–Neville scheme for the evaluation of univariate interpolation polynomials, avoiding the explicit computation of the coefficients of the interpolating polynomial, is a classical topic in almost any standard textbook on Numerical Analysis. It has been introduced first by A. C. Aitken [1] in 1932, based on earlier work by Ch. Jordan [11], and later, in a modified form, by E. H. Neville [14] and 1934, respectively, see the footnote on p. 96 of [14] for details. Given any distinct nodes $x_j \in \mathbb{R}$, $j \in \mathbb{Z}_{n+1} := \{0, 1, \ldots, n\}$, and prescribed values f_j, $j \in \mathbb{Z}_{n+1}$, the Aitken–Neville scheme computes, for any $x \in \mathbb{R}$,

Approximation Theory X: Abstract and Classical Analysis
Charles K. Chui, Larry L. Schumaker, and Joachim Stöckler (eds.), pp. 353–366.
Copyright ℗ 2002 by Vanderbilt University Press, Nashville, TN.
ISBN 0-8265-1415-4.

the value $p(x)$ of the interpolation polynomial p at the location x by an
iterated linear interpolation process

$$\varphi_j^0(x) := f_j, \qquad j \in \mathbb{Z}_{n+1}$$

$$\varphi_j^k(x) := \frac{x_{j+k} - x}{x_{j+k} - x_j} \, \varphi_j^{k-1}(x) + \frac{x - x_j}{x_{j+k} - x_j} \, \varphi_{j+1}^{k-1}(x), \qquad (1)$$

$$j \in \mathbb{Z}_{n-k+1}, \, k - 1 \in \mathbb{Z}_n.$$

For these polynomials φ_j^k, $j \in \mathbb{Z}_{n-k+1}$, $k \in \mathbb{Z}_{n+1}$, of degree at most k it
can be easily shown by induction that

$$\varphi_j^k(x_\ell) = f_\ell, \qquad \ell - j \in \mathbb{Z}_{k+1}, \qquad (2)$$

and therefore, in particular, that $\varphi_0^n(x_\ell) = f_\ell$, $\ell \in \mathbb{Z}_{n+1}$.

We remark that the recurrence formula (1) has an appealing geometric interpretation: the weights in the recursive rule are nothing but
the **barycentric coordinates** of the point x with respect to the line segment
$[x_j, x_{j+k}]$, $j \in \mathbb{Z}_{n-k+1}$, $k \in \mathbb{Z}_{n+1}$. Our aim is to extend this geometric
approach to the multivariate case and, in particular, to show its limita-
tions: the availability of an evaluation algorithm in terms of barycentric
coordinates imposes severe restrictions on the geometry of the interpolation nodes.

There were various (non–geometric) approaches to extend the Aitken–
Neville formula to higher dimensions, for example by G. Mühlbach [13]
(see also [4]) or by M. Gasca and his collaborators [7, 8]. For general
information on polynomial interpolation in several variables and its history
we refer to the survey articles [6,9,10].

In Section 2 of this paper we will develop and study the aforemen-
tioned extension of the Aitken–Neville scheme and use it in Section 3
to develop a recursive formula for divided differences associated to this
Aitken–Neville scheme.

§2. The Multivariate Case

We now consider the vector space $\Pi = \mathbb{R}[x] = \mathbb{R}[x_1, \ldots, x_d]$ of polynomials in d variables with real coefficients and denote by Π_n the subspace of
all polynomials of total degree at most n. For the set of all homogenized
multiindices of length n we write

$$\Gamma_n^0 := \left\{ \alpha = (\alpha_0, \ldots, \alpha_d) \in \mathbb{N}_0^{d+1} : \alpha_0 + \cdots + \alpha_d = n \right\}, \qquad n \in \mathbb{N}_0. \quad (3)$$

We now assume that we are given points $x_\alpha \in \mathbb{R}^d$, $\alpha \in \Gamma_n^0$, and values f_α,
$\alpha \in \Gamma_n^0$. Writing $[X]$ for the convex hull of a set $X \subset \mathbb{R}^d$ and denoting

the canonical "unit" multiindices by $\epsilon_j \in \Gamma_1^0$, we define, for $k \in \mathbb{Z}_{n+1}$ and $\alpha \in \Gamma_{n-k}^0$, the simplices

$$\Delta_\alpha^k := \left[x_{\alpha + k\epsilon_j} : j \in \mathbb{Z}_{d+1} \right]. \tag{4}$$

We will always assume throughout this paper that all these simplices are nondegenerate, i.e., that

$$\operatorname{vol}_d \Delta_\alpha^k > 0. \tag{5}$$

Let $\Delta = [v_j : j \in \mathbb{Z}_{d+1}]$, $v_j \in \mathbb{R}^d$, be a non–degenerate simplex and let

$$V := [v_0 \cdots v_d] \in \mathbb{R}^{d \times d+1},$$

denote the arrangement of these vertices according to the (arbitrary) order acording to their indexing, which is conveniently written as a matrix. The barycentric coordinates

$$u(x \,|\, V) = (u_j(x \,|\, V) : j \in \mathbb{Z}_{d+1})$$

of a point $x \in \mathbb{R}^d$ with respect to V are the *unique* coefficients such that

$$x = \sum_{\in \mathbb{Z}_{d+1}} u_j(x \,|\, V)\, v_j \quad \text{and} \quad \sum_{j \in \mathbb{Z}_{d+1}} u_j(x \,|\, V) = 1. \tag{6}$$

Note that the barycentric coordinates obviously depend on the indexing of the vertices $\{v_j : j \in \mathbb{Z}_{d+1}\}$ of the simplex Δ, but that a permutation of the vertices results in the respective permutation of the barycentric coordinates. Moreover, defining for $v \in \mathbb{R}^d$ the vector

$$\widehat{v} := \begin{bmatrix} 1 \\ v \end{bmatrix} \in \mathbb{R}^{d+1},$$

it can easily be seen by Cramer's rule that

$$u_j(x \,|\, V) = \frac{\tau_j(x \,|\, V)}{\tau(V)}, \tag{7}$$

where

$$\tau(V) := \det \widehat{V} := \det [\widehat{v}_0 \cdots \widehat{v}_d]$$

and

$$\tau_j(x \,|\, V) = \det [\widehat{v}_0 \cdots \widehat{v}_{j-1}\, \widehat{x}\, \widehat{v}_{j+1} \cdots \widehat{v}_d],$$

the latter one being the matrix obtained by replacing the j–th column of $\widehat{V}$, that is $\widehat{v}_j$, by $\widehat{x}$. Now, the multivariate analog of the Aitken–Neville scheme works as follows:

$$\varphi_\alpha^0(x) := f_\alpha, \qquad \alpha \in \Gamma_n^0,$$

$$\varphi_\alpha^k(x) := \sum_{j \in \mathbb{Z}_{d+1}} \varphi_{\alpha+\epsilon_j}^{k-1}(x)\, u_j(x \,|\, V_\alpha^k), \qquad \alpha \in \Gamma_{n-k}^0, \quad k-1 \in \mathbb{Z}_n, \tag{8}$$

where V_α^k denotes the vertices of Δ_α^k, indexed as described in (4). Since the barycentric coordinates are affine polynomials with respect to x by (7), we can immediately conclude that

$$\varphi_\alpha^k \in \Pi_k, \qquad \alpha \in \Gamma_{n-k}^0, \quad k \in \mathbb{Z}_{n+1}. \tag{9}$$

To formulate the main result on the validity of (2) in the multivariate case, we define the affine hull of a finite set $\{v_j : j \in J\} \subset \mathbb{R}^d$ as

$$\langle v_j : j \in J \rangle = \left\{ \sum_{j \in J} u_j v_j : u_j \in \mathbb{R}, \sum_{j \in J} u_j = 1 \right\}. \tag{10}$$

Theorem 1. *Let* $x_\alpha \in \mathbb{R}^d$, $\alpha \in \Gamma_n^0$, *be given. Then the following two statements are equivalent:*

1) *For any data* f_α, $\alpha \in \Gamma_n^0$, *all* $k \in \mathbb{Z}_{n+1}$ *and* $\beta \in \Gamma_{n-k}^0$ *we have*

$$\varphi_\beta^k (x_\gamma) = f_\gamma, \qquad \gamma \in \beta + \Gamma_k^0. \tag{11}$$

2) *For all* $J \subset \mathbb{Z}_{d+1}$, $k \in \mathbb{Z}_{n+1}$, *and* $\beta \in \Gamma_{n-k}^0$ *we have that*

$$\gamma \in [\beta + k\epsilon_j \,:\, j \in J] \quad \Longrightarrow \quad x_\gamma \in \langle x_{\beta+k\epsilon_j} \,:\, j \in J \rangle. \tag{12}$$

Before we prove this result, we briefly comment on the geometric condition (12). This condition says that, whenever an index is a convex combination of other indices, i.e., lies on the simplex generated by these indices, then the associated point lies on the affine variety generated by the points associated to these indices. In other words, the geometry of the indices carries over to the associated points.

To illustrate these geometric effects, let us consider two bivariate examples, namely the construction of *all* Aitken–Neville configurations of order 2 and 3 – clearly, order 1 is more or less trivial since it just requests three points in general position.

For the order $n = 2$, the geometric constraints come from

$$\epsilon_j + \epsilon_k = \frac{1}{2} (2\epsilon_j + 2\epsilon_k), \qquad j \neq k \in \mathbb{Z}_3,$$

hence, the respective points $x_{\epsilon_j+\epsilon_k}$ have to lie on the face of the triangle which is generated by the two vertices $x_{2\epsilon_j}$ and $x_{2\epsilon_k}$. Note, however, that this also permits that these points lie *outside* the simplex generated by $x_{(2,0,0)}$, $x_{(0,2,0)}$ and $x_{(0,0,2)}$. This situation is depicted in Fig. 1.

The situation already becomes more intricate when we consider the order $n = 3$. Besides the already familiar situation that the points $x_{2\epsilon_j+\epsilon_k}$

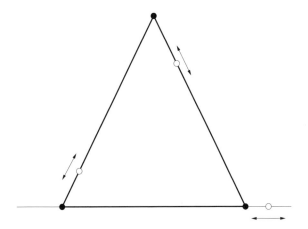

Fig. 1. Degrees of freedom for Aitken–Neville points of order 2.

and $x_{\epsilon_j + 2\epsilon_k}$ must lie on the straight line passing through the two vertices $x_{3\epsilon_j}$ and x_{ϵ_k}, we now have the additional identity of multiindices

$$(1,1,1) = \frac{1}{2}\left((2,1,0) + (0,1,2)\right) = \frac{1}{2}\left((1,2,0) + (1,0,2)\right)$$
$$= \frac{1}{2}\left((2,0,1) + (0,2,1)\right),$$

which means that the point $x_{(1,1,1)}$ lies on the intersection of the three lines determined by the above pairs of multiindices, a condition which also limits the possibilities of choices for the points $x_{2\epsilon_j + \epsilon_k}$, $j \neq k \in \mathbb{Z}_3$. As depicted in Fig. 2 we can, however, use this property to construct such configurations by choosing $x_{(1,1,1)}$ anywhere except on the boundary of the simplex and then we choose any three lines passing through this point and choose the additional points appropriately from the intersection of these lines with the faces of the simplex or their extensions.

Proof of Theorem 1. We begin with proving the sufficiency of (12) which we will do by induction on k. The case $k = 0$ is trivial and the case $k = 1$ is due to the fact that barycentric coordinates are fundamental polynomials with respect to (linear) interpolation, i.e.,

$$u_j\left(v_k \mid V\right) = \delta_{j,k}, \qquad j, k \in \mathbb{Z}_{n+1}.$$

Hence, suppose that (11) holds true for some $k \geq 1$. Choose $\beta \in \Gamma^0_{n-k-1}$ and $\gamma \in \beta + \Gamma^0_{k+1} \subseteq \Gamma^0_n$. This means that there exists $\eta \in \Gamma^0_{k+1}$ such that $\gamma = \beta + \eta$. Define $J \subset \mathbb{Z}_{d+1}$ as the support of η, i.e.

$$J := \left\{ j \in \mathbb{Z}_{d+1} : \eta_j > 0 \right\}.$$

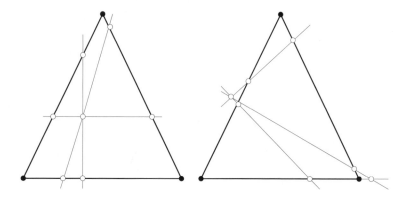

Fig. 2. The construction for Aitken–Neville points of order 3.

Then

$$\gamma \in \beta + \epsilon_j + \Gamma_k^0 \qquad \Longleftrightarrow \qquad \eta_j > 0 \qquad \Longleftrightarrow \qquad j \in J. \qquad (13)$$

Thus, since $\eta \in \Gamma_{k+1}^0$,

$$\gamma = \beta + \eta = \beta + \sum_{j \in J} \eta_j \epsilon_j$$
$$= \beta + \sum_{j \in J} \frac{\eta_j}{k+1} \epsilon_j (k+1) = \sum_{j \in J} \frac{\eta_j}{k+1} (\beta + (k+1)\epsilon_j),$$

i.e., $\gamma \in [\beta + (k+1)\epsilon_j : j \in J]$. By (12), we have that

$$x_\gamma = x_{\beta+\eta} \in \left\langle x_{\beta+(k+1)\epsilon_j} : j \in J \right\rangle,$$

which is the affine hull of a $(\#J-1)$–dimensional face of V_β^{k+1}. Therefore,

$$u_j \left(x_\gamma \mid V_\beta^{k+1} \right) = 0, \quad j \notin J, \qquad \text{and} \qquad \sum_{j \in J} u_j \left(x_\gamma \mid V_\beta^{k+1} \right) = 1. \quad (14)$$

Moreover, from (13) we conclude, for any $j \in J$, that $\gamma \in (\beta + \epsilon_j) + \Gamma_k^0$, and therefore, by the induction hypothesis, that $\varphi_{\beta+\epsilon_j}^k (x_\gamma) = f_\gamma$ for any $j \in J$. Hence, using (14) we obtain that

$$\varphi_\beta^{k+1} (x_\gamma) = \sum_{j \in \mathbb{Z}_{d+1}} \varphi_{\beta+\epsilon_j}^k (x_\gamma) \, u_j \left(x_\gamma \mid V_\beta^{k+1} \right)$$
$$= \sum_{j \in J} \varphi_{\beta+\epsilon_j}^k (x_\gamma) \, u_j \left(x_\gamma \mid V_\beta^{k+1} \right) = f_\gamma \sum_{j \in J} u_j \left(x_\gamma \mid V_\beta^{k+1} \right)$$
$$= f_\gamma,$$

which advances the induction hypothesis and proves the sufficiency of (12).

For the converse, we assume that (12) is violated, that is, that there exist $J \subseteq \mathbb{Z}_{d+1}$, $k \in \mathbb{N}$, $\alpha \in \Gamma_n^0$ and $\beta \in \Gamma_{n-k}^0$ such that

$$\alpha \in [\beta + k\epsilon_j : j \in J] \qquad \text{but} \qquad x_\alpha \notin \langle x_{\beta+k\epsilon_j} : j \in J \rangle. \tag{15}$$

Note that this implies that $J \neq \mathbb{Z}_{d+1}$ since V_β^k is nondegenerate, and therefore

$$\langle x_{\beta+k\epsilon_j} : j \in \mathbb{Z}_{d+1} \rangle = \langle V_\beta^k \rangle = \mathbb{R}^d.$$

In addition, if there are several values for k such that (15) holds, we choose the *minimal* one. We will now construct values f_γ such that (11) will be violated. Since α belongs to a $(\#J - 1)$–dimensional face of V_β^k, we again have that $\alpha = \beta + \eta$, where $\eta_j = 0$, $j \notin J$. Now, set $f_\gamma = \delta_{\alpha,\gamma}$, $\gamma \in \Gamma_n^0$, and consider

$$\varphi_\beta^k (x_\alpha) = \sum_{j \in \mathbb{Z}_{d+1}} \varphi_{\beta+\epsilon_j}^{k-1} (x_\alpha) \, u_j \left(x_\alpha \,|\, V_\beta^k \right).$$

Since k is the minimal index for which (12) is violated, we can use the sufficiency of (12) proved above to conclude that, for $\beta \in \Gamma_{n-k}^0$ and $j \in \mathbb{Z}_{d+1}$, the polynomials $\varphi_{\beta+\epsilon_j}^{k-1}$ belong to Π_{k-1} and interpolate f_γ at x_γ for any $\gamma \in \beta + \epsilon_j + \Gamma_{k-1}^0$. Since for $j \notin J$ we have that $\alpha_j = \beta_j$, we obtain that $\alpha \notin \beta + \epsilon_j + \Gamma_{k-1}^0$ whenever $j \notin J$. By definition of f_γ, we therefore conclude that $\varphi_{\beta+\epsilon_j}^{k-1} = 0$ if $j \notin J$, and therefore

$$\varphi_{\beta+\epsilon_j}^{k-1} (x_\alpha) = \begin{cases} 0, & j \notin J, \\ 1, & j \in J. \end{cases} \tag{16}$$

On the other hand, since $x_\alpha \notin \langle \Delta_\beta^k \rangle$, we have that

$$\sum_{j \in J} u_j \left(x_\alpha \,|\, V_\beta^k \right) \neq 1.$$

Hence,

$$\varphi_\beta^k (x_\alpha) = \sum_{j \in \mathbb{Z}_{d+1}} \varphi_{\beta+\epsilon_j}^{k-1} (x_\alpha) \, u_j \left(x_\alpha \,|\, V_\beta^k \right) = \sum_{j \in J} \varphi_{\beta+\epsilon_j}^{k-1} (x_\alpha) \, u_j \left(x_\alpha \,|\, V_\beta^k \right)$$

$$= \sum_{j \in J} u_j \left(x_\alpha \,|\, V_\beta^k \right) \neq 1 = f_\alpha,$$

and therefore (11) is also violated. $\square$

In view of Theorem 1 we make the following definition.

Definition 2. *A set of points* $X = \left\{ x_\alpha \in \mathbb{R}^d : \alpha \in \Gamma_n^0 \right\}$ *is called an* Aitken–Neville configuration *of order* n *if for all* $J \subset \mathbb{Z}_{d+1}$, $k \in \mathbb{Z}_{n+1}$, *and* $\beta \in \Gamma_{n-k}^0$

$$\gamma \in [\beta + k\epsilon_j : j \in J] \quad \Longrightarrow \quad x_\gamma \in \langle x_{\beta + k\epsilon_j} : j \in J \rangle. \tag{17}$$

Since for any Aitken–Neville configuration one can effectively construct the interpolation polynomial by the recurrence formula (8), we can immediately draw the following conclusion.

Corollary 3. *Any Aitken–Neville configuration* $X = \left\{ x_\alpha : \alpha \in \Gamma_n^0 \right\} \subset \mathbb{R}^d$ *defines a* poised *or* unisolvent *or* correct *interpolation problem, that is, for any data* $f_\alpha \in \mathbb{R}$, $\alpha \in \Gamma_n^0$, *there exists a unique polynomial* $p \in \Pi_n$ *such that* $p(x_\alpha) = f_\alpha$, $\alpha \in \Gamma_n^0$.

The fact that Aitken–Neville configurations can be characterized by interpolation properties on *all* intermediate levels allows us to make the following conclusion.

Corollary 4. *If* $X = \left\{ x_\alpha : \alpha \in \Gamma_n^0 \right\}$ *is an Aitken–Neville configuration, then all subsets of the form*

$$x_{\alpha + \Gamma_k^0} := \left\{ x_{\alpha + \beta} : \beta \in \Gamma_k^0 \right\}, \qquad \alpha \in \Gamma_{n-k}^0, \, k \in \mathbb{Z}_{n+1}, \tag{18}$$

are Aitken–Neville configurations as well.

§3. A Divided Difference for Aitken–Neville Configurations

There are various generalizations of divided differences to the multivariate case, depending on which property one wishes to extend, see for example [2,3,16] and also the section on Salzer's work in [10]. Another point of view, which we are going to follow here and which has been pursued recently in [15], is to use the name "divided difference" for the coefficients with respect to the monomials of maximal possible degree in the interpolation polynomial. For this purpose, we introduce some more notation. Let

$$\Gamma_n = \left\{ \alpha = (\alpha_1, \ldots, \alpha_d) \in \mathbb{N}_0^d : \alpha_1 + \cdots + \alpha_d = n \right\}$$

denote the set of all multiindices of length n. We also write

$$\boldsymbol{x}^n = (x^\alpha : \alpha \in \Gamma_n) \in (\Pi_n)^{\#\Gamma_n}, \qquad n \in \mathbb{N}_0,$$

for the vector of all monomials of total degree n and $\boldsymbol{c}_n$ for any vector of the form

$$\boldsymbol{c}_n = (c_\alpha^n : \alpha \in \Gamma_n) \in \mathbb{R}^{\#\Gamma_n}, \qquad n \in \mathbb{N}.$$

With this notation we can conveniently write any polynomial $p \in \Pi_n$ as

$$p(x) = \sum_{j=0}^{n} c_j \cdot x^j, \qquad x \in \mathbb{R}^d,$$

where "$\cdot$" denotes the usual inner product. Now, suppose that an Aitken–Neville configuration $X = \{x_\alpha : \alpha \in \Gamma_n^0\}$ is given. Clearly, for any $k \in \mathbb{Z}_{n+1}$ and $\beta \in \Gamma_{n-k}^0$ there exist coefficients $c_j(\beta)$ such that

$$\varphi_\beta^k(x) = \sum_{j=0}^{n-|\beta|} c_j(\beta) \cdot x^j. \tag{19}$$

We will call these vector valued coefficients the **divided difference** with respect to the Aitken–Neville configuration X. Our first goal is to give a recurrence relation for those coefficients which correspond to the terms of maximal possible degree $n - |\beta|$; note however, that this need not be the **homogeneous leading term** of these polynomials since the case $c_{n-|\beta|}(\beta) = 0$ is *not* excluded – this happens if $f \in \Pi_{n-|\beta|-1}$. For this purpose, for a simplex $\Delta = [v_j : j \in \mathbb{Z}_{d+1}]$ and $j, k \in \mathbb{Z}_{d+1}$, we let $\tau_{j,k}(V)$ denote the determinant obtained from $\tau(V)$ by deleting its jth column and kth row.

Lemma 5. *The vectors* $c_k(\alpha)$, $\alpha \in \Gamma_{n-k}^0$, $k \in \mathbb{Z}_{n+1}$, *satisfy the recurrence relation*

$$c_0(\alpha) = f_\alpha, \qquad \alpha \in \Gamma_n^0, \tag{20}$$

and, for $\alpha \in \Gamma_{n-k}$, $k - 1 \in \mathbb{Z}_n$,

$$c_k(\alpha) = \left(\sum_{\ell=0}^{d} \sum_{j=1}^{d} (-1)^{\ell+j} \frac{\tau_{\ell,j}(V_\alpha^k)}{\tau(V_\alpha^k)} (c_k(\alpha + \epsilon_\ell))_{\beta - \epsilon_j} : \beta \in \Gamma_k^0 \right), \tag{21}$$

with the convention that $(c_k(\alpha))_\beta = 0$ *if* $\beta \in \mathbb{Z}^{d+1} \setminus \Gamma_k^0$.

Proof: Using the recurrence relation (8) we obtain that

$$\sum_{j=0}^{k} c_j(\alpha) \cdot x^j = \varphi_\alpha^k(x) = \sum_{\ell=0}^{d} \varphi_{\alpha+\epsilon_\ell}^{k-1}(x) \, u_\ell(x \mid V_\alpha^k)$$

$$= \sum_{m=0}^{k-1} \sum_{\ell=0}^{d} (c_m(\alpha + \epsilon_\ell) \cdot x^m) \, u_\ell(x \mid V_\alpha^k).$$

For $\ell \in \mathbb{Z}_{d+1}$, we expand the determinant in the numerator of (7) and write

$$u_\ell(x \mid V_\alpha^k) = \frac{(-1)^\ell}{\tau(V_\alpha^k)} \left(\tau_{\ell,0}(V_\alpha^k) + \sum_{j=1}^{d} (-1)^j x_j \, \tau_{\ell,j}(V_\alpha^k) \right).$$

Substituting this equation into the right hand-side of the previous one yields

$$\sum_{j=0}^{k} c_j(\alpha) \cdot x^j = \sum_{m=0}^{k-1}\sum_{\ell=0}^{d} (c_m(\alpha + \epsilon_\ell) \cdot x^m)(-1)^\ell \frac{\tau_{\ell,0}\left(V_\alpha^k\right)}{\tau\left(V_\alpha^k\right)}$$

$$+ \sum_{m=0}^{k-1}\sum_{\ell=0}^{d}\sum_{j=1}^{d}(-1)^{\ell+j}\frac{\tau_{\ell,j}\left(V_\alpha^k\right)}{\tau\left(V_\alpha^k\right)}c_m(\alpha+\epsilon_\ell)\cdot x^m \cdot x_j$$

$$=: \Sigma_1 + \Sigma_2,$$

where $\Sigma_1 \in \Pi_{k-1}$. Moreover,

$$\Sigma_2 = \sum_{m=0}^{k-1}\sum_{l=0}^{d}\sum_{j=1}^{d}\sum_{\gamma\in\Gamma_m}(-1)^{l+j}\frac{\tau_{l,j}\left(V_\alpha^k\right)}{\tau\left(V_\alpha^k\right)}c_m(\alpha+\epsilon_l)_\gamma\, x^{\gamma+\epsilon_j}$$

$$= \sum_{m=1}^{k}\sum_{l=0}^{d}\sum_{j=1}^{d}\sum_{\gamma\in\Gamma_m}(-1)^{l+j}\frac{\tau_{l,j}\left(V_\alpha^k\right)}{\tau\left(V_\alpha^k\right)}c_m(\alpha+\epsilon_l)_{\gamma-\epsilon_j}\, x^\gamma,$$

and a comparison of leading terms yields (21). $\square$

We now give a recursively defined divided difference which realizes the above coefficients. For this purpose we introduce, for a given simplex $\Delta = [v_j : j \in \mathbb{Z}_{d+1}]$, $j - 1 \in \mathbb{Z}_d$ and $0 \le k < \ell \le d$ the determinant $\widehat{\tau}_{k,\ell}^j(V)$ which is obtained from $\tau(V)$ by deleting its rows number 0 and j and columns number k and ℓ. The kth divided difference of $f : \mathbb{R}^d \to \mathbb{R}$ with respect to the Aitken–Neville points x_α, $\alpha \in \Gamma_n^0$, denoted by

$$\left[x_{\alpha+\beta} : \beta \in \Gamma_k^0\right]f = \left[x_{\alpha+\Gamma_k^0}\right]f = \left(\left[x_{\alpha+\Gamma_k^0}\right]_\beta f : \beta \in \Gamma_k^0\right), \qquad \alpha \in \Gamma_{n-k}^0,$$

using the abbreviation from (18), is recursively defined as

$$[x]f = [x]_0 f := f(x),$$

$$\left[x_{\alpha \mid \Gamma_k^0}\right]_\beta f := \sum_{j=1}^{d}\sum_{0\le m<l\le d}\widehat{\tau}_{l,m}^j\left(V_\alpha^k\right)(-1)^{j+l+m}\times$$

$$\times \frac{\left[x_{\alpha+\epsilon_l+\Gamma_{k-1}^0}\right]_{\beta-\epsilon_j}f - \left[x_{\alpha+\epsilon_m+\Gamma_{k-1}^0}\right]_{\beta-\epsilon_j}f}{\tau\left(V_\alpha^k\right)}, \qquad (22)$$

$$\beta \in \Gamma_k^0, \ \alpha \in \Gamma_{n-k}^0.$$

Before we prove that this recursion realizes the leading coefficients in the interpolation polynomials, let us first point out that the above recurrence

(22) indeed defines *the* divided difference if $d = 1$. If $n \in \mathbb{N}$, then we are considering a point set

$$x_{n,0} = x_0, \; x_{n-1,1} = x_1, \ldots, x_{0,n} = x_n$$

and

$$\tau\left(\left[x_{n\epsilon_j} : j \in \mathbb{Z}_{d+1}\right]\right) = \det \begin{bmatrix} 1 & 1 \\ x_0 & x_n \end{bmatrix} = x_n - x_0,$$

while, trivially, $\widehat{\tau}^{\,j}_{k,l}\left(\left[x_{n\epsilon_j} : j \in \mathbb{Z}_{d+1}\right]\right) = 1$. Therefore, (22) becomes

$$[x_0, \ldots, x_n]\, f = \frac{[x_1, \ldots, x_n]\, f - [x_0, \ldots, x_{n-1}]\, f}{x_n - x_0},$$

which is the usual divided difference recurrence relation.

Theorem 6. *Let* $\left\{x_\alpha : \alpha \in \Gamma^0_n\right\}$ *be an Aitken–Neville configuration of points. Then the vectors* $c_k\,(\alpha)$, $k \in \mathbb{Z}_{n+1}$, $\alpha \in \Gamma^0_{n-k}$, *satisfy*

$$c_k\,(\alpha) = \left[x_{\alpha+\Gamma^0_k}\right] f. \tag{23}$$

Proof: We prove Theorem 6 by induction on k, where the case $k = 0$ is immediate from (20) and (22). So, we assume that (23) is valid for some $k \geq 0$. We first note that for any $\Delta = [v_j : j \in \mathbb{Z}_{d+1}]$ an expansion of the determinant $\tau_{k,j}(V)$ with respect to the first row $(1, \ldots, 1)^T$ yields that

$$\tau_{k,j}(V) = \sum_{\ell=0}^{k-1}(-1)^\ell \widehat{\tau}^{\,j}_{k,\ell}(V) - \sum_{\ell=k+1}^{d}(-1)^\ell \widehat{\tau}^{\,j}_{k,\ell}(V), \qquad k \in \mathbb{Z}_{d+1}, \tag{24}$$

and, obviously, for $k, \ell \in \mathbb{Z}_{d+1}$ that

$$\widehat{\tau}^{\,j}_{k,\ell}(V) = \widehat{\tau}^{\,j}_{\ell,k}(V). \tag{25}$$

Using (21) and the induction hypothesis we now get, for $\alpha \in \Gamma^0_{n-k-1}$ and $\beta \in \Gamma^0_{k+1}$, that

$$c_{k+1}\,(\alpha)_\beta = \sum_{\ell=0}^{d}\sum_{j=1}^{d}(-1)^{\ell+j}\frac{\tau_{\ell,j}\left(V^k_\alpha\right)}{\tau\left(V^k_\alpha\right)}\left(c_k(\alpha + \epsilon_\ell)\right)_{\beta-\epsilon_j}$$

$$= \sum_{\ell=0}^{d}\sum_{j=1}^{d}(-1)^{\ell+j}\frac{\tau_{\ell,j}\left(V^k_\alpha\right)}{\tau\left(V^k_\alpha\right)}\left[x_{\alpha+\epsilon_\ell+\Gamma^0_{n-k}}\right]_{\beta-\epsilon_j} f.$$

To simplify the presentation of the computations to follow, we introduce the notation

$$f\,(\ell, j, \beta) := \left[x_{\alpha+\epsilon_\ell+\Gamma^0_{n-k}}\right]_{\beta-\epsilon_j} f, \qquad \ell \in \mathbb{Z}_{d+1}, \; j-1 \in \mathbb{Z}_d, \; \beta \in \Gamma^0_{k+1}.$$

Hence, it follows from (24) that

$$c_{k+1}\left(\alpha\right)_\beta = \frac{1}{\tau\left(V_\alpha^k\right)} \sum_{\ell=0}^{d}\sum_{j=1}^{d}(-1)^{\ell+j}\times$$

$$\times\left(\sum_{m=0}^{\ell-1}(-1)^m\widehat{\tau}_{\ell,m}^j\left(V_\alpha^k\right) - \sum_{m=\ell+1}^{d}(-1)^m\widehat{\tau}_{\ell,m}^j\left(V_\alpha^k\right)\right)f\left(\ell,j,\beta\right).$$

Now, we conclude that

$$c_{k+1}\left(\alpha\right)_\beta = \frac{1}{\tau\left(V_\alpha^k\right)} \sum_{\ell=1}^{d}\sum_{j=1}^{d}\sum_{m=0}^{\ell-1}(-1)^{\ell+j+m}\,\widehat{\tau}_{\ell,m}^j\left(V_\alpha^k\right)f\left(\ell,j,\beta\right)$$

$$-\frac{1}{\tau\left(V_\alpha^k\right)} \sum_{\ell=0}^{d-1}\sum_{j=1}^{d}\sum_{m=\ell+1}^{d}(-1)^{\ell+j+m}\,\widehat{\tau}_{\ell,m}^j\left(V_\alpha^k\right)f\left(\ell,j,\beta\right)$$

$$=\frac{1}{\tau\left(V_\alpha^k\right)} \sum_{j=1}^{d}\sum_{0\leq m<\ell\leq d}(-1)^{\ell+j+m}\,\widehat{\tau}_{\ell,m}^j\left(V_\alpha^k\right)f\left(\ell,j,\beta\right)$$

$$-\frac{1}{\tau\left(V_\alpha^k\right)} \sum_{j=1}^{d}\sum_{0\leq \ell<m\leq d}(-1)^{\ell+j+m}\,\widehat{\tau}_{\ell,m}^j\left(V_\alpha^k\right)f\left(\ell,j,\beta\right).$$

Exchanging ℓ and m in the second sum and taking into account (25), we obtain that the second sum takes the value

$$\sum_{j=1}^{d}\sum_{0\leq m<\ell\leq d}(-1)^{\ell+j+m}\,\widehat{\tau}_{\ell,m}^j\left(V_\alpha^k\right)f\left(m,j,\beta\right)$$

which yields that

$$c_{k+1}\left(\alpha\right)_\beta = \sum_{j=1}^{d}\sum_{0\leq m<\ell\leq d}(-1)^{\ell+j+m}\widehat{\tau}_{\ell,m}^j\left(V_\alpha^k\right)\times$$

$$\times\frac{\left[x_{\alpha+\epsilon_\ell+\Gamma_{n-k}^0}\right]_{\beta-\epsilon_j}f - \left[x_{\alpha+\epsilon_m+\Gamma_{n-k}^0}\right]_{\beta-\epsilon_j}f}{\tau\left(V_\alpha^k\right)}$$

$$=\left[x_{\alpha+\Gamma_{n-k-1}^0}\right]_\beta f.$$

This completes the induction step and our proof. $\square$

To end the paper, we remark another property of the univariate divided difference which is preserved by the multivariate generalization.

Corollary 7. *Let X be an Aitken–Neville configuration of order n and let $f \in \Pi_\ell$, $\ell \leq n$. Then*

$$\left[x_{\alpha + \Gamma_k^0} \right] f = 0, \qquad \alpha \in \Gamma_{n-k}^0, \, \ell < k \leq n. \tag{26}$$

Proof: From Corollary 4 and the fact that $f \in \Pi_\ell$ it follows that for every $k > \ell$ and $\alpha \in \Gamma_{n-k}^0$ the polynomials φ_α^k satisfy

$$\varphi_\alpha^k \left(x_{\alpha + \Gamma_k^0} \right) = f \left(x_{\alpha + \Gamma_k^0} \right),$$

and consequently $\varphi_\alpha^k = f \in \Pi_\ell$. Since the divided difference consists of the homogeneous coefficients of order $k \leq \ell$ by Theorem 6, equation (26) follows readily. $\square$

Acknowledgments. The second author is supported in part by the US National Science Foundation under grant DMS-9973427 and by the Chinese Academy of Sciences under program "One Hundred Distinguished Young Scientists".

References

1. A. C. Aitken, On interpolation by iteration of proportional parts, without the use of differences, Proc. Edinburgh Math. Soc. **3** (1932), 56–76.

2. C. de Boor, A multivariate divided difference, in *Approximation Theory VIII, Vol. 1: Approximation and Interpolation*, C. K. Chui and L. L. Schumaker (eds.), World Scientific Publishing Co., 1995, 87–96.

3. C. de Boor, On the Sauer–Xu formula in multivariate polynomial interpolation, Math. Comp. **65** (1996), 1231–1234.

4. C. Brezinski, The Mühlbach–Neville–Aitken algorithm and some extensions, BIT **80** (1980), 444–451.

5. K. C. Chung and T. H. Yao, On lattices admitting unique Lagrange interpolation, SIAM J. Numer. Anal. **14** (1977), 735–743.

6. M. Gasca, Multivariate polynomial interpolation, in *Computation of Curves and Surfaces*, W. Dahmen, M. Gasca, and C. A. Micchelli (eds.), Kluver Academic Publishers, 1990, 215–236.

7. M. Gasca and E. Lebrón, On Aitken–Neville formulae for multivariate interpolation, in *Numerical approximation of partial differential equations*, E. L. Ortiz (ed.), North Holland, 1987, 133–140.

8. M. Gasca and A. López-Carmona, A general recurrence interpolation formula and its applications to multivariate interpolation, J. Approx. Theory **34** (1982), 361–374.

9. M. Gasca and T. Sauer, Polynomial interpolation in several variables, Advances Comput. Math. **12** (2000), 377–410.

10. M. Gasca and T. Sauer, On the history of multivariate polynomial interpolation, J. Comput. Appl. Math. **122** (2000), 23–35.

11. Ch. Jordan, Sur une formule d'interpolation dérivée de la formule d'Everett, Metron **7** (1928), 47–51.

12. C. A. Micchelli, T. Sauer and Yuesheng Xu, A construction of refinable sets for interpolating wavelets, Result. Math. **34** (1998), 359-372.

13. G. Mühlbach, The general Neville–Aitken algorithm and some applications, Numer. Math. **31** (1978), 97–110.

14. E. H. Neville, Iterative interpolation, J. Indian Math. Soc. **20** (1933), 87–120.

15. C. Rabut, Generalized divided difference with simple knots, SIAM J. Numer. Anal. (2001), to appear.

16. T. Sauer and Yuan Xu, On multivariate Lagrange interpolation, Math. Comp. **64** (1995), 1146-1170.

Thomas Sauer
Lehrstuhl für Numerische Mathematik
Justus–Liebig–Universität Gießen
Heinrich–Buff–Ring 44
D–35092 Gießen, Germany
Tomas.Sauer@math.uni-giessen.de

Yuesheng Xu
Department of Mathematics
West Virginia University
Morgantown, WV 26506, USA
and
Institute of Mathematics
The Chinese Academy of Sciences
Beijing 100080, P. R. China
yxu@math.wvu.edu

Interpolation by Polynomials in Several Variables

Boris Shekhtman

Abstract. A subspace $H \subset C(\mathbb{R}_n)$ is called k-interpolating if for any choice of distinct points $x_1, \ldots, x_k \in \mathbb{R}_n$ and for any choice $\alpha_1, \ldots, \alpha_k \in \mathbb{R}$ there exists $h \in H$ such that $h(x_j) = \alpha_j, j = 1, \ldots, k$. Let Π_{k-1}^n be the space of polynomials of n variables of degree at most $k - 1$. We show that for almost all choices $h_1, \ldots, h_N \in \Pi_{k-1}^n$ with $N = nk + k$ the span $[h_1, \ldots, h_N]$ is k-interpolating.

§1. Preliminaries

Let $C(\mathbb{R}_n)$ be the space of continuous functions on $\mathbb{R}_n$. Let $H \subset C(\mathbb{R}_n)$ be a linear subspace.

Definition 1. *The space H is called k–interpolating if for any k distinct points $x_1, x_2, \ldots, x_k \in \mathbb{R}_n$ and for any k values $\alpha_1, \ldots, \alpha_k \in \mathbb{R}$ there exists (not necessarily unique) $h \in H$ such that*

$$h(x_j) = \alpha_j, \quad j = 1, \ldots, k.$$

Definition 2. *Define*

$$i(n, k) = \min\{\dim H : H \text{ is } k \text{ interpolating}\}.$$

Examples. *Obviously, we have $i(n, k) \geq k$, and it follows from Mairhuber's Theorem that $i(n, k) > k$ if $n > 1$.*

1) Let Π_{k-1} be the space of polynomials of degree at most $k - 1$. Then Π_{k-1} is a k–interpolating subspace of $C(\mathbb{R})$ and $i(n, k) = k$.

Approximation Theory X: Abstract and Classical Analysis
Charles K. Chui, Larry L. Schumaker, and Joachim Stöckler (eds.), pp. 367–372.
Copyright © 2002 by Vanderbilt University Press, Nashville, TN.
ISBN 0-8265-1415-4.

2) Let $H = \text{span}\{1, \text{Re}z^j, \text{Im}z^j\}_{j=1}^{k-1}$. Then $\dim H = 2k - 1$ and H is
   k–interpolating in $C(\mathbb{R}_2)$. Hence $i(2, k) \leq 2k - 1$. An amazing result
   discovered by D. Handel and F. Cohen [1] and then rediscovered by
   V. Vasiliev [5] states that

$$2k - \eta(k) \leq i(2, k) \leq 2k - 1,$$

where $\eta(k)$ is the number of ones in the binary expansion of the integer
k.

3) Let $x = (t_1, \ldots, t_n) \in \mathbb{R}_n$ and let $\pi_j(x) = t_j$. Then $[1] \subset C(\mathbb{R}_n)$
   is a one–interpolating subspace of $C(\mathbb{R}_n)$; $\text{span}\{1, \pi_1, \pi_2, \ldots, \pi_n\}$ is
   2–interpolating in $C(\mathbb{R}_n)$; and $\text{span}\left\{1, \pi_1, \pi_2, \ldots, \pi_n, \sum_{j=1}^{n} \pi_j^2\right\}$ is 3–
   interpolating in $C(\mathbb{R}_n)$. As was shown by the author [4] and by D.
   Wulbert [6] these are the "extremal" subspaces in the sense that

$$i(n, 1) = 1; \quad i(n, 2) = n + 1; \quad i(n, 3) = n + 2.$$

The purpose of this note is to obtain an upper bound on the quantity
$i(n, k)$. The following conjecture is reasonable.

Conjecture. $i(n, k) \leq n(k - 1) + 1$.

Let Π_{k-1}^n be the space of polynomials of n variables of degree at
most $k - 1$. It is easy to see that Π_{k-1}^n is k-interpolating in $C(\mathbb{R}_n)$ and
$\dim \Pi_{k-1}^n = \binom{k-1+n}{n}$.

Theorem 1. *Let $N = nk + k$. Then generic (almost all) subspaces $H \subset$
Π_{k-1}^n with $\dim H = N$ are k–interpolating. In particular $i(n, k) \leq N$.*

The proof of this theorem is based on the notion of transversality.
We will finish this section with the discussion of transversality. All the
relevant details can be found in [2].

Definition 3. *Let A and B be two differentiable manifolds and F be a
differentiable map*

$$F : A \to B.$$

*Let C be a submanifold of B and $x \in A$. We say that F is transversal to
C at the point x if*

$$F(x) \notin C$$

or

$$T_{F(x)}B = T_{F(x)}C + (DF)_x A.$$

F is transversal to C if it is transversal to C at every point $x \in A$. We use the symbol

$$F \pitchfork C$$

to denote the transversality of F to C.

Here $T_z D$ denotes the tangent plane to the manifold D at the point z; and $(DF)_x$ is the derivative of the function F at the point x.

Let G be another differential manifold and Φ be a differentiable mapping

$$\Phi : G \times A \to B.$$

For any $g \in G$,

$$\Phi(g) : A \to B$$

is defined by

$$\Phi(g)(a) = \Phi(g, a).$$

The main ingredient of our proof is the following theorem.

Transversality Theorem. *Let*

$$\Phi : G \times A \to B$$

be as above. Let $C \subset B$ be a submanifold and

$$\Phi \pitchfork C.$$

Then for almost all (in the sense of category) $g \in G$

$$\Phi(g) \pitchfork C.$$

§2. Proof of Theorem 1

As before we let $N = (n+1)k$. Further, we use $G = \left(\Pi_{k-1}^n\right)^N$ for a differential manifold (vector space) which is a direct product of N copies of Π_{k-1}^n. Let $A = A(n, k)$ be a configuration manifold, i.e., the set of all distinct k-tuples of points $x_1, \ldots, x_k \in \mathbb{R}_n$. Finally, let $B = B(N, k)$ be the space of all k by N matrices and let $C \subset B$ be a submanifold of matrices of rank less than k.

We define a mapping

$$\Phi : G \times A \to B$$

by

$$\Phi\left(h_1, \ldots, h_N, x_1, \ldots, x_k\right) = \left[h_j\left(x_m\right)\right] \in B, \quad j = 1, \ldots, N, \quad m = 1, \ldots, k,$$

where

$$g = (h_j), \quad h_j \in \Pi_{k-1}^n$$

and

$$x_1, \ldots, x_k \in \mathbb{R}_n$$

are distinct. Our immediate goal is to show that $\Phi \pitchfork C$.

In order to do so we need to verify that

$$T_{\Phi(g,x)}B = T_{\Phi(g,x)}C + (D\Phi)_{(g,x)}(G \times A). \tag{2.1}$$

Since B is a linear space, its tangent space is itself, and to prove (2.1) it is sufficient to show that

$$\operatorname{rank}(D\Phi)_{(g,x)} = \dim\left(T_{\Phi(g,x)}B\right) = \dim B = kN. \tag{2.2}$$

The Jacobian $(D\Phi)_{(g,x)}$ is an augmented matrix

$$L = [L_1 \mid L_2],$$

where $L_1 = \frac{\partial \Phi}{\partial g}(h_1, \ldots, h_N, x_1, \ldots, x_k)$ and $L_2 = \frac{\partial \Phi}{\partial x}(g, x)$. To show (2.2) we will verify that rank $L_1 = kN$. Let $p_1, \ldots, p_d$ be a basis for Π_{k-1}^n where $d = \dim\left(\Pi_{k-1}^n\right)$. Then $g = (h_j)_{j=1}^N$ and

$$h_j = \sum_{m=1}^d a_m^{(j)} p_m.$$

Hence L_1 can be represented as

$$L_1 = \begin{bmatrix} \tilde{H} & 0 & \cdots & 0 \\ 0 & \tilde{H} & & \vdots \\ \vdots & & \ddots & \\ 0 & \cdots & & \tilde{H} \end{bmatrix}$$

where $\tilde{H} = \left[\frac{\partial}{\partial a_m^{(j)}} h_j\right](x_1, \ldots, x_k)$. This is so since $\frac{\partial}{\partial a_m^{(j)}} h_s = 0$ if $s \neq j$. Finally observe that

$$\tilde{H} = [p_s(x_m)], \quad s = 1, \ldots, d \text{ and } m = 1, \ldots, k.$$

Since Π_{k-1}^n is k-interpolating it follows that rank $\left(\tilde{H}\right) = k$ and since there are N of them we conclude that rank $L_1 = kN$ and $\Phi \pitchfork C$. By the Transversality Theorem we conclude that for almost all N-tuples of polynomials

$$h_1, \ldots, h_N \in \Pi_{k-1}^n,$$

the mapping

$$\Phi(h_1, \ldots, h_N) : A \to B$$

is transversal to C. Thus either

$$\Phi(h_1, \ldots, h_N)(x_1, \ldots, x_k) = [h_j(x_m)] \notin C \tag{2.3}$$

or

$$T_{\Phi(h)(x)}B = T_{\Phi(h)(x)}C + (D\Phi(h))_x A. \tag{2.4}$$

A simple dimension count shows that (2.4) is not possible for any choice of $x = (x_1, \ldots, x_k)$. Indeed

$$\dim TB = \dim B = Nk$$

$$\dim(D\Phi(h))_x A \le \dim A = kn$$

$$\dim TC = \dim C.$$

So if (2.4) were to be true, then

$$Nk \le \dim C + kn$$

gives $\operatorname{codim} C \le kn$. But it is well known (see [3]) that

$$\operatorname{codim} C = N - k + 1.$$

Since $N = nk + k$ we obtain a contradiction. The remaining viable alternative is (2.3); *i.e.*, we have

$$[h_j(x_m)] \notin C \text{ for all distinct } x_1, \ldots, x_k \in \mathbb{R}_n.$$

Since $B \setminus C$ is the set of all matrices of full rank, we conclude that

$$H := \operatorname{span}[h_1, \ldots, h_N]$$

is k–interpolating. $\square$

Remark. *The disparity between the conjectured bound of $n(k-1)+1$ and the proven bound $(n+1)k$ can be partially explained by the "generic nature" of Theorem 1.*

Indeed $i(1,1) = 1$, yet the set of all polynomials (say of degree 10) which do not vanish at any point of the real line, are not dense among polynomials. However, the set of all pairs of polynomials (p_1, p_2) of degree 10 that do not vanish simultaneously at any point is dense in $\Pi_{10} \times \Pi_{10}$. It also suggests that generic subspaces of polynomials do not form interpolating spaces of minimal dimensions.

Acknowledgments. I would like to thank my friend, Laurent Barachart, who patiently introduced me to the beauty of transversality.

References

1. Cohen, F. R. and D. Handel, k-regular embeddings of the plane, Proc. AMS **72** (1978), 202–204.

2. Hirsch, M. W., *Differential Topology*, Graduate Texts in Mathematics, Springer-Verlag, 1976.

3. Milnor, J., *Differential Topology*, Princeton University, 1958.

4. Shekhtman, B., Interpolating subspaces in $\mathbb{R}_n$, in *Approximation Theory, Wavelets and Applications*, S. P. Singh (ed.), Kluwer, 1995, 465–471.

5. Vasil'ev, V. A., On function spaces that are interpolating at any k nodes, Funct. Anal. Appl. **26** (1992), 209–210.

6. Wulbert, D., Interpolation at a few points, J. Approx. Theory **96** (1999), no. 1, 139–148.

Boris Shekhtman
Department of Mathematics
University of South Florida
Tampa, FL 33620-5700
boris@math.usf.edu

Nonlinear Approximation with Regard to Bases

V. N. Temlyakov

Abstract. The paper is a survey on greedy type approximation with regard to bases. We have included in this paper some of the basic recent results in this area with complete proofs. For understanding those proofs the reader does not need an extensive background in approximation theory, they can be taught in a graduate course. The paper also contains a discussion of other results in the area which are given without proofs.

§1. Introduction

Let a Banach space X with a basis $\Psi = \{\psi_k\}_{k=1}^\infty$ be given. We consider the following theoretical **greedy algorithm**. For a given element $f \in X$ we consider the expansion

$$f = \sum_{k=1}^\infty c_k(f, \Psi)\psi_k.$$

Let an element $f \in X$ be given. We call a permutation ρ, $\rho(j) = k_j$, $j = 1, 2, \ldots$, of the positive integers **decreasing** and write $\rho \in D(f)$ if

$$\|c_{k_1}(f, \Psi)\psi_{k_1}\|_X \geq \|c_{k_2}(f, \Psi)\psi_{k_2}\|_X \geq \cdots \quad .$$

In the case of strict inequalities, $D(f)$ consists of only one permutation. We define the m-th **greedy approximant** of f with regard to the basis Ψ corresponding to a permutation $\rho \in D(f)$ by the formula

$$G_m(f, \Psi) := G_m^X(f, \Psi) := G_m(f, \Psi, \rho) := \sum_{j=1}^m c_{k_j}(f, \Psi)\psi_{k_j}.$$

Approximation Theory X: Abstract and Classical Analysis
Charles K. Chui, Larry L. Schumaker, and Joachim Stöckler (eds.), pp. 373–402.
Copyright © 2002 by Vanderbilt University Press, Nashville, TN.
ISBN 0-8265-1415-4.

In the case $X = L_p$ we will write G_m^p. The above formula gives a simple algorithm which describes a theoretical (not computationally ready) scheme for m-term approximation of an element f. We will call this algorithm the **Thresholding Greedy Algorithm** (TGA). In order to understand the efficiency of this algorithm, we compare its accuracy with the best possible when an approximant is a linear combination of m terms from Ψ. We define the best m-term approximation with regard to Ψ as

$$\sigma_m(f, \Psi)_X := \inf_{c_k, \Lambda} \left\| f - \sum_{k \in \Lambda} c_k \psi_k \right\|_X,$$

where inf is taken over coefficients c_k and sets of indices Λ with cardinality $|\Lambda| = m$. The best we can achieve with the algorithm G_m is

$$\|f - G_m(f, \Psi, \rho)\|_X = \sigma_m(f, \Psi)_X,$$

or a little weaker

$$\|f - G_m(f, \Psi, \rho)\|_X \leq G\sigma_m(f, \Psi)_X \tag{1.1}$$

for all elements $f \in X$ with a constant $G = C(X, \Psi)$ independent of f and m. It is clear that if $X = H$ is a Hilbert space and Ψ is an orthonormal basis, we have

$$\|f - G_m(f, \Psi, \rho)\|_H = \sigma_m(f, \Psi)_H.$$

In Section 2 we consider the case $X = L_p(T^d)$, $1 \leq p \leq \infty$ and $\Psi = \mathcal{T} := \{e^{i(k,x)}\}_{k \in \mathbb{Z}^d}$. For $f \in L_p(T)$ we write

$$E_n(f, \mathcal{T})_p := \inf_{c_k, |k| \leq n} \left\| f(x) - \sum_{|k| \leq n} c_k e^{ikx} \right\|_p.$$

de la Vallée Poussin (1908) and S. N. Bernstein (1912) proved that

$$E_n(|\sin x|, \mathcal{T})_\infty \asymp n^{-1}.$$

R. S. Ismagilov [11] (1974) proved that

$$\sigma_n(|\sin x|, \mathcal{T})_\infty \leq C_\epsilon n^{-6/5+\epsilon}$$

with arbitrary $\epsilon > 0$. A little later V. E. Maiorov [15] (1978) proved that

$$\sigma_n(|\sin x|, \mathcal{T})_\infty \asymp n^{-3/2}. \tag{1.2}$$

These results showed an advantage of nonlinear approximation over linear approximation for typical individual functions. We now compare these methods for the simplest well known function classes, namely, for the Sobolev classes of functions of a single variable

$$W_q^r := \{f \ : \ \|f^{(r)}\|_q \le 1\}.$$

We assume in this definition that $f^{(r-1)}$ is absolutely continuous. For a function class F we define

$$E_n(F, \mathcal{T})_p := \sup_{f \in F} E_n(f, \mathcal{T})_p,$$

$$\sigma_m(F, \mathcal{T})_p := \sup_{f \in F} \sigma_m(f, \mathcal{T})_p.$$

The following relation is well known:

$$E_n(W_q^r, \mathcal{T})_p \asymp n^{-r+(1/q-1/p)_+}, \quad 1 \le q, p \le \infty, \tag{1.3}$$

with $r > 1$ for $q = 1$, $p = \infty$.

We proved in [8] that

$$\sigma_m(W_q^r, \mathcal{T})_p \asymp m^{-r+(1/q-\max(1/2,1/p))_+}, \quad 1 \le q, p \le \infty, \tag{1.4}$$

with $r > 1$ for $q = 1$, $p = \infty$.

We note that (1.4) is better than (1.3) in the case $p > \max(2, q)$. Thus the upper estimates in (1.4) are nontrivial (do not follow from (1.3)) only in this case ($p > \max(2, q)$). The most difficult case $p = \infty$ was studied in [8] using Gluskin's theorem for finite dimensional Banach spaces. The case $2 < p < \infty$ was studied in [16] using a probabilistic approach. The most difficult case of the lower estimate, $p = 1$, $q = \infty$, was studied in [8] using the volume estimates of finite dimensional bodies. For the further development of this method, see [21].

Comparing (1.3) and (1.4), we conclude that nonlinear approximation is better than linear approximation in the case $p > \max(2, q)$. Now, when we know that efficiency of the m-term best approximation is good, the following important problem arises. Construct an algorithm which realizes good m-term approximation. In Section 2 we concentrate on efficiency of the Thresholding Greedy Algorithm. We prove the following theorem [22].

Theorem 2.1. *For each $f \in L_p(T^d)$ we have*

$$\|f - G_m(f, \mathcal{T}^d)\|_p \le (1 + 3m^{h(p)})\sigma_m(f, \mathcal{T}^d)_p, \quad 1 \le p \le \infty,$$

where $h(p) := |1/2 - 1/p|$.

In Section 3 we discuss another important class of bases – wavelet type bases. We discuss in detail the simplest representative of wavelet bases, the Haar basis. Let $\mathcal{H} := \{H_k\}_{k=1}^{\infty}$ be the Haar basis on $[0,1)$ normalized in $L_2(0,1)$: $H_1 = 1$ on $[0,1)$ and for $k = 2^n + l$, $n = 0,1,\ldots$, $l = 1,2,\ldots,2^n$

$$H_k = \begin{cases} 2^{n/2}, & x \in [(2l-2)2^{-n-1},(2l-1)2^{-n-1}) \\ -2^{n/2}, & x \in [(2l-1)2^{-n-1},2l2^{-n-1}) \\ 0, & \text{otherwise.} \end{cases}$$

We denote by $\mathcal{H}_p := \{H_{k,p}\}_{k=1}^{\infty}$ the Haar basis $\mathcal{H}$ renormalized in $L_p(0,1)$. We will use the following definition of the L_p-equivalence of bases. We say that $\Psi = \{\psi_k\}_{k=1}^{\infty}$ is L_p-equivalent to $\Phi = \{\phi_k\}_{k=1}^{\infty}$ if for any finite set Λ and any coefficients c_k, $k \in \Lambda$, we have

$$C_1(p,\Psi,\Phi)\left\|\sum_{k\in\Lambda} c_k\phi_k\right\|_p \leq \left\|\sum_{k\in\Lambda} c_k\psi_k\right\|_p \leq C_2(p,\Psi,\Phi)\left\|\sum_{k\in\Lambda} c_k\phi_k\right\|_p \quad (1.5)$$

with two positive constants $C_1(p,\Psi,\Phi)$, $C_2(p,\Psi,\Phi)$ which may depend on p, Ψ, and Φ. For sufficient conditions on Ψ to be L_p-equivalent to $\mathcal{H}$, see [10] and [6]. We prove the following theorem in Section 3.

Theorem 3.1. *Fix $1 < p < \infty$, and let Ψ be a basis L_p-equivalent to the Haar basis $\mathcal{H}$. Then for any $f \in L_p(0,1)$ and any $\rho \in D(f)$, we have*

$$\|f - G_m(f,\Psi,\rho)\|_p \leq C(p,\Psi)\sigma_m(f,\Psi)_p \quad (1.6)$$

with a constant $C(p,\Psi)$ independent of f, ρ, and m.

In Section 4 we consider the general setting of greedy approximation in Banach spaces. We will concentrate on studying bases which satisfy (1.6) for all individual functions.

Definition 1.1. *We call a basis Ψ a greedy basis if for every $f \in X$ there exists a permutation $\rho \in D(f)$ such that*

$$\|f - G_m(f,\Psi,\rho)\|_X \leq G\sigma_m(f,\Psi)_X \quad (1.7)$$

with a constant independent of f, m.

The following proposition has been proved in [13].

Proposition 1.1. *If Ψ is a greedy basis, then (1.7) holds for any permutation $\rho \in D(f)$.*

Theorem 3.1 shows that each basis Ψ which is L_p-equivalent to the univariate Haar basis $\mathcal{H}$ is a greedy basis for $L_p(0,1)$, $1 < p < \infty$. We note that in the case of Hilbert space, each orthonormal basis is a greedy basis with a constant $G = 1$ (see (1.7)). We now give definitions of unconditional and democratic bases.

Definition 1.2. *A basis* $\Psi = \{\psi_k\}_{k=1}^{\infty}$ *of a Banach space* X *is said to be* unconditional *if for every choice of signs* $\theta = \{\theta_k\}_{k=1}^{\infty}$, $\theta_k = 1$ *or* -1, $k = 1, 2, \ldots$, *the linear operator* M_θ *defined by*

$$M_\theta\Big(\sum_{k=1}^{\infty} a_k\psi_k\Big) = \sum_{k=1}^{\infty} a_k\theta_k\psi_k$$

is a bounded operator from X *into* X.

Definition 1.3. *We say that a normalized basis* $\Psi = \{\psi_k\}_{k=1}^{\infty}$ *is a* democratic basis *for* X *if there exists a constant* $D := D(X, \Psi)$ *such that for any two finite sets of indices* P *and* Q *with the same cardinality* $|P| = |Q|$, *we have*

$$\Big\|\sum_{k \in P} \psi_k\Big\| \le D\Big\|\sum_{k \in Q} \psi_k\Big\|.$$

We proved in [13] the following theorem.

Theorem 4.1. *A normalized basis is greedy if and only if it is unconditional and democratic.*

Section 5 contains a discussion (see [13]) of notions close to the notion of greedy basis. In Section 6 we give some results on direct and inverse theorems for m-term approximation with regard to bases. The technique developed in Sections 3 and 4 provides a simple and straightforward way to get the equivalence relation between appropriate Lorenz spaces norms of the sequences of coefficients and best m-term approximations. We also discuss an interesting generalization of m-term approximation (restricted approximation) from [2].

§2. The Trigonometric System

In this section we prove Theorem 2.1 of the introduction.

Theorem 2.1. *For each* $f \in L_p(T^d)$,

$$\|f - G_m(f, T^d)\|_p \le (1 + 3m^{h(p)})\sigma_m(f, T^d)_p, \quad 1 \le p \le \infty,$$

where $h(p) := |1/2 - 1/p|$.

Proof: We treat separately the two cases $1 \le p \le 2$ and $2 \le p \le \infty$. Before splitting into these two cases, we prove an auxiliary statement for $1 \le p \le \infty$, using the notation

$$\hat{f}(k) := (2\pi)^{-d}\int_{T^d} f(x)e^{-i(k,x)}dx.$$

Lemma 2.1. Let $\Lambda \subset \mathbb{Z}^d$ be a finite subset with cardinality $|\Lambda| = m$. Then for the operator S_Λ defined on $L_1(T^d)$ by

$$S_\Lambda(f) := \sum_{k \in \Lambda} \hat{f}(k) e^{i(k,x)},$$

we have for all $1 \leq p \leq \infty$,

$$\|S_\Lambda(f)\|_p \leq m^{h(p)} \|f\|_p. \tag{2.1}$$

Proof: For a given linear operator A, denote by $\|A\|_{a \to b}$ the norm of this operator as an operator from $L_a(T^d)$ to $L_b(T^d)$. Then it is obvious that

$$\|S_\Lambda\|_{2 \to 2} = 1. \tag{2.2}$$

Consider

$$\mathcal{D}_\Lambda(x) := \sum_{k \in \Lambda} e^{i(k,x)}. \tag{2.3}$$

Then

$$S_\Lambda(f) = f * \mathcal{D}_\Lambda := (2\pi)^{-d} \int_{T^d} f(x-y)\mathcal{D}_\Lambda(y)dy,$$

and for $p = 1$, or ∞ we have

$$\|S_\Lambda\|_{p \to p} \leq \|\mathcal{D}_\Lambda\|_1 \leq \|\mathcal{D}_\Lambda\|_2 = m^{1/2}. \tag{2.4}$$

The relations (2.2) and (2.4) and the Riesz-Thorin theorem imply (2.1).
$\square$

We return now to the proof of Theorem 2.1.

Case 1: $2 \leq p \leq \infty$.

Take any function $f \in L_p(T^d)$. Let t_m be a trigonometric polynomial which realizes the best m-term approximation to f in $L_p(T^d)$. For the existence of t_m, see [1] and also [22]. Denote by Λ the set of frequencies of t_m, i.e., $\Lambda := \{k : \hat{t}_m(k) \neq 0\}$. Then $|\Lambda| \leq m$. Denote by Λ' the set of frequencies of $G_m(f) := G_m(f, T^d)$. Then $|\Lambda'| = m$. We use the representation

$$f - G_m(f) = f - S_{\Lambda'}(f) = f - S_\Lambda(f) + S_\Lambda(f) - S_{\Lambda'}(f).$$

From this representation we derive

$$\|f - G_m(f)\|_p \leq \|f - S_\Lambda(f)\|_p + \|S_\Lambda(f) - S_{\Lambda'}(f)\|_p. \tag{2.5}$$

We use Lemma 2.1 to estimate the first term in the right hand side of (2.5):

$$\|f - S_\Lambda(f)\|_p = \|f - t_m - S_\Lambda(f - t_m)\|_p \leq (1 + m^{h(p)})\sigma_m(f, T^d)_p. \tag{2.6}$$

In estimating the second term in (2.5), we use the well-known inequality $\|f\|_2 \leq \|f\|_p$ for $2 \leq p \leq \infty$ and the following lemma.

Lemma 2.2. Let $\Lambda \subset \mathbb{Z}^d$ be a finite subset with cardinality $|\Lambda| = n$. Then, for $2 \leq p \leq \infty$, we have

$$\|S_\Lambda(f)\|_p \leq n^{h(p)} \|f\|_2. \tag{2.7}$$

Proof: For $p = \infty$ we have

$$\|S_\Lambda(f)\|_\infty \leq \sum_{k \in \Lambda} |\hat{f}(k)| \leq n^{1/2} \Big(\sum_{k \in \Lambda} |\hat{f}(k)|^2 \Big)^{1/2} \leq n^{1/2} \|f\|_2. \tag{2.8}$$

For $2 < p < \infty$ we use (2.8) and the following well-known inequality

$$\|g\|_p \leq \|g\|_2^{2/p} \|g\|_\infty^{1-2/p}. \quad \square$$

We continue now estimating $\|S_\Lambda(f) - S_{\Lambda'}(f)\|_p$. Using Lemma 2.2 we get

$$\begin{aligned}
\|S_\Lambda(f) - S_{\Lambda'}(f)\|_p &= \|S_{\Lambda \setminus \Lambda'}(f) - S_{\Lambda' \setminus \Lambda}(f)\|_p \\
&\leq \|S_{\Lambda \setminus \Lambda'}(f)\|_p + \|S_{\Lambda' \setminus \Lambda}(f)\|_p \\
&\leq m^{h(p)} (\|S_{\Lambda \setminus \Lambda'}(f)\|_2 + \|S_{\Lambda' \setminus \Lambda}(f)\|_2).
\end{aligned} \tag{2.9}$$

The definition of Λ' and the relations $|\Lambda'| = m$, $|\Lambda| \leq m$ imply

$$\|S_{\Lambda \setminus \Lambda'}(f)\|_2 \leq \|S_{\Lambda' \setminus \Lambda}(f)\|_2. \tag{2.10}$$

Finally, we have

$$\|S_{\Lambda' \setminus \Lambda}(f)\|_2 \leq \|f - S_\Lambda(f)\|_2 \leq \|f - t_m\|_2 \leq \|f - t_m\|_p = \sigma_m(f, \mathcal{T}^d)_p. \tag{2.11}$$

Combining the relations (2.9)–(2.11) we get

$$\|S_\Lambda(f) - S_{\Lambda'}(f)\|_p \leq 2m^{h(p)} \sigma_m(f, \mathcal{T}^d)_p. \tag{2.12}$$

The relations (2.5), (2.6) and (2.12) result in

$$\|f - G_m(f)\|_p \leq (1 + 3m^{h(p)}) \sigma_m(f, \mathcal{T}^d)_p.$$

This completes the proof of Theorem 2.1 in the case $2 \leq p \leq \infty$.

Case 2: $1 \leq p \leq 2$.

We keep the notation of Case 1. We start again with the inequality (2.5). Now, the inequality (2.6) holds also for $1 \leq p \leq 2$ because it is based on Lemma 2.1 which covers the whole range $1 \leq p \leq \infty$ of the parameter p. Thus, it remains to estimate $\|S_\Lambda(f) - S_{\Lambda'}(f)\|_p$. Using the inequality $\|f\|_p \leq \|f\|_2$ we get

$$
\begin{aligned}
\|S_\Lambda(f) - S_{\Lambda'}(f)\|_p &= \|S_{\Lambda \setminus \Lambda'}(f) - S_{\Lambda' \setminus \Lambda}(f)\|_p \\
&\leq \|S_{\Lambda \setminus \Lambda'}(f)\|_p + \|S_{\Lambda' \setminus \Lambda}(f)\|_p \qquad (2.13) \\
&\leq \|S_{\Lambda \setminus \Lambda'}(f)\|_2 + \|S_{\Lambda' \setminus \Lambda}(f)\|_2.
\end{aligned}
$$

In order to estimate $\|S_{\Lambda' \setminus \Lambda}(f)\|_2$, we use the part of the Hausdorff-Young theorem (see [26, Ch.12,§2]) which states that

$$
\|(\hat{f}(k))_{k \in \mathbb{Z}^d}\|_{l_{p'}} \leq \|f\|_p, \quad 1 \leq p \leq 2, \quad p' := \frac{p}{p-1}.
$$

We have

$$
\begin{aligned}
\|S_{\Lambda' \setminus \Lambda}(f)\|_2 &= \|(\hat{f}(k))_{k \in \Lambda' \setminus \Lambda}\|_{l_2} \\
&\leq |\Lambda' \setminus \Lambda|^{1/p - 1/2} \|(\hat{f}(k))_{k \in \Lambda' \setminus \Lambda}\|_{l_{p'}} \\
&\leq m^{h(p)} \|(\hat{f}(k) - \hat{t}_m(k))_{k \in \mathbb{Z}^d}\|_{l_{p'}} \qquad (2.14) \\
&\leq m^{h(p)} \|f - t_m\|_p = m^{h(p)} \sigma_m(f, \mathcal{T}^d)_p.
\end{aligned}
$$

Gathering (2.5), (2.6), (2.10), (2.13) and (2.14), we get

$$
\|f - G_m(f)\|_p \leq (1 + 3 m^{h(p)}) \sigma_m(f, \mathcal{T}^d)_p,
$$

which completes the proof of Theorem 2.1. $\square$

Remark 2.1. *Lemma 2.1 implies for all* $1 \leq p \leq \infty$

$$
\|G_m(f)\|_p \leq m^{h(p)} \|f\|_p. \qquad (2.15)
$$

We will give without proof some remarks from [22].

Remark 2.2. *There is a positive absolute constant C such that for each m and $1 \leq p \leq \infty$, there exists a function $f \neq 0$ with the property*

$$
\|G_m(f)\|_p \geq C m^{h(p)} \|f\|_p.
$$

Remark 2.3. *The trivial inequality* $\sigma_m(f, T^d)_p \leq \|f\|_p$ *and Remark 2.2 show that the factor* $m^{h(p)}$ *in Theorem 2.1 is sharp in the sense of order.*

Remark 2.4. *Using Remark 2.2 it is easy to construct for each* $p \neq 2$ *a function* $f \in L_p(T)$ *such that the sequence* $\{\|G_m(f)\|_p\}_{m=1}^{\infty}$ *is not bounded.*

We now make some remarks about possible generalizations of Theorem 2.1. Reviewing the proof of Theorem 2.1, one verifies that all arguments hold for any orthonormal system $\{\phi_j\}_{j=1}^{\infty}$ of uniformly bounded functions $\|\phi_j\|_{\infty} \leq M, j = 1, 2, \ldots$. The only difference is that instead of the Hausdorff–Young theorem, we shall use the F. Riesz theorem and the constants in Lemmas 2.1 and 2.2 will depend on M. Let us formulate the corresponding analogue of Theorem 2.1. Let $\Phi := \{\phi_j\}_{j=1}^{\infty}$ be an orthonormal system in $L_2(T^d)$ such that $\|\phi_j\|_{\infty} \leq M, j = 1, 2, \ldots$.

Theorem 2.2. *For any orthonormal system* $\Phi = \{\phi_j\}_{j=1}^{\infty}$ *of uniformly bounded functions* $\|\phi_j\|_{\infty} \leq M$, *there exists a constant* $C(M)$ *such that*

$$\|f - G_m(f, \Phi)\|_p \leq C(M)m^{h(p)}\sigma_m(f, \Phi)_p, \quad 1 \leq p \leq \infty,$$

where $h(p) := |1/2 - 1/p|$.

§3. The Wavelet Bases

In this section it will be convenient for us to index elements of bases by dyadic intervals: $\psi_1 =: \psi_{[0,1]}$ and

$$\psi_{2^n + l} =: \psi_I, \quad I = [(l-1)2^{-n}, l2^{-n}).$$

We begin by proving Theorem 3.1 of the introduction.

Theorem 3.1. *Let* $1 < p < \infty$ *and a basis* $\Psi := \{\psi_I\}_I$ *be* L_p*-equivalent to* $\mathcal{H}$. *Then for any* $f \in L_p$ *we have*

$$\|f - G_m^p(f, \Psi)\|_p \leq C(p)\sigma_m(f, \Psi)_p.$$

Proof: Let us take a parameter $0 < t \leq 1$ and consider the following greedy type algorithm $G^{p,t}$ with regard to the Haar system. For the Haar basis $\mathcal{H}$ we define

$$c_I(f) := \langle f, H_I \rangle = \int_0^1 f(x)H_I(x)dx.$$

Let $\Lambda_m(t)$ be any set of m dyadic intervals such that

$$\min_{I \in \Lambda_m(t)} \|c_I(f)H_I\|_p \geq t \max_{J \notin \Lambda_m(t)} \|c_J(f)H_J\|_p, \tag{3.1}$$

and define

$$G_m^{p,t}(f) := G_m^{p,t}(f, \mathcal{H}) := \sum_{I \in \Lambda_m(t)} c_I(f)H_I. \tag{3.2}$$

For a given function $f \in L_p$ of the form

$$f = \sum_I c_I(f, \Psi)\psi_I,$$

we define

$$g(f) := \sum_I c_I(f, \Psi)H_I. \tag{3.3}$$

It is clear that $g(f) \in L_p$ and

$$\sigma_m(g(f), \mathcal{H})_p \leq C_1(p)^{-1}\sigma_m(f, \Psi)_p, \tag{3.4}$$

here and later on we use brief notation $C_i(p) := C_i(p, \Psi, \mathcal{H})$, $i = 1, 2$, for the constants from (1.5). Let

$$G_m^p(f, \Psi) = \sum_{I \in \Lambda_m} c_I(f, \Psi)\psi_I.$$

For any two intervals $I \in \Lambda_m$, $J \notin \Lambda_m$, by the definition of Λ_m we have

$$\|c_I(f, \Psi)\psi_I\|_p \geq \|c_J(f, \Psi)\psi_J\|.$$

Using (1.5) we get

$$\begin{aligned}
\|c_I(g(f))H_I\|_p &= \|c_I(f, \Psi)H_I\|_p \\
&\geq C_2(p)^{-1}\|c_I(f, \Psi)\psi_I\|_p \\
&\geq C_2(p)^{-1}\|c_J(f, \Psi)\psi_J\|_p \\
&\geq C_1(p)C_2(p)^{-1}\|c_J(g(f))H_J\|_p.
\end{aligned} \tag{3.5}$$

This inequality implies that for any m, we can find a set $\Lambda_m(t)$, where $t = C_1(p)C_2(p)^{-1}$, such that $\Lambda_m(t) = \Lambda_m$ and, therefore,

$$\|f - G_m^p(f, \Psi)\|_p \leq C_2(p)\|g(f) - G_m^{p,t}(g(f))\|_p. \tag{3.6}$$

The relations (3.4) and (3.6) show that Theorem 3.1 will follow from Theorem 3.2. $\square$

Theorem 3.2. *Let $1 < p < \infty$ and $0 < t \leq 1$. Then for any $g \in L_p$, we have*

$$\|g - G_m^{p,t}(g, \mathcal{H})\|_p \leq C(p,t)\sigma_m(g, \mathcal{H})_p.$$

Proof: The Littlewood-Paley Theorem for the Haar system (see for instance [12]) gives for $1 < p < \infty$

$$C_3(p)\left\|\left(\sum_I |c_I(g)H_I|^2\right)^{1/2}\right\|_p \leq \|g\|_p \leq C_4(p)\left\|\left(\sum_I |c_I(g)H_I|^2\right)^{1/2}\right\|_p.$$
(3.7)

We formulate first two simple corollaries from (3.7) :

$$\|g\|_p \leq C_5(p)\left(\sum_I \|c_I(g)H_I\|_p^p\right)^{1/p}, \qquad 1 < p \leq 2,$$
(3.8)

$$\|g\|_p \leq C_6(p)\left(\sum_I \|c_I(g)H_I\|_p^2\right)^{1/2}, \qquad 2 \leq p < \infty.$$
(3.9)

Analogs of these inequalities for the trigonometric system are known (see, for instance, [24, p.37]). The same proof gives (3.8) and (3.9).

The dual inequalities to (3.8) and (3.9) are

$$\|g\|_p \geq C_7(p)\left(\sum_I \|c_I(g)H_I\|_p^2\right)^{1/2}, \qquad 1 < p \leq 2,$$
(3.10)

$$\|g\|_p \geq C_8(p)\left(\sum_I \|c_I(g)H_I\|_p^p\right)^{1/p}, \qquad 2 \leq p < \infty.$$
(3.11)

Let T_m be an m-term Haar polynomial of best m-term approximation to g in L_p (for existence see [1] and also [9]):

$$T_m = \sum_{I \in \Lambda} a_I H_I, \qquad |\Lambda| = m.$$

For any finite set Q of dyadic intervals, we denote by S_Q the projector

$$S_Q(f) := \sum_{I \in Q} c_I(f)H_I.$$

From (3.7) we get

$$\begin{aligned}
\|g - S_\Lambda(g)\|_p &= \|g - T_m - S_\Lambda(g - T_m)\|_p \\
&\leq \|Id - S_\Lambda\|_{p \to p}\sigma_m(g, \mathcal{H})_p \\
&\leq C_4(p)C_3(p)^{-1}\sigma_m(g, \mathcal{H})_p,
\end{aligned}$$
(3.12)

where Id denotes the identity operator. Further, we have

$$G_m^{p,t}(g) = S_{\Lambda_m(t)}(g),$$

and

$$\|g - G_m^{p,t}(g)\|_p \le \|g - S_\Lambda(g)\|_p + \|S_\Lambda(g) - S_{\Lambda_m(t)}(g)\|_p. \qquad (3.13)$$

The first term in the right side of (3.13) has been estimated in (3.12). We estimate now the second term. We represent it in the form

$$S_\Lambda(g) - S_{\Lambda_m(t)}(g) = S_{\Lambda \setminus \Lambda_m(t)}(g) - S_{\Lambda_m(t) \setminus \Lambda}(g),$$

and remark that similarly to (3.12) we get

$$\|S_{\Lambda_m(t) \setminus \Lambda}(g)\|_p \le C_9(p)\sigma_m(g, \mathcal{H})_p. \qquad (3.14)$$

The key point of the proof of Theorem 3.2 is the estimate

$$\|S_{\Lambda \setminus \Lambda_m(t)}(g)\|_p \le C(p,t)\|S_{\Lambda_m(t) \setminus \Lambda}(g)\|_p \qquad (3.15)$$

which will be derived from the following two lemmas.

Lemma 3.1. *Consider*

$$f = \sum_{I \in Q} c_I H_I, \qquad |Q| = N.$$

Let $1 \le p < \infty$. *Assume*

$$\|c_I H_I\|_p \le 1, \qquad I \in Q. \qquad (3.16)$$

Then

$$\|f\|_p \le C_{10}(p)N^{1/p}.$$

Lemma 3.2. *Consider*

$$f = \sum_{I \in Q} c_I H_I, \qquad |Q| = N.$$

Let $1 < p \le \infty$. *Assume*

$$\|c_I H_I\|_p \ge 1, \qquad I \in Q.$$

Then

$$\|f\|_p \ge C_{11}(p)N^{1/p}.$$

Proof: We begin with the proof of Lemma 3.1. We note that in the case $1 < p \le 2$, the statement of Lemma 3.1 follows from (3.8). We will give a proof of this lemma for all $1 \le p < \infty$. We have

$$\|c_I H_I\|_p = |c_I| |I|^{1/p - 1/2}.$$

The assumption (3.16) implies

$$|c_I| \le |I|^{1/2 - 1/p}.$$

Next, we have

$$\|f\|_p \le \left\| \sum_{I \in Q} |c_I H_I| \right\|_p \le \left\| \sum_{I \in Q} |I|^{-1/p} \chi_I(x) \right\|_p, \tag{3.17}$$

where $\chi_I(x)$ is a characteristic function of the interval I

$$\chi_I(x) = \begin{cases} 1, & x \in I \\ 0, & x \notin I. \end{cases}$$

In order to proceed further we need a lemma.

Lemma 3.3. *Let* $n_1 < n_2 < \cdots < n_s$ *be integers, and let* $E_j \subset [0,1]$ *be measurable sets,* $j = 1, \ldots, s$. *Then for any* $0 < q < \infty$, *we have*

$$\int_0^1 \left(\sum_{j=1}^s 2^{n_j/q} \chi_{E_j}(x) \right)^q dx \le C_{12}(q) \sum_{j=1}^s 2^{n_j} |E_j|.$$

Proof: Denote

$$F(x) := \sum_{j=1}^s 2^{n_j/q} \chi_{E_j}(x)$$

and estimate it on the sets

$$E_l^- := E_l \setminus \cup_{k=l+1}^s E_k, \quad l = 1, \ldots, s-1; \quad E_s^- := E_s.$$

We have for $x \in E_l^-$

$$F(x) \le \sum_{j=1}^l 2^{n_j/q} \le C(q) 2^{n_l/q}.$$

Therefore,

$$\int_0^1 F(x)^q dx \le C(q)^q \sum_{l=1}^s 2^{n_l} |E_l^-| \le C(q)^q \sum_{l=1}^s 2^{n_l} |E_l|,$$

which proves the lemma. $\square$

We return to the proof of Lemma 3.1. Denote by $n_1 < n_2 < \cdots < n_s$ all integers such that there is $I \in Q$ with $|I| = 2^{-n_j}$. Introduce the sets

$$E_j := \cup_{I \in Q; |I| = 2^{-n_j}} I.$$

Then the number N of elements in Q can be written in the form

$$N = \sum_{j=1}^{s} |E_j| 2^{n_j}. \tag{3.18}$$

Using these notations, the right hand side of (3.17) can be rewritten as

$$Y := \left(\int_0^1 \left(\sum_{j=1}^{s} 2^{n_j/p} \chi_{E_j}(x) \right)^p dx \right)^{1/p}.$$

Applying Lemma 3.3 with $q = p$, we get

$$\|f\|_p \leq Y \leq C_{13}(p) \left(\sum_{j=1}^{s} |E_j| 2^{n_j} \right)^{1/p} = C_{13}(p) N^{1/p}.$$

In the last step we used (3.18). Lemma 3.1 is now proved. $\square$

Proof: We now derive Lemma 3.2 from Lemma 3.1. Define

$$u := \sum_{I \in Q} \bar{c}_I |c_I|^{-1} |I|^{1/p-1/2} H_I,$$

where the bar means complex conjugate. Then for $p' = \frac{p}{p-1}$, we have

$$\|\bar{c}_I |c_I|^{-1} |I|^{1/p-1/2} H_I\|_{p'} = 1,$$

and by Lemma 3.1

$$\|u\|_{p'} \leq C_{10}(p) N^{1/p'}. \tag{3.19}$$

Consider $\langle f, u \rangle$. We have on one hand

$$\langle f, u \rangle = \sum_{I \in Q} |c_I| |I|^{1/p-1/2} = \sum_{I \in Q} \|c_I H_I\|_p \geq N, \tag{3.20}$$

and on the other hand

$$\langle f, u \rangle \leq \|f\|_p \|u\|_{p'}. \tag{3.21}$$

Combining (3.19) – (3.21) we get the statement of Lemma 3.2. $\square$

We now complete the proof of Theorem 3.2. It remained to prove the inequality (3.15). Let

$$A := \max_{I \in \Lambda \setminus \Lambda_m(t)} \|c_I(g) H_I\|_p,$$

and

$$B := \min_{I \in \Lambda_m(t) \setminus \Lambda} \|c_I(g) H_I\|_p.$$

Then by the definition of $\Lambda_m(t)$, we have

$$B \geq tA. \tag{3.22}$$

Using Lemma 3.1 we get

$$\|S_{\Lambda \setminus \Lambda_m(t)}(g)\|_p \leq AC_{10}(p)|\Lambda \setminus \Lambda_m(t)|^{1/p} \leq t^{-1}BC_{10}(p)|\Lambda \setminus \Lambda_m(t)|^{1/p}. \tag{3.23}$$

Using Lemma 3.2 we get

$$\|S_{\Lambda_m(t) \setminus \Lambda}(g)\|_p \geq BC_{11}(p)|\Lambda_m(t) \setminus \Lambda|^{1/p}. \tag{3.24}$$

Taking into account that $|\Lambda_m(t) \setminus \Lambda| = |\Lambda \setminus \Lambda_m(t)|$, we get the relation (3.15) from (3.23) and (3.24). This completes the proof of Theorem 3.2. $\square$

We now discuss the multivariate analog of Theorem 3.1. There are several natural generalizations of the Haar system to the d-dimensional case. We describe first the one for which the statement of Theorem 3.1 and its proof coincide with the one-dimensional version. First of all, we include in the system the constant function

$$H_{[0,1]^d}(x) = 1, \quad x \in [0,1)^d.$$

Next we define $2^d - 1$ functions with support $[0,1)^d$. Take any combination of intervals $Q_1, \ldots, Q_d$, where $Q_i = [0,1]$ or $Q_i = [0,1)$ with at least one $Q_j = [0,1)$, and define for $Q = Q_1 \times \ldots \times Q_d$, $x = (x_1, \ldots, x_d)$,

$$H_Q(x) := \prod_{i=1}^{d} H_{Q_i}(x_i).$$

We shall also denote these functions by $H_{[0,1)^d}^k(x)$, $k = 1, \ldots, 2^d - 1$. We define the basis of Haar functions with supports on dyadic cubes of the form

$$J = [(j_1 - 1)2^{-n}, j_1 2^{-n}) \times \ldots \times [(j_d - 1)2^{-n}, j_d 2^{-n}),$$
$$j_i = 1, \ldots, 2^n; \quad n = 0, 1, \ldots \quad . \tag{3.25}$$

For each dyadic cube of the form (3.25), we define $2^d - 1$ basis functions

$$H_J^k(x) := 2^{nd/2} H_{[0,1)^d}^k(2^n(x - (j_1 - 1, \ldots, j_d - 1)2^{-n})), \quad k = 1, \ldots, 2^d - 1.$$

We can also use another enumeration of these functions. Let $H_{[0,1)^d}^k(x) = H_Q(x)$ with

$$Q = Q_1 \times \ldots \times Q_d, \quad Q_i = [0,1), \quad i \in E,$$

$$Q_i = [0,1], \quad i \in \{1,d\} \setminus E, \quad E \neq \emptyset.$$

Consider a dyadic interval I of the form

$$I = I_1 \times \ldots \times I_d, \quad I_i = [(j_i - 1)2^{-n}, j_i 2^{-n}), \quad i \in E, \tag{3.26}$$

$$I_i = [(j_i - 1)2^{-n}, j_i 2^{-n}], \quad i \in \{1,d\} \setminus E, \quad E \neq \emptyset,$$

and define $H_I(x) := H_J^k(x)$. Denoting the set of dyadic intervals D as the set of all dyadic cubes of the form (3.26) amended by the cube $[0,1]^d$, and denoting by $\mathcal{H}$ the corresponding basis $\{H_I\}_{I \in D}$, we get the multivariate Haar system.

Remark 3.5. *Theorem 3.1 holds for the multivariate Haar system $\mathcal{H}$ with the constant $C(p)$ allowed to depend also on d.*

In this section we studied approximation in $L_p([0,1])$ and made a remark about approximation in $L_p([0,1]^d)$. In the same way we can treat approximation in $L_p(\mathbb{R}^d)$.

Remark 3.6. *Theorem 3.1 holds for approximation in $L_p(\mathbb{R}^d)$.*

Results on approximation of function classes using the multivariate greedy algorithm $G_m^p(\cdot, \Psi)$ can be found in [5].

Let us discuss now another multivariate Haar basis $\mathcal{H}^d := \mathcal{H} \times \cdots \times \mathcal{H}$ which is obtained from the univariate one by tensor product. It is known (see [23]) that the tensor product structure of multivariate wavelet bases makes them universal for approximation of anisotropic smoothness classes with different anisotropy. We have the following theorem in this case.

Theorem 3.3. *Let $1 < p < \infty$. Then for any $f \in L_p([0,1]^d)$ we have*

$$\|f - G_m^p(f, \mathcal{H}^d)\|_p \leq C(p,d)(\log(m+1))^{(d-1)|1/2-1/p|}\sigma_m(f, \mathcal{H}^d)_p.$$

This theorem was conjectured in [20] and was proved there in the particular case $d = 2$, $4/3 \leq p \leq 4$. In the general case, Theorem 3.3 was proved in [25].

Theorem 3.4. *For any $1 < p < \infty$, there exists $C(p,d) > 0$ such that for any m there is $f_m \in L_p$, $f_m \neq 0$, with the property*

$$\|f_m - G_m^p(f_m, \mathcal{H}^d)\|_p \geq C(p,d)(\log(m+1))^{(d-1)|1/2 - 1/p|}\sigma_m(f, \mathcal{H}^d)_p.$$

Proof: This theorem was proved by R. Hochmuth. We will give a proof of it from [20]. For a set Λ of indices, we define

$$g_{\Lambda,p} := \sum_{I \in \Lambda} |I|^{1/2 - 1/p} H_I.$$

For each $n \in \mathbb{N}$, we define two sets A and B of dyadic intervals I as follows

$$A := \{I : \quad |I| = 2^{-n}\};$$

$$B := \{I : \quad I \notin A, \quad \forall I \neq I' \quad I \cap I' = \emptyset; \quad |B| = |A|\}.$$

Let $2 \leq p < \infty$ be given. Denote $m = |A|$ and consider

$$f = g_{A,p} + 2g_{B,p}.$$

Then we have on one hand

$$G_m^p(f, \mathcal{H}^d) = 2g_{B,p}$$

and

$$\|f - G_m^p(f, \mathcal{H}^d)\|_p = \|g_{A,p}\|_p \gg m^{1/p}(\log m)^{(1/2 - 1/p)(d-1)}. \tag{3.27}$$

On the other hand, we have

$$\sigma_m(f, \mathcal{H}^d)_p \leq \|2g_{B,p}\|_p \ll m^{1/p}. \tag{3.28}$$

The relations (3.27) and (3.28) imply the required lower estimate in the case $2 \leq p < \infty$. The remaining case $1 < p \leq 2$ can be handled in the same way considering the function $f = 2g_{A,p} + g_{B,p}$. $\square$

§4. Greedy Bases

We begin with proving Theorem 4.1 of the introduction.

Theorem 4.1. *A basis is greedy if and only if it is unconditional and democratic.*

Proof: In fact we prove a little more, namely, that (1.7) holds for any $\rho \in D(f)$. This combined with Theorem 4.1 implies Proposition 1.1. The following theorem is a well known fact about unconditional bases (see [14], p.19).

Theorem 4.2. *Let Ψ be an unconditional basis for X. Then for every choice of bounded scalars $\{\lambda_k\}_{k=1}^{\infty}$,*

$$\left\| \sum_{k=1}^{\infty} \lambda_k a_k \psi_k \right\| \le 2K \sup_k |\lambda_k| \left\| \sum_{k=1}^{\infty} a_k \psi_k \right\|$$

(in the case of real Banach space X we can take K instead of $2K$) where $K := \sup_{\theta} \|M_{\theta}\|$ is the unconditional constant of Ψ.

Take any $\epsilon > 0$, and find

$$p_m(f) := \sum_{k \in P} b_k \psi_k$$

such that $|P| = m$ and

$$\|f - p_m(f)\| \le \sigma_m(f, \Psi) + \epsilon. \tag{4.1}$$

For any finite set of indices Λ, let S_{Λ} be the projector

$$S_{\Lambda}(f) := \sum_{k \in \Lambda} c_k(f) \psi_k.$$

The assumption that Ψ is an unconditional basis implies that

$$\|f - S_P(f)\| \le K(\sigma_m(f, \Psi) + \epsilon). \tag{4.2}$$

Let $\rho \in D(f)$ and

$$G_m(f, \Psi, \rho) = \sum_{k \in Q} c_k(f) \psi_k = S_Q(f).$$

Then

$$\|f - G_m(f, \Psi, \rho)\| \le \|f - S_P(f)\| + \|S_P(f) - S_Q(f)\|. \tag{4.3}$$

The first term in the right side of (4.3) has been estimated in (4.2). We estimate now the second term. We have

$$S_P(f) - S_Q(f) = S_{P \setminus Q}(f) - S_{Q \setminus P}(f). \tag{4.4}$$

Similarly to (4.2) we have

$$\|S_{Q \setminus P}(f)\| \le K(\sigma_m(f, \Psi) + \epsilon). \tag{4.5}$$

We now estimate $\|S_{P\backslash Q}(f)\|$. By the definition of the greedy algorithm G_m, we have

$$A := \max_{k\in P\backslash Q} |c_k(f)| \le \min_{k\in Q\backslash P} |c_k(f)| =: B. \tag{4.6}$$

By Theorem 4.2 we have

$$\|S_{P\backslash Q}(f)\| \le 2KA\Big\|\sum_{k\in P\backslash Q} \psi_k\Big\| \tag{4.7}$$

and

$$\|S_{Q\backslash P}(f)\| \ge (2K)^{-1}B\Big\|\sum_{k\in Q\backslash P} \psi_k\Big\|. \tag{4.8}$$

By the assumption that Ψ is democratic, we get

$$\Big\|\sum_{k\in P\backslash Q} \psi_k\Big\| \le D\Big\|\sum_{k\in Q\backslash P} \psi_k\Big\|. \tag{4.9}$$

Combining (4.7)–(4.9), we obtain

$$\|S_{P\backslash Q}(f)\| \le 4DK^2\|S_{Q\backslash P}(f)\|. \tag{4.10}$$

Using (4.5) and (4.10), we derive from (4.4) and (4.3) that

$$\|f - G_m(f,\Psi,\rho)\| \le 4DK^3(\sigma_m(f,\Psi) + \epsilon)$$

and, therefore, the inequality

$$\|f - G_m(f,\Psi,\rho)\| \le 4DK^3\sigma_m(f,\Psi)$$

holds.

We prove now the inverse part of the theorem that a greedy basis is always unconditional and democratic. Assume that a basis Ψ satisfies (1.7) for all $f \in X$. We begin with the unconditionality. We shall prove that for each function $f \in X$ and any finite set Λ, we have

$$\|S_\Lambda(f)\| \le (G+1)\|f\|, \tag{4.11}$$

where G is from (1.7). It is well known that (4.11) implies that Ψ is an unconditional basis. Take a number N such that

$$N > \max_k |c_k(f)|,$$

and consider a new function

$$g := f - S_\Lambda(f) + N \sum_{k \in \Lambda} \psi_k.$$

Then we obviously have

$$\sigma_m(g) \leq \|f\|, \tag{4.12}$$

and

$$G_m(g) := G_m(g, \Psi, \rho) = N \sum_{k \in \Lambda} \psi_k. \tag{4.13}$$

Thus, by our assumption that Ψ is a greedy basis, we get

$$\|f - S_\Lambda(f)\| = \|g - G_m(g)\| \leq G\sigma_m(g) \leq G\|f\|.$$

This implies (4.11).

We proceed now to proving that Ψ is democratic. Let two finite sets P and Q, $|P| = |Q| = m$, be given. Take a third one Y such that $|Y| = m$ and $Y \cap P = \emptyset$, $Y \cap Q = \emptyset$. For a finite set Λ, denote

$$\psi_\Lambda := \sum_{k \in \Lambda} \psi_k.$$

Fix any $\epsilon > 0$ and consider the function

$$f := (1 + \epsilon)\psi_Q + \psi_Y.$$

Then

$$\sigma_m(f) \leq (1 + \epsilon)\|\psi_Q\|$$

and

$$\|f - G_m(f)\| = \|\psi_Y\|.$$

Therefore, by the assumption that Ψ is greedy we get

$$\|\psi_Y\| \leq G(1 + \epsilon)\|\psi_Q\|. \tag{4.14}$$

Similarly, we obtain

$$\|\psi_P\| \leq G(1 + \epsilon)\|\psi_Y\|. \tag{4.15}$$

Combining the above two inequalities and taking into account that ϵ is arbitrarily small, we obtain

$$\|\psi_P\| \leq G^2 \|\psi_Q\|.$$

This completes the proof of Theorem 4.1. $\quad\square$

§5. Some Examples

5.1 Unconditionality does not imply democracy

This follows from properties of the multivariate Haar system $\mathcal{H}^2 = \mathcal{H} \times \mathcal{H}$ defined as the tensor product of the univariate Haar systems $\mathcal{H}$ (see Theorem 3.4).

5.2 Democracy does not imply unconditionality

Let X be the set of all real sequences $x = (x_1, x_2, \ldots)$ such that

$$\|x\|_X = \sup_{N \in \mathbb{N}} \left| \sum_{n=1}^{N} x_n \right|$$

is finite. Clearly, X equipped with the norm $\|\cdot\|_X$ is a Banach space. Let $\psi_k \in X$, $k = 1, 2, \ldots$, be defined as

$$(\psi_k)_n = \begin{cases} 1, & n = k, \\ 0, & n \neq k. \end{cases}$$

Let X_0 be the subspace of X generated by the elements ψ_k. It is easy to see that $\{\psi_k\}$ is a democratic basis in X_0. However, it is not an unconditional basis since

$$\left\| \sum_{k=1}^{m} \psi_k \right\|_X = m,$$

but

$$\left\| \sum_{k=1}^{m} (-1)^k \psi_k \right\|_X = 1.$$

5.3 Superdemocracy does not imply unconditionality

It is clear that an unconditional and democratic basis Ψ satisfies the following inequality

$$\left\| \sum_{k \in P} \theta_k \psi_k \right\| \leq D_S \left\| \sum_{k \in Q} \epsilon_k \psi_k \right\| \tag{5.1}$$

for any two finite sets P and Q, $|P| = |Q|$, and any choices of signs $\theta_k = \pm 1$, $k \in P$, and $\epsilon_k = \pm 1$, $k \in Q$.

Definition 5.1. *We say that a basis Ψ is a* superdemocratic basis *if it satisfies (5.1).*

Theorem 4.1 implies that a greedy basis is a superdemocratic one. Now we will construct an example of superdemocratic basis which is not an unconditional basis and, therefore, by Theorem 4.1 is not a greedy basis.

Let X be the set of all real sequences $x = (x_1, x_2, \ldots) \in l_2$ such that

$$\|x\|_u = \sup_{N \in \mathbf{N}} \left| \sum_{n=1}^{N} x_n / \sqrt{n} \right|$$

is finite. Clearly, X equipped with the norm

$$\|\cdot\| = \max(\|\cdot\|_{l_2}, \|\cdot\|_u)$$

is a Banach space. Let $\psi_k \in X$, $k = 1, 2, \ldots$, be defined as

$$(\psi_k)_n = \begin{cases} 1, & n = k, \\ 0, & n \neq k. \end{cases}$$

By X_0 denote the subspace of X generated by the elements ψ_k. It is easy to see that $\Psi = \{\psi_k\}$ is a democratic basis in X_0. Moreover, it is superdemocratic: for any $k_1, \ldots, k_m$ and for any choice of signs

$$\sqrt{m} \leq \left\| \sum_{j=1}^{m} \pm \psi_{k_j} \right\| < 2\sqrt{m}. \tag{5.2}$$

Indeed, we have

$$\left\| \sum_{j=1}^{m} \pm \psi_{k_j} \right\|_{l_2} = \sqrt{m},$$

$$\left\| \sum_{j=1}^{m} \pm \psi_{k_j} \right\|_u \leq \sum_{j=1}^{m} 1/\sqrt{j} < 2\sqrt{m},$$

and (5.2) follows. However, Ψ is not an unconditional basis since for $m \geq 2$

$$\left\| \sum_{k=1}^{m} \psi_k / \sqrt{k} \right\| \geq \sum_{k=1}^{m} 1/k \asymp \log m,$$

but

$$\left\| \sum_{k=1}^{m} (-1)^k \psi_k / \sqrt{k} \right\| \asymp \sqrt{\log m}.$$

5.4 A quasigreedy basis is not necessarily an unconditional basis

It follows from the definition of greedy basis (see (1.7)) that the inequality

$$\|G_m(f, \Psi, \rho)\| \leq (G+1)\|f\| \tag{5.3}$$

holds for all m and all $f \in X$, with some $\rho \in D(f)$.

Definition 5.2. *We say that a basis Ψ is* quasigreedy *if there exists a constant C_Q such that for any $f \in X$ and any finite set of indices Λ, having the property*

$$\min_{k \in \Lambda} \|c_k(f)\psi_k\| \geq \max_{k \notin \Lambda} \|c_k(f)\psi_k\|, \tag{5.4}$$

we have

$$\|S_\Lambda(f, \Psi)\| = \left\| \sum_{k \in \Lambda} c_k(f)\psi_k \right\| \leq C_Q \|f\|. \tag{5.5}$$

It is clear that for elements f with the unique decreasing rearrangement of coefficients ($\#D(f) = 1$), the inequalities (5.3) and (5.5) are equivalent. By slightly modifying the coefficients and using the continuity argument, we get that (5.3) and (5.5) are equivalent.

We shall prove now that the basis Ψ constructed in Subsection 5.3 is a quasigreedy basis. Combining this with the result from 5.3 that Ψ is not unconditional, we get the required statement of this subsection.

Assume $\|f\| = 1$. Then by definition of $\|\cdot\|$ we have

$$\sum_{k=1}^{\infty} |c_k(f)|^2 \leq 1, \tag{5.6}$$

and for any M

$$\left| \sum_{k=1}^{M} c_k(f)k^{-1/2} \right| \leq 1. \tag{5.7}$$

It is clear that for any Λ we have

$$\|S_\Lambda(f, \Psi)\|_{l_2} \leq \|f\|_{l_2} \leq 1. \tag{5.8}$$

We now estimate $\|S_\Lambda(f, \Psi)\|_u$. Let Λ satisfy (5.4), and let

$$\alpha := \min_{k \in \Lambda} |c_k(f)|.$$

If $\alpha = 0$ we get $S_\Lambda(f, \Psi) = f$ and (5.5) holds. Let $\alpha > 0$. Denote for any N

$$\Lambda^+(N) := \{k \in \Lambda : k > N\}, \qquad \Lambda^-(N) := \{k \in \Lambda : k \leq N\}.$$

We have for any N

$$\sum_{k\in\Lambda^+(N)} |c_k(f)|k^{-1/2} \le \Big(\sum_{k\in\Lambda^+(N)} |c_k(f)|^{3/2} \Big)^{2/3} \Big(\sum_{k>N} k^{-3/2} \Big)^{1/3} \ll$$

$$\ll N^{-1/6} \Big(\sum_{k\in\Lambda^+(N)} |c_k(f)|^{3/2}(|c_k(f)|/\alpha)^{1/2} \Big)^{2/3} \ll (\alpha^2 N)^{-1/6}. \qquad (5.9)$$

Choose $N_\alpha := [\alpha^{-2}] + 1$. Then for any $M \le N_\alpha$ we have by (5.7) that

$$\Big| \sum_{k\in\Lambda^-(M)} c_k(f)k^{-1/2} \Big| \le \Big| \sum_{k=1}^M c_k(f)k^{-1/2} \Big| + \Big| \sum_{k\notin\Lambda^-(M),k\le M} c_k(f)k^{-1/2} \Big| \le$$

$$\le 1 + \alpha \sum_{k=1}^M k^{-1/2} \le 1 + 2\alpha M^{1/2} \ll 1. \qquad (5.10)$$

For $M > N_\alpha$ we get using (5.9) and (5.10)

$$\Big| \sum_{k\in\Lambda^-(M)} c_k(f)k^{-1/2} \Big| \le \Big| \sum_{k\in\Lambda^-(N_\alpha)} c_k(f)k^{-1/2} \Big| + \sum_{k\in\Lambda^+(N_\alpha)} |c_k(f)|k^{-1/2} \ll 1.$$

Thus

$$\|S_\Lambda(f,\Psi)\|_u \le C,$$

which completes the proof. $\square$

The above example and Theorem 4.1 show that a quasigreedy basis is not necessarily a greedy basis. Further results on quasigreedy bases can be found in [25].

§6. Further Results

6.1 Greedy bases. Direct and inverse theorems

Theorem 3.1 points out the importance of bases which are L_p-equivalent to the Haar basis. We will now discuss necessary and sufficient conditions for f to have a prescribed decay of $\{\sigma_m(f,\Psi)_p\}$ under the assumption that Ψ is L_p-equivalent to the Haar basis $\mathcal{H}$, $1 < p < \infty$. We will express these conditions in terms of coefficients $\{c_n(f)\}$ of the expansion

$$f = \sum_{n=1}^\infty c_n(f)\psi_n.$$

We present results from [20] here. The following lemma from [19] (see Lemmas 3.1 and 3.2 from Section 3 of this paper) plays the key role in this consideration.

Lemma 6.1. *Let a basis Ψ be L_p-equivalent to $\mathcal{H}_p$, $1 < p < \infty$. Then for any finite Λ and $a \le |c_n| \le b$, $n \in \Lambda$, we have*

$$C_1(p, \Psi)a(|\Lambda|)^{1/p} \le \left\| \sum_{n \in \Lambda} c_n \psi_n \right\|_p \le C_2(p, \Psi)b(|\Lambda|)^{1/p}.$$

We formulate a general statement and then consider several important particular examples of rate of decrease of $\{\sigma_m(f, \Psi)_p\}$. We begin by introducing some notation. For a sequence $\mathcal{E} = \{\epsilon_k\}_{k=0}^{\infty}$ of positive numbers (we write $\mathcal{E} \in MDP$) monotonically decreasing to zero we define inductively a sequence $\{N_s\}_{s=0}^{\infty}$ of nonnegative integers: $N_0 = 0$; N_s is the smallest integer satisfying

$$\epsilon_{N_s} < 2^{-s}; \qquad n_s := \max(N_{s+1} - N_s, 1). \tag{6.1}$$

We are going to consider the following examples of sequences.

Example 6.1. *Take $\epsilon_0 = 1$ and $\epsilon_k = k^{-r}$, $r > 0$, $k = 1, 2, \ldots$. Then*

$$N_s \asymp 2^{s/r}, \qquad n_s \asymp 2^{s/r}.$$

Example 6.2. *Fix $0 < b < 1$ and take $\epsilon_k = 2^{-k^b}$, $k = 0, 1, 2, \ldots$. Then*

$$N_s = s^{1/b} + O(1), \qquad n_s \asymp s^{1/b-1}.$$

Let $f \in L_p$. Rearrange the sequence $\|c_n(f)\psi_n\|_p$ in decreasing order

$$\|c_{n_1}(f)\psi_{n_1}\|_p \ge \|c_{n_2}(f)\psi_{n_2}\|_p \ge \cdots$$

and denote

$$a_k(f, p) := \|c_{n_k}(f)\psi_{n_k}\|_p.$$

We now give some inequalities for $a_k(f, p)$ and $\sigma_m(f, \Psi)_p$. We will use the simplified notation $\sigma_m(f)_p := \sigma_m(f, \Psi)_p$ and $\sigma_0(f)_p := \|f\|_p$.

Lemma 6.2. *For any two positive integers $N < M$, we have*

$$a_M(f, p) \le C(p, \Psi)\sigma_N(f)_p(M - N)^{-1/p}.$$

Lemma 6.3. *For any sequence $m_0 < m_1 < m_2 < \cdots$ of nonnegative integers, we have*

$$\sigma_{m_s}(f)_p \le C(p, \Psi) \sum_{l=s}^{\infty} a_{m_l}(f, p)(m_{l+1} - m_l)^{1/p}.$$

Theorem 6.1. *Assume a given sequence $\mathcal{E} \in MDP$ satisfies the conditions*

$$\epsilon_{N_s} \geq C_1 2^{-s}, \qquad n_{s+1} \leq C_2 n_s, \qquad s = 0, 1, 2, \ldots.$$

Then we have the equivalence

$$\sigma_n(f)_p \ll \epsilon_n \quad \Longleftrightarrow \quad a_{N_s}(f, p) \ll 2^{-s} n_s^{-1/p}.$$

Corollary 6.1. *Theorem 6.1 applied to Examples 6.1, 6.2 gives the following relations:*

$$\sigma_m(f)_p \ll (m+1)^{-r} \quad \Longleftrightarrow \quad a_n(f, p) \ll n^{-r-1/p}, \qquad (6.2)$$

$$\sigma_m(f)_p \ll 2^{-m^b} \quad \Longleftrightarrow \quad a_n(f, p) \ll 2^{-n^b} n^{(1-1/b)/p}. \qquad (6.3)$$

Remark 6.1. *Making use of Lemmas 6.2 and 6.3, we can prove a version of Corollary 6.1 with the sign $\ll$ replaced by $\asymp$.*

Theorem 6.1 and Corollary 6.1 are in the spirit of classical Jackson-Bernstein direct and inverse theorems in linear approximation theory, where conditions on the corresponding sequences of approximating characteristics are imposed in the form

$$E_n(f)_p \ll \epsilon_n,$$

or

$$\|E_n(f)_p/\epsilon_n\|_{l_\infty} < \infty. \qquad (6.4)$$

It is well known (see [3]) that in studying many questions of approximation theory, it is convenient to consider along with restriction (6.4) the following generalization

$$\|E_n(f)_p/\epsilon_n\|_{l_q} < \infty. \qquad (6.5)$$

Lemmas 6.2 and 6.3 are also useful in considering this more general case. For instance, in the particular case of Example 6.1 one gets the following statement.

Theorem 6.2. *Let $1 < p < \infty$ and $0 < q < \infty$. Then for any positive r we have the equivalence relation*

$$\sum_m \sigma_m(f)_p^q m^{rq-1} < \infty \quad \Longleftrightarrow \quad \sum_n a_n(f, p)^q n^{rq-1+q/p} < \infty.$$

Remark 6.2. *The condition*

$$\sum_n a_n(f, p)^q n^{rq-1+q/p} < \infty$$

with $q = \beta := (r + 1/p)^{-1}$ takes a very simple form

$$\sum_n a_n(f,p)^\beta = \sum_n \|c_n(f)\psi_n\|_p^\beta < \infty. \tag{6.6}$$

In the case $\Psi = \mathcal{H}_p$ the condition (6.6) is equivalent to f being in the Besov space $B_\beta^r(L_\beta)$.

Corollary 6.2. *Theorem 6.2 implies the following relation*

$$\sum_m \sigma_m(f, \mathcal{H})_p^\beta m^{r\beta-1} < \infty \quad \Longleftrightarrow \quad f \in B_\beta^r(L_\beta),$$

where $\beta := (r + 1/p)^{-1}$.

The statement similar to Corollary 6.2 for free knot spline approximation was proved in [18]. Corollary 6.2 and further results in this direction can be found in [7] and [5]. We want to remark here that conditions in terms of $a_n(f,p)$ are convenient in applications. For instance, the relation (6.2) can be rewritten using the idea of thresholding. For a given $f \in L_p$ denote

$$T(\epsilon) := \#\{a_k(f,p) \quad : \quad a_k(f,p) \geq \epsilon\}.$$

Then (6.2) is equivalent to

$$\sigma_m(f)_p \ll (m+1)^{-r} \quad \Longleftrightarrow \quad T(\epsilon) \ll \epsilon^{-(r+1/p)^{-1}}.$$

For further results in this direction see [3, 2, 17].

An interesting generalization of m-term approximation was considered in [2]. Let $\Psi = \{\psi_I\}_I$ be a basis indexed by dyadic intervals. Take an α and assign to each index set Λ the following measure

$$\Phi_\alpha(\Lambda) := \sum_{I \in \Lambda} |I|^\alpha.$$

In the case $\alpha = 0$ we get $\Phi_0(\Lambda) = |\Lambda|$. An analog of best m-term approximation is the following

$$\inf_{\Lambda : \Phi_\alpha(\Lambda) \leq m} \quad \inf_{c_I, I \in \Lambda} \left\| f - \sum_{I \in \Lambda} c_I \psi_I \right\|_p.$$

A detailed study of this type of approximation (restricted approximation) can be found in [2]. It is proved in [2] that the technique developed for m-term approximation can be generalized for restricted approximation.

6.2 Stability

In this section we assume that a basis $\Psi = \{\psi_k\}_{k=1}^{\infty}$ is an unconditional normalized ($\|\psi_k\| = 1, k = 1, 2, \ldots$) basis for X (see Definition 1.2).

In numerical implementation of nonlinear m-term approximation one usually prefers to employ the strategy known as **thresholding** (see [3, §7.8]) instead of greedy algorithm. We define and study here **soft thresholding** (see [4]). Suppose a real function $v(x)$ defined for $x \geq 0$ satisfies the following relations

$$v(x) = \begin{cases} 1, & \text{for } x \geq 1 \\ 0, & \text{for } 0 \leq x \leq 1/2, \end{cases} \tag{6.7}$$

$$|v(x)| \leq A, \quad x \in [0, 1]; \tag{6.8}$$

and there is a constant C_L such that for any $x, y \in [0, \infty)$ we have

$$|v(x) - v(y)| \leq C_L|x - y|. \tag{6.9}$$

Let

$$f = \sum_{k=1}^{\infty} c_k(f)\psi_k.$$

We define a soft thresholding mapping $T_{\epsilon,v}$ as follows. Take $\epsilon > 0$ and set

$$T_{\epsilon,v}(f) := \sum_k v(|c_k(f)|/\epsilon)c_k(f)\psi_k.$$

Theorem 4.2 implies that

$$\|T_{\epsilon,v}(f)\| \leq 2KA\|f\|. \tag{6.10}$$

It was proved in [23] that the mapping $T_{\epsilon,v}$ satisfies the Lipschitz condition with a constant independent of ϵ.

Theorem 6.3. *For any ϵ and any functions $f, g \in X$,*

$$\|T_{\epsilon,v}(f) - T_{\epsilon,v}(g)\| \leq (3A + 2C_L)2K\|f - g\|.$$

Acknowledgments. This research was supported by the National Science Foundation Grant DMS 9970326 and by ONR Grant N00014-91-J1343.

References

1. B. M. Baishanski, Approximation by polynomials of given length, Illinois J. Math. **27** (1983), 449–458.

2. A. Cohen, R. A. DeVore, and R. Hochmuth, Restricted nonlinear approximation, Constructive Approx. **16** (2000), 85–113.

3. R. A. DeVore, Nonlinear approximation, Acta Numerica (1998), 51–150.

4. D. Donoho and I. Johnstone, Ideal spatial adaptation via wavelet shrinkage, Biometrica **81** (1994), 425–455.

5. R. DeVore, B. Jawerth, and V. Popov, Compression of wavelet decompositions, American Journal of Mathematics **114** (1992), 737–785.

6. R. A. DeVore, S. V. Konyagin, and V. N. Temlyakov, Hyperbolic wavelet approximation, Constructive Approximation **14** (1998), 1–26.

7. R. A. DeVore and V. A. Popov, Interpolation spaces and non-linear approximation, Lecture Notes in Mathematics **1302** (1988), 191–205.

8. R. A. DeVore and V. N. Temlyakov, Nonlinear approximation by trigonometric sums, J. Fourier Analysis and Applications **2** (1995), 29–48.

9. V. V. Dubinin, *Greedy Algorithms and Applications*, Ph.D. Thesis, University of South Carolina, 1997.

10. M. Frazier and B. Jawerth, A discrete transform and decomposition of distribution spaces, J. Functional Analysis **93** (1990), 34-170.

11. R. S. Ismagilov, Widths of sets in normed linear spaces and the approximation of functions by trigonometric polynomials, Uspekhi Mat. Nauk **29** (1974), 161–178; English transl. in Russian Math. Surveys **29** (1974).

12. B. S. Kashin and A. A. Saakyan, *Orthogonal Series*, American Math. Soc., Providence, R.I., 1989.

13. S. V. Konyagin and V. N. Temlyakov, A remark on greedy approximation in Banach spaces, East J. Approx. **5** (1999), 365–379.

14. J. Lindenstrauss and L. Tzafriri, *Classical Banach Spaces I*, Springer-Verlag, Berlin, 1977.

15. V. E. Maiorov, Trigonometric diameters of the Sobolev classes W_p^r in the space L_q, Math. Notes **40** (1986), 590–597.

16. Y. Makovoz, On trigonometric n-widths and their generalization, J. Approx. Theory **41** (1984), 361–366.

17. P. Oswald, Greedy algorithms and best m-term approximation with respect to biorthogonal systems, Preprint (2000), 1–22.

18. P. Petrushev, Direct and converse theorems for spline and rational approximation and Besov spaces, Lecture Notes in Mathematics **1302** (1988), 363–377.

19. V. N. Temlyakov, The best m-term approximation and greedy algorithms, Advances in Comp. Math. **8** (1998), 249–265.

20. V. N. Temlyakov, Nonlinear m-term approximation with regard to the multivariate Haar system, East J. Approx. **4** (1998), 87–106.

21. V. N. Temlyakov, Nonlinear Kolmogorov's widths, Matem. Zametki **63** (1998), 891–902.

22. V. N. Temlyakov, Greedy algorithm and m-term trigonometric approximation, Constructive Approx. **14** (1998), 569–587.

23. V.N. Temlyakov, Universal bases and greedy algorithms, IMI-Preprint series **8** (1999), 1–20.

24. V. Temlyakov, *Approximation of Functions with Bounded Mixed Derivative*, Proc. Steklov Institute, 1989, Issue 1.

25. P. Wojtaszczyk, Greedy algorithms for general systems, J. Approx. Theory **107** (2000), 293–314.

26. A. Zygmund, *Trigonometric Series*, University Press, Cambridge, 1959.

V. Temlyakov
Department of Mathematics
University of South Carolina
Columbia, SC 29208
USA
temlyak@math.sc.edu